TALLAHASSEE
LIBRARY
COMMUNITY COLLEGE

ENVIRONMENTAL **EFFECTS** of TRANSGENIC **PLANTS**

THE SCOPE AND ADEQUACY OF REGULATION

Committee on Environmental Impacts Associated with
Commercialization of Transgenic Plants

Board on Agriculture and Natural Resources
Division on Earth and Life Studies
National Research Council

NATIONAL ACADEMY PRESS
Washington, D.C.

NATIONAL ACADEMY PRESS 2101 Constitution Avenue, NW Washington, D.C. 20418

NOTICE: The project that is the subject of this report was approved by the Governing Board of the National Research Council, whose members are drawn from the councils of the National Academy of Sciences, the National Academy of Engineering, and the Institute of Medicine. The members of the committee responsible for the report were chosen for their special competences and with regard for appropriate balance.

This study was supported by Cooperative Agreements No. 59-0790-0-173 and No. 99-1001-0229-GR between the National Academy of Sciences and the U.S. Department of Agriculture. Any opinions, findings, conclusions, or recommendations expressed in this publication are those of the author(s) and do not necessarily reflect the views of the organizations or agencies that provided support for the project.

Library of Congress Cataloging-in-Publication Data

Environmental effects of transgenic plants : the scope and adequacy of regulation / Committee on Environmental Impacts associated with Commercialization of Transgenic Plants Board on Agriculture and Natural Resources Division on Earth and Life Studies, National Research Council.

p. cm.

Includes bibliographical references (p.).

ISBN 0-309-08263-3 (hardcover)

1. Transgenic plants—Risk assessment. 2. Agricultural biotechnology—Environmental aspects. I. National Research Council. Committee on Environmental Impacts.

SB123.57 .E58 2002

631.5′233—dc21

2001008715

Environmental Effects of Transgenic Plants: The Scope and Adequacy of Regulation is available from the National Academy Press, 2101 Constitution Avenue, N.W., Lockbox 285, Washington, DC 20055; (800) 624-6242 or (202) 334-3313 (in the Washington metropolitan area); *http://www.nap.edu*.

Printed in the United States of America

THE NATIONAL ACADEMIES

National Academy of Sciences
National Academy of Engineering
Institute of Medicine
National Research Council

The **National Academy of Sciences** is a private, nonprofit, self-perpetuating society of distinguished scholars engaged in scientific and engineering research, dedicated to the furtherance of science and technology and to their use for the general welfare. Upon the authority of the charter granted to it by the Congress in 1863, the Academy has a mandate that requires it to advise the federal government on scientific and technical matters. Dr. Bruce M. Alberts is president of the National Academy of Sciences.

The **National Academy of Engineering** was established in 1964, under the charter of the National Academy of Sciences, as a parallel organization of outstanding engineers. It is autonomous in its administration and in the selection of its members, sharing with the National Academy of Sciences the responsibility for advising the federal government. The National Academy of Engineering also sponsors engineering programs aimed at meeting national needs, encourages education and research, and recognizes the superior achievements of engineers. Dr. Wm. A. Wulf is president of the National Academy of Engineering.

The **Institute of Medicine** was established in 1970 by the National Academy of Sciences to secure the services of eminent members of appropriate professions in the examination of policy matters pertaining to the health of the public. The Institute acts under the responsibility given to the National Academy of Sciences by its congressional charter to be an adviser to the federal government and, upon its own initiative, to identify issues of medical care, research, and education. Dr. Kenneth I. Shine is president of the Institute of Medicine.

The **National Research Council** was organized by the National Academy of Sciences in 1916 to associate the broad community of science and technology with the Academy's purposes of furthering knowledge and advising the federal government. Functioning in accordance with general policies determined by the Academy, the Council has become the principal operating agency of both the National Academy of Sciences and the National Academy of Engineering in providing services to the government, the public, and the scientific and engineering communities. The Council is administered jointly by both Academies and the Institute of Medicine. Dr. Bruce M. Alberts and Dr. Wm. A. Wulf are chairman and vice chairman, respectively, of the National Research Council.

COMMITTEE* ON ENVIRONMENTAL IMPACTS ASSOCIATED WITH COMMERCIALIZATION OF TRANSGENIC PLANTS

FRED L. GOULD, *Chair*, North Carolina State University, Raleigh
DAVID A. ANDOW, University of Minnesota, St. Paul
BERND BLOSSEY, Cornell University, Ithaca, New York
IGNACIO CHAPELA, University of California, Berkeley
NORMAN C. ELLSTRAND, University of California, Riverside
NICHOLAS JORDAN, University of Minnesota, St. Paul
KENDALL R. LAMKEY, Iowa State University, Ames
BRIAN A. LARKINS, University of Arizona, Tucson
DEBORAH K. LETOURNEAU, University of California, Santa Cruz
ALAN McHUGHEN, University of Saskatchewan, Saskatoon
RONALD L. PHILLIPS, University of Minnesota, St. Paul
PAUL B. THOMPSON, Purdue University, West Lafayette, Indiana

Staff

KIM WADDELL, *Study Director*
HEATHER CHRISTIANSEN, *Research Associate*
KAREN L. IMHOF, *Project Assistant* (from January 2000 to March 2001)
MICHAEL R. KISIELEWSKI, *Research Assistant* (since March 2001)
BARBARA BODLING, *Editor*

---

*The work of this committee was overseen by the Committee on Agricultural Biotechnology, Health, and the Environment, of the Board on Agriculture and Natural Resources and the Board on Life Sciences.

COMMITTEE ON AGRICULTURAL BIOTECHNOLOGY, HEALTH, AND THE ENVIRONMENT

BARBARA A. SCHAAL, *Co-Chair*, Washington University, St. Louis
HAROLD E. VARMUS, *Co-Chair*, Memorial Sloan-Kettering Cancer Center, New York, New York
DAVID A. ANDOW, University of Minnesota, St. Paul
FREDERICK M. AUSUBEL, Harvard Medical School, Boston, Massachusetts
NEAL L. FIRST, University of Wisconsin, Madison
LYNN J. FREWER, Institute of Food Research, Norwich, England
HENRY L. GHOLZ, National Science Foundation, Arlington, Virginia
ERIC M. HALLERMAN, Virginia Polytechnic Institute and State University, Blacksburg
CALESTOUS JUMA, Harvard University
NOEL T. KEEN, University of California, Riverside
SAMUEL B. LEHRER, Tulane University School of Medicine, New Orleans, Louisiana
J. MICHAEL McGINNIS, Robert Wood Johnson Foundation, Princeton, New Jersey
SANFORD A. MILLER, Georgetown University
PER PINSTRUP-ANDERSON, International Food Policy Research Institute, Washington, D.C.
VERNON W. RUTTAN, University of Minnesota, St. Paul
ELLEN K. SILBERGELD, University of Maryland Medical School, Baltimore
ROBERT E. SMITH, R.E. Smith Consulting, Inc., Newport, Vermont
ALLISON A. SNOW, Ohio State University, Columbus
DIANA H. WALL, Colorado State University, Fort Collins

Staff

JENNIFER KUZMA, *Senior Program Officer*
LAURA HOLLIDAY, *Research Assistant*

BOARD ON LIFE SCIENCES

COREY S. GOODMAN, *Chair*, University of California, Berkeley
DAVID EISENBERG, University of California, Los Angeles
DAVID J. GALAS, Keck Graduate Institute of Applied Life Science, Claremont, California
BARBARA GASTEL, Texas A&M University, College Station
JAMES M. GENTILE, Hope College, Holland, Michigan
ROBERT T. PAINE, University of Washington, Seattle
STUART L. PIMM, Columbia University
JOAN B. ROSE, University of South Florida, St. Petersburg
GERALD M. RUBIN, Howard Hughes Medical Institute, Chevy Chase, Maryland
RAYMOND L. WHITE, University of Utah, Salt Lake City

Staff

WARREN R. MUIR, *Executive Director* (since June 1999)
FRANCES SHARPLES, *Director*
BRIDGET AVILA, *Senior Project Assistant*

Preface

In assessing the conclusions of any report on a subject as controversial as agricultural biotechnology, I certainly would want to know about the background of the individuals who wrote the report, and the process used to write it. So before you delve into the contents of this report I would like to tell you about our committee and the process that we used in writing this report. "About the Authors" provides background information on each of the 12 committee members who wrote the report. The committee followed the general National Research Council guidelines for report writing, with more specific steps in the process determined by the committee members. This report is a consensus document. Therefore, every member of the committee had an opportunity to question the content of each page, and in the end had to determine that he or she could consent to all of the report findings and recommendations. Had any committee member written the report alone, the conclusions would have been different. Some view this as a weakness of the consensus process—too much compromise. Based on my experience with this specific report, I strongly disagree with that perspective. What I saw in our consensus process was that logic and detailed information prevailed. It was easy for us to come to consensus on some issues but in other cases there were lengthy debates. In the approximately 15 months from the time of our first meeting until we finally signed off on the report, members of the committee had time to present specific arguments on multiple occasions with the opportunity to collect data to back up their arguments in between meetings or conference calls. Evidence to me of the success of our specific consensus process is that a report written by a single member of our committee would have been substantially different before and after he or she had gone through the study process. We learned a lot from each other, and the report reflects this enhanced pool of knowledge.

The report clearly was the product of the committee, but there were a number of other important inputs. A workshop was convened by the committee to obtain input from scientists working on novel plant traits, from individuals with special expertise in regulation of transgenic plants, and from members of public interest groups (Appendix A). We also sent a letter to nearly 400 selected individuals and groups to solicit input (Appendix C). The letter specifically probed for unique perspectives on potential environmental impacts of transgenic plants. We received 35 useful, individual responses to this letter (copies available from NRC). In addition, members of the committee met with APHIS personnel and representatives from industry and public interest groups. All of these meetings were followed up by written communications to ensure that the information gathered from these meetings was accurate. The draft of our report was reviewed in detail by 12 individuals approved by the NRC's Report Review Committee in order to provide distinct expertise and perspectives on the topics covered. The comments from the reviewers were given thorough consideration by the committee, and the Report Review Committee of the NRC assessed the revised draft before our report was accepted for publication.

You will read many findings and recommendations in this report. I would like to highlight a few of them that reflect on the nature of the issues addressed, and on the study process. During the initial meetings of our committee it became apparent that there was a need to examine the environmental risks of transgenic plants within the context of environmental risks posed by the entire modern agricultural enterprise. Our assessment confirmed the general findings of others that many agricultural practices have substantial negative environmental impacts. Additionally, we found that the current standards used by the federal government to assure environmental safety of transgenic plants were higher than the standards used in assuring safety of other agricultural practices and technologies. After much discussion of this finding we did not conclude that the standards for transgenics were too high. We found that over the past 70 years there has been growing concern about the impacts of agriculture on the environment and that, in general, agricultural technologies introduced many years ago have not been as carefully scrutinized as newer technologies. Therefore, in the future, it will be important to reconsider the standards that are being used to examine environmental effects of older technologies such as conventional plant breeding.

Deliberations among our committee members—most of whom were biologists—led to a consensus that effective environmental risk analysis and management must consider both biological and social factors. While risk of environmental effects can be defined simply as a multiple of hazard and exposure, the measurement of both hazard and exposure involves a complex blend of ecological and social factors. This is in part because the value of every organism and habitat is based on its ecosystem, economic, and cultural functions. The recent assessment of risks to monarch butterflies from transgenic corn exemplifies these interactions. From a purely ecological perspective, decline in monarch butterfly populations is not, a priori, expected to be more environmentally disruptive than the decline in a randomly selected species of ground beetle. However, appropriate risk analyses for these two species should differ because of the role of the monarch butterfly in American culture. While ecologists must insist on careful examination of environmental risk to all species, decision makers cannot ignore other factors.

One general finding of the committee was that a rigorous scientific risk analysis has two roles: 1) it offers essential technical information to the agencies charged with making decisions about commercializing a transgenic plant; 2) it also serves as evidence to the public that the decision-making agencies are deserving of their trust. This second role is not fully appreciated in many cases. The more clearly an agency can explain the rigor of its methods, and the more engaged it becomes in responding to the public, the more likely it is to gain the public's confidence.

The report of our committee does not paint a simple black and white picture of transgenic plant regulation by USDA-APHIS personnel. As stated in the report, our committee took on the role of searching for problems, and recommended changes "as a means to help improve a functioning system." I hope that members of the press and other organizations will not yield to the temptation of focusing only on our finding that environmental standards for transgenic plants are higher than those for other agricultural technologies, or only on our findings that suggest the need for improvement in environmental regulation of transgenic plants.

I want to thank the entire committee for their diligence and perseverance in examining mountains of background documents, and for writing and rewriting the pieces of this report. I am proud of the committee members for their willingness to argue forcefully, and for their ability to listen carefully to the perspectives of others. Without this combination of traits it would have been impossible to develop this consensus document. Special thanks go to Drs. Norman Ellstrand, David Andow, Bernd Blossey, and Paul Thompson for their leadership roles with the major organizing and writing responsibilities. External reviewers substantially improved the content of the report, and our technical editor, Barbara Bodling, improved the prose. Karen Imhof and Mike Kisielewski offered valuable technical and organizational expertise in setting up meetings and in pulling the report together. Heather Christiansen's research efforts gave us access to essential information from both the public and private sectors. The study process and the writing of this report could not have been accomplished without the hard work, insight, and diplomacy of our study director, Dr. Kim Waddell.

Fred Gould
Chair
Committee on Environmental Impacts Associated with Commercialization of Transgenic Crops

Acknowledgments

This study was enhanced by the contributions of many individuals who graciously offered their time, expertise, and knowledge. The committee thanks all who attended and/or participated in its public workshop:

STANELY ABRAMSON, Arent, Fox, Kintner, Plotkin, & Kahn, Washington, D.C.
DAVID E. ADELMAN, Natural Resources Defense Council, Washington, D.C.
FAITH CAMPBELL, American Lands Alliance, Washington, D.C.
THOMAS CORS, Dynamics Technology, Arlington, Virginia
DEAN DELLAPENNA, Michigan State University, East Lansing
SHARON FRIEDMAN, Office of Science and Technology Policy, Executive Office of the President, Washington, D.C.
ELIOT HERMAN, Climate Stress Laboratory, Beltsville, Maryland
MAUREEN K. HINKLE, National Audubon Society, Bethesda, Maryland
SHIRLEY INGEBRITSEN, U.S. Department of Agriculture, Animal and Plant Health Inspection Service, Riverdale, Maryland
GANESH KISHORE, Monsanto Company (formerly), St. Louis, Missouri
WARREN LEON, Northeast Sustainable Energy Association, Greenfield, Massachusetts
TERRY L. MEDLEY, Dupont BioSolutions Enterprise, Wilmington, Delaware

DEBORAH OLSTER, National Science Foundation, Arlington, Virginia
CRAIG ROSELAND, U.S. Department of Agriculture, Animal and Plant Health Inspection Service, Riverdale, Maryland
JOAN ROTHENBERG, Institute of Food Technology, Washington, D.C.
ALLISON SNOW, Ohio State University, Columbus
ED SOULE, McDonough School of Business, Washington, D.C.
JOHN TURNER, U.S. Department of Agriculture, Animal and Plant Health Inspection Service, Riverdale, Maryland
MICHAEL F. THOMASHOW, Michigan State University, East Lansing
LAREESA WOLFENBERGER, U.S. Environmental Protection Agency, Washington, D.C.

The committee extends its appreciation to the staff members of the National Research Council's (NRC) Division on Earth and Life Studies and Board on Agriculture and Natural Resources for their commitment to the study process and their efforts in preparing this report.

This report has been reviewed in draft form by individuals chosen for their diverse perpectives and technical expertise, in accordance with procedures approved by the NRC's Report Review Committee. The purpose of this independent review is to provide candid and critical comments that will assist the institution in making its published report as sound as possible and to ensure that the report meets institutional standards for objectivity, evidence, and responsiveness to the study charge. The review comments and draft manuscript remain confidential to protect the integrity of the deliberative process. We thank the following individuals for their review of this report:

STEVEN LINDOW, University of California, Berkeley
ROGER BEACHY, Donald Danforth Plant Science Center
MAY BERENBAUM, University of Illinois
ALLISON SNOW, The Ohio State University
TERRY L. MEDLEY, DuPont BioSolutions Enterprise
JAMES PRATT, Portland State University
DANIEL SIMBERLOFF, The University of Tennessee
JANE RISSLER, Union of Concerned Scientists
THOMAS E. NICKSON, Monsanto Company, St. Louis, Missouri
WYATT ANDERSON, University of Georgia
DOUGLASS GURIAN-SHERMAN, Center for Science in the Public Interest
FREDERICK BUTTEL, University of Wisconsin

Although the reviewers above have provided many constructive comments and suggestions, they were not asked to endorse the conclusions or recommendations nor did they see the final draft of the report before its release. The review of this report was overseen by Michael T. Clegg, University of California, Riverside, and John E. Dowling, Harvard University. Appointed by the National Research Council, they were responsible for making certain that an independent examination of this report was carried out in accordance with institutional procedures and that all review comments were carefully considered. Responsibility for the final content of this report rests entirely with the authoring committee and the institution.

Contents

TABLES, FIGURES, AND BOXES

Tables

Figures

Boxes

Executive Summary

Before transgenic plants can be grown outside the laboratory, approval must be obtained from the Animal and Plant Health Inspection Service (APHIS) of the United States Department of Agriculture (USDA). APHIS derives its authority for regulating transgenic plants from the Federal Plant Pest Act (FPPA) and the Federal Plant Quarantine Act (FPQA). As a participant in the U.S. Coordinated Framework for the Regulation of Biotechnology, APHIS developed its first formal procedures for assessing potential environmental effects of transgenic plants in 1987. Currently, APHIS's Biotechnology, Biologics, and Environmental Protection unit (BBEP) reviews approximately 1,000 applications for field testing and deregulation of transgenic plants each year.

TASK OF THE COMMITTEE

In January 2000 the USDA requested that the National Academy of Sciences examine the scientific basis for and the operation of APHIS regulatory oversight. The specific task set before this committee by the USDA and the NRC (National Research Council) Committee on Agricultural Biotechnology, Health, and the Environment (CABHE) was as follows:

> "The committee will review the scientific basis that supports the scope and adequacy of USDA's oversight of environmental issues related to current and anticipated transgenic plants and their products. In order to address these issues, the committee will:
>
> 1. Evaluate the scientific premises and assumptions underpinning the environmental regulation and oversight of transgenic plants. This evaluation will include a comparison of the processes and products of genetic engineering with those of conventional plant breeding as they pertain

to environmental risks. This evaluation may result in recommendations for research relevant to environmental oversight and effects of transgenic plants.

2. Assess the relevant scientific and regulatory literature in order to evaluate the scope and adequacy of APHIS's environmental review regarding the process of notification and determination of non-regulated status. The committee will focus on the identification of effects of transgenic plants on non-target organisms and the Environmental Assessments (EA) of those effects. The study will also provide guidance on the assessment of non-target effects, appropriate tests for environmental evaluation, and assessment of cumulative effects on agricultural and non-agricultural environments.

3. Evaluate the need for and approaches to environmental monitoring and validation processes."

Previous National Research Council (NRC) committees have examined a number of issues related to the safety of genetically engineered organisms (NRC 1982, 1989, 2000c) but none specifically examined APHIS oversight or how commercial use of genetically engineered crops with non-pesticidal traits could affect agricultural and nonagricultural environments.

The task of this committee specifically included provision of guidance for assessment of the cumulative effects of commercialization of engineered crops on the environment. Therefore, the committee examined the potential effects on the environment that could result from the use of engineered crops on large spatial scales over many years. In addition to evaluating the potential direct environmental impacts of single engineered traits within existing agricultural systems, the committee also examined how commercialization of engineered crops with single and multiple traits could actually change farming and thereby impact agricultural and nonagricultural landscapes of the United States. As part of its task, the committee conducted a detailed study of the relevant scientific and regulatory literature and used its findings from this study in developing what the committee considers an appropriate framework for assessing the environmental effects of transgenic plants. The committee used this framework to evaluate the scope and adequacy of the APHIS review process.

COMPARISON OF ENVIRONMENTAL ASSESSMENT OF TRANSGENIC PLANTS WITH ASSESSMENT OF OTHER AGRICULTURAL TECHNOLOGIES

Risk assessment literature and history demonstrate that environmental regulation of agricultural practices and technologies involves an interplay of ecological and social factors. Therefore, any analysis of the scope and adequacy of an environmental assessment program must address interaction of these two factors. At the outset of the twentieth century environmental issues were not a dominant public concern, and U.S. farmers were, in general, free to use whatever agricultural practices best suited their needs. As the twentieth century progressed, the impacts of agricultural practices on human health and the environment became a focus of public attention. Regulations and incentive programs were developed for agriculture and now have a major influence on farming, ranging from the choice of tillage practices to the choice of pest control techniques.

In the era when the scientific foundation of conventional plant breeding was developed and put in practice, potential for non-target effects and gene flow were not a concern. In contrast, transgenic technology is coming of age during a time when environmental assessments are much more sophisticated and when a growing segment of the public is voicing concern about environmental degradation and lack of faith in government agencies to prevent that degradation. Concern over the impact of transgenic plants on the environment has led governments in a number of countries to raise the standards by which they judge what constitutes a significant negative effect of agriculture on the environment. Environmental standards being developed for transgenic plant varieties consider impacts that were rarely even measured when novel conventional crop varieties or synthetic chemicals were introduced in the 1960s and 1970s. Today, government agencies charged with the regulation of transgenic plants find themselves in the difficult position of enforcing a higher environmental standard for transgenic plants than the standards currently used to regulate the impacts of other agricultural technologies and practices. It is possible that the higher standards being developed for transgenic plants will, in the future, be applied in some fashion to other agricultural technologies and practices. Because decisions that are now being made with regard to transgenic plants could set a precedent for evaluating all agriculture, government agencies and the public must keep this in mind as regulations for transgenic plants evolve.

ENVIRONMENTAL EFFECTS OF AGRICULTURAL PRACTICES, NOVEL GENETIC MATERIAL, AND THE PROCESSES USED IN PLANT IMPROVEMENT

From a scientific perspective, the committee saw a need to place potential impacts of transgenic crops within the context of environmental effects caused by other agricultural practices and technologies. There is substantial evidence that ecological effects of farming practices exert simplifying and destabilizing effects on neighboring natural ecosystems.

These effects are of concern because they appear to weaken or destroy ecosystems' capacity for resilience—that is, an ecosystem's ability to return to its initial state despite disturbance. Potential ecological effects of transgenic crops, and other crops bearing novel traits, may be heightened in this destabilized ecological milieu. This argues for a cautious approach to the release of any crop that bears a novel trait. Equally, it is an argument for a cautious approach to any extensive change in agricultural practices.

The two plant pest statutes (Federal Plant Pest Act and the Federal Plant Quarantine Act), which are used by APHIS to regulate transgenic plants, were originally developed to regulate the introduction of nonindigenous plant species. Because the amount of novel genetic information added to an ecosystem by the introduction of a new species is much greater than that added by a single transgene, the use of these plant pest statutes to regulate transgenic plants has been criticized. On the other hand, the use of these statutes for regulating transgenic but not conventionally improved plants has been defended because the introduced genes in transgenic plants can have a much more distant taxonomic origin (for example, bacterial genes transferred to plants). These arguments are based on a general assumption that the risks associated with the introduction of genetic novelty are related to the number of genetic changes and the origin of the novel genes.

The committee compared empirical evidence of environmental impacts involving small to large amounts of genetic novelty from taxonomically related and unrelated sources and found no general support for this assumption. **More specifically, it was found that (1) small and large genetic changes have had substantial environmental consequences; (2) the consequences of biological novelty depend strongly on the specific environment, including the genomic, physical, and biological environments into which they are introduced; (3) the significance of the consequences of biological novelty depend on societal values; (4) introduction of biological novelty can have unintended and unpredicted effects on the recipient community and ecosystem; (5) *a priori* there is no strict dichotomy between the possibility of environmental hazard associated with releases of cultivated plants with novel traits and the introduction of nonindigenous plant species. However, the highly domesticated characteristics of many cultivated plants decrease the potential of certain hazards.**

The conventional development of semi-dwarf, short-season varieties of rice and wheat that propelled the Green Revolution of the twentieth century clearly exemplifies how a small number of genetic changes to a crop can impact the environment. In the case of rice, a single gene for short stature made rice much more responsive to fertilizer, while a few genes for more rapid maturation allowed farmers to grow one or two extra crops per year. These small genetic changes enabled massive changes in agricultural practices that increased production. However, these changes have increased soil salinity, lowered water tables, and altered wetlands in some regions. Thus, there are tradeoffs between the long-term positive and negative effects of these varieties. The environmental impacts of the genes introduced to Green Revolution varieties and other crops are often indirect, which makes their assessment more complex but no less important.

In comparing conventional and transgenic approaches to crop improvement, **the committee is in agreement with a previous NRC report (2000c) which found that both transgenic and conventional approaches (for example, hybridization, mutagenesis) for adding genetic variation to crops can cause changes in the plant genome that result in unintended effects on crop traits.** Genetic improvement of crops by both approaches typically involves the addition of genetic variation to existing varieties, followed by screening for individuals that have only desirable traits. The screening component will remove many but not all of the unanticipated physical and ecological traits that could adversely affect the environment.

Based on a detailed evaluation of the intended and unintended traits produced by the two approaches to crop improvement, **the committee finds that the transgenic process presents no new categories of risk compared to conventional methods of crop improvement but that specific traits introduced by both approaches can pose unique risks.** There is currently no formal environmental regulation of most conventionally improved crops, so it is clear that the standards being set for transgenic crops are much higher than for their conventional counterparts. **The committee finds that the scientific justification for regulation of transgenic plants is not dependent on historically set precedents for not regulating conventionally modified plants. While there is a need to reevaluate the potential environmental effects of conventionally improved crops, for practical reasons, the committee does not recommend immediate regulation of conventional crops.** A previous NRC report (2000c) also raised issues related to the lack of rigorous examination of conventionally produced crop varieties. Transgenic and conventional approaches are in a period of rapid change. This makes it difficult to assess the potential risks of specific traits that each approach will be able to alter in the future.

While it is not possible to assess the risks of any genetically modified plant without empirical examination, **the committee finds that it should be possible to relatively quickly screen modified plants for potential environmental risk and then conduct detailed tests on only the subset of plants for which preliminary screening indicates potential risk.**

RISK ANALYSIS AND THE REGULATION OF TRANSGENIC PLANTS: SCIENTIFIC ASSUMPTIONS AND PREMISES

The committee reviewed a number of models for risk analysis, with special attention to those outlined in two previous reports (NRC 1983, 1996). Risk analysis often involves the use of scientific information to provide technical guidance to decision makers about the management of those risks (referred to as the "decision support" role of risk analysis). This decision support role presumes that the decision maker (an individual, a group, or an organization) is well defined and has the legitimate and uncontested authority to make a decision. Traditionally, officials in government agencies have viewed risk analyses as decision support for the exercise of their legislatively mandated authority. However, the assumption that "mandated authority" provides "uncontested authority" does not hold in all cases. Indeed there are many situations in which scientifically rigorous risk analysis and involvement of interested and affected parties in the risk analysis process perform a second, well-recognized role in regulation—that of maintaining the legitimacy of regulatory agencies to exercise such authority.

The committee discussed a number of points of tension that arise between the use of risk analysis to create and maintain legitimacy and its use as a decision support tool. It is clear that democracy is best served when people affected by regulatory decision making can be significantly involved in the decision making, and that inclusion of diverse interests in the risk analysis process can be a powerful force to garner legitimacy of a decision. This is especially true because the significance of environmental effects of novel genetic material depends on societal values. However, especially when the decision options under consideration are not well defined, broad public involvement in risk analysis can result in risk management decisions that lack scientific rigor.

In the analysis of risks from transgenic plants, APHIS has concentrated on the decision support role of risk analysis. However, it is clear that risk analysis of transgenic plants has played an important role in maintaining the legitimacy of regulatory decision making concerning environmental and food safety in the United States. **The committee concludes that risk analysis of transgenic plants must continue to fulfill two distinct roles: (1) technical support for regulatory decision making and (2) establishment and maintenance of regulatory legitimacy.**

The use of more rigorous methods in decision support are likely to help risk analysis fulfill its role of establishing and maintaining regulatory authority. At least five standards of evidence can be used in a risk assessment for decision support. The scientifically rigorous methods include epidemiological, modeling, and experimental methods. Other methods include the judgments of external scientific panels with specific technical expertise, and judgment of experienced regulatory personnel. A consensus of multiple external scientific experts is likely to be more rigorous than regulatory judgments because disagreements among external experts are likely to lead to more robust risk assessments.

As indicated above, this committee agrees with previous NRC committees (NRC, 1989, 2000c) that there are no new categories of risk associated with transgenic plants. The categories of risks from transgenic plants include those associated with the movement of the transgenes, impacts of the whole plant through escape, and through impacts on agricultural practices, non-target organism effects, and resistance evolution. For this reason, the process of producing new plant varieties should not enter into the assessment. However, the committee's analysis indicates that specific traits introduced by *either* of the two approaches can pose unique risks. For example, within the general category of "risks to non-target organisms," production of *Bt* toxins in corn pollen could pose a unique airborne toxin-exposure that was never found in conventional corn varieties. **For purposes of decision support this committee agrees with previous NRC reports which conclude that risks must be assessed on a case-by-case basis with consideration for the organism, trait, and environment.**

Typically there are a number of comparisons that are appropriate for assessing the risks of transgenic crops. For example, the environmental effects of a transgenic crop could be compared to chemically intensive farming practices and to farming practices developed to be more ecologically sustainable. Another obvious comparison is that of a crop variety with a transgenic trait to a similar variety (that is, a near isoline) lacking that trait. Therefore, the maintenance of such varieties is critical for appropriate testing.

The committee recognizes that in any attempt to mitigate environmental risk there is a need to be mindful of the fact that avoiding one risk can sometimes inadvertently cause another greater risk. For example, a regulation that discouraged research on pest-specific, plant-produced compounds could in some cases lead to continued use of environmentally disruptive synthetic pesticides.

ANALYSIS OF THE APHIS REGULATORY PROCESS

The major focus of the committee's work was on analysis of the scope and adequacy of the APHIS environmental review process for transgenic crops. There were three phases to the analysis. First, the committee examined the general statutes and rules used by APHIS to regulate transgenic plants and the documents that APHIS has developed as guidance for applicants. Next, APHIS assessments of specific applications for testing and commercialization were examined in detail (case studies). During these two phases the committee communicated with APHIS personnel to avoid missing any crucial unpublished information and to learn more about the day-to-day operations of the APHIS-BBEP. Finally, the information gathered was used to determine how well APHIS oversight is meeting the two general roles of risk assessment, and to develop recommendations for specific changes in that oversight.

The committee finds that APHIS and other regulatory agencies charged with assessing the safety of transgenic plants face a daunting task. This is so in part because environmental risk assessment of transgenic plants is new and in part because the social context in which regulatory decisions about transgenic organisms must now be made is dramatically different from the one in which these agencies have been accustomed to work. **The committee finds that the APHIS regulatory system has improved substantially since it was initiated.** For example, in two *Bt* corn petitions for nonregulated status, one completed in 1994 and one in 1997, the breadth of environmental issues addressed and the degree of rigor with which they were addressed improved with time. Furthermore, **the development of a notification process that utilizes ecologically-based performance standards was an important step in effectively streamlining the field-testing process.** The learning process at APHIS has not come without missteps, but the agency seems to use them as opportunities for further improvement. In its role of analyzing APHIS environmental reviews the committee mostly searched for problem areas as a means to help improve a functioning system.

APHIS has been criticized for regulating transgenic crops with statutes that do not cover all transgenic plants. **The committee finds that APHIS currently has the authority to base regulatory scrutiny on potential plant pest status, regardless of the process of derivation, and therefore can theoretically regulate any transgenic plant. However, the only practical trigger used by APHIS is the presence of a previously identified plant pest or genes from a plant pest in the transformed plant. Other operational triggers are needed for transgenic plants that may have associated risks but lack the above characteristics.**

APHIS jurisdiction and the focus of its Environmental Assessments are confined to the United States, but some APHIS assessments discuss potential environmental effects of specific transgenic plants outside the United States. There is a need to clarify this discrepancy. If APHIS jurisdiction is to remain confined to the United States, Environmental Assessments should clearly state that they do not consider risks beyond United States borders.

APHIS documents reviewed by the committee also are inconsistent regarding APHIS authority to deregulate transgenic plants on a limited geographic basis within the United States. This inconsistency is important because in one case the perceived inability of APHIS to set geographic limits led it and the Environmental Protection Agency (EPA) to make different decisions on the planting range of transgenic cotton in the United States. There may be future situations where at least temporary geographic limits would be beneficial.

Current APHIS oversight involves three processes: notification, permitting, and petitioning for nonregulated status. It is possible to commercialize the nonliving products of transgenic plants through each of these processes, but petitioning for nonregulated status is currently the predominant mode for commercialization of all transgenic plant products and is the only process for commercialization of living transgenic plants. Initially, all field testing of transgenic plants needed to be approved by the permitting process (see Chapter 3). However, APHIS determined that for some transgenic plants, safety could be assured through a more streamlined approach of having the applicant notify the agency in advance of planting. The notification process was first used for a limited set of crops, but currently almost all field testing is conducted through the notification process that requires APHIS to complete its decision making in less than 30 days. Within this time frame, one APHIS staff member typically determines if the notification process is sufficient for the particular transgenic plant. The applicant must follow general guidelines to ensure that there are no environmental effects from the planting, but the process involves no public or external scientific input. Plants that cannot be grown in the field, based on notification, include those that produce substances intended for use as pharmaceuticals and those that could affect non-target organisms. Some plant products have been commercialized using the notification process, and there is no limit to the acreage that can be planted under the notification system. Commercialization of certain plant products through notification could result in large plantings and increased risks through scale effects. In the committee's examination of specific cases where commercialization involved only oversight through the notification process, one case was found where it appears that a transgenic plant with toxic properties (avidin-producing corn) was grown under the notification process. **The committee finds that the notification process is conceptually appropriate, but there is a need to reexamine which transgenic plants should be tested and commercialized through the notification process.**

In comparison with the notification process the permitting process requires more detail from the applicant, and if APHIS determines that there is a need for a formal Environmental Assessment of the plant, a description of the application is published in the *Federal Register* and is open to public comment. The permitting process is not commonly used at present, but as more pharmaceutical-producing plants are developed, it may be used more frequently.

The dominant path toward commercialization—petitioning for nonregulated status—is in essence a request for APHIS to determine that there is no plant pest risk (or as commonly understood, no environmental risk) associated with the specific transgenic plant. If APHIS makes this determination, it agrees that the plant no longer needs regulation. As currently implemented, APHIS deregulation is absolute. Once deregulated, the agency does not assume further oversight of the plant or its progeny and descendants. As part of the petitioning process APHIS always conducts a formal Environmental Assessment and publishes this assessment in the *Federal Register*, providing the public with a 60-day comment period. APHIS personnel are required to respond to each comment received.

The committee examined six individual petitions for nonregulated status conducted over a period of four years. Based on these detailed assessments as well as an examination of the general process of APHIS oversight, the committee finds a number of places where APHIS could improve its technical risk assessments and the manner in which it involves the public in policy development and decision making. In general, **the committee finds that the APHIS process should be made significantly more transparent and rigorous by enhanced scientific peer review, solicitation of public input, and development of determination documents with more explicit presentation of data, methods, analyses, and interpretations.** Such changes are likely to improve the agency's risk analyses at both the level of decision support and the level of maintenance of regulatory authority.

To improve the rigor of decision support, **the committee recommends that, whenever changes in regulatory policy are being considered, APHIS should convene a scientific advisory group.** This is a common practice of the EPA. Before APHIS first introduced the notification procedure, it formally requested input from the then-active USDA Agricultural Biotechnology Research Advisory Committee (ABRAC). Such formal input has not been sought since that time. **The committee recommends that before making specific, precedent-setting decisions, APHIS should solicit broad external scientific review well beyond the use of *Federal Register* notices.**

Specific attributes of APHIS's environmental assessments require comment. **The committee recommends that APHIS should not use the term "no evidence" in its environmental assessments.** The term "no evidence" can mean either that no one has looked for evidence or that the examination provides contrary evidence. Lack of evidence is not typically useful in making regulatory decisions about risk. **The committee also recommends that APHIS not use general weediness characteristics in its assessments because these characteristics have no predictive value.** APHIS must instead use criteria specific to the regulated article and the environments to which it could be exposed. Until recently it was difficult or impossible to determine the full sequence of an inserted gene. Therefore, APHIS's past acceptance of partial sequence data was reasonable. **At this time the agency should require reporting of full DNA sequences of transgenes as they are integrated into the plant genome unless the applicant can provide scientific justification not to do so.** Data on flanking sequences also would be useful to determine the exact insertion site of the transgene

APHIS's environmental assessments of transgenic plants with pesticidal properties include assessment of effects on non-target organisms as well as assessment of the risk posed by the potential of pests to evolve resistance to the pesticidal substance. The treatment of these two issues in APHIS's Environmental Assessment documents is generally superficial. **The committee recommends that for pesticidal plants APHIS should either increase the rigor of assessments of resistance risk and non-target impacts, or it should completely defer to the EPA, which also assesses these risks.**

The committee commends APHIS for developing and making available guidelines for applicants who are using any of the three APHIS processes. These guidelines clearly are helpful, especially to small companies and scientists who are generally not familiar with regulatory processes. One way in which these guidelines could be improved would be for APHIS to provide information about what types of evidence it considers necessary for each of the characteristics listed in the guidelines. Without such information it is difficult for applicants to determine the degree of rigor required by the agency in making its assessments. The committee recognizes that APHIS staff are open to personal interaction with applicants, but more detailed published guidance still would be useful. All of these changes would increase the utility of APHIS risk assessments in decision support. The increased rigor provided by these changes also could increase public confidence.

There are a number of aspects of APHIS oversight that bear directly on public confidence. **The committee finds that the extent of confidential business information (CBI) in registrant documents sent to APHIS hampers external review and transparency of the decision-making process.** Indeed, the committee often found it difficult to gather the information needed to write this report due to inaccessible CBI. It is not clear that APHIS has the power to decrease the unwarranted use of CBI. However, regulatory agencies of other countries receive documents with less CBI than does APHIS. A previous NRC report (NRC 2000c) raised similar concerns about CBI.

In the committee's review of public participation in the review process it was apparent that the number of comments on *Federal Register* notices has declined almost to zero. Committee discussions with representatives of public interest groups indicate that this decline in responses to APHIS-BBEP *Federal Register* notices is at least in part due to a perception that APHIS is only superficially responsive to comments. **The committee finds that there is a need for APHIS to actively involve more groups of interested and affected parties in the risk analysis process while maintaining a scientific basis for decisions.** As indicated above, there is a tension between use of the risk analysis process for decision support and maintenance of authority. APHIS could benefit from more attention to maintaining a balance between these two roles of its risk analyses.

In examining its day-to-day operations the committee finds that APHIS-BBEP is understaffed and questions the match between the scientific areas of staff training and their responsibilities. **The committee specifically noted understaffing in the area of ecology. The committee recommends that APHIS improve the balance between the scientific areas of staff training and job responsibilities of the unit by increasing staff and making appropriate hires.** In making this recommendation the committee is aware that APHIS needs help in making its hiring practices and salary ranges more flexible. Because of the large number of applications for field testing, more resources are needed in order to maintain a suitable number of well-trained APHIS officers for field inspection.

As pointed out above, **the committee commends APHIS for maintaining an environment in which the decision-making process can be adjusted based on knowledge gained from past risk assessments and regulatory decisions. The committee thinks APHIS would profit from formalizing its learning process.** A fault-tree analysis is one approach to such a formalized learning process.

POSTCOMMERCIALIZATION TESTING AND MONITORING

Environmental testing of transgenic plants prior to commercialization can be effective in screening plants for many types of risks, but **the committee finds there are several compelling arguments for validation-testing and ecological monitoring after commercialization of these plants**.

Because APHIS has considered deregulation absolute, it does not currently conduct postcommercialization monitoring unless commercialization has been conducted through the process of notification (or permitting) and the extent of monitoring for notifications is dependent on APHIS staffing levels and priorities. The committee finds that APHIS assessments of petitions for deregulation are largely based on environmental effects considered at small spatial scales. Potential effects from "scale-up" associated with commercialization are rarely considered. A good example of this is the use of toxicology-type testing of transgenic plants with pesticidal traits. These toxicology tests on a limited set of organisms are certainly helpful in preliminary assessment of hazards from effects on non-target organisms, but these acute toxicity tests might miss potential chronic effects that occur within field environments or effects that are not possible with the model organisms tested. **The committee recommends that postcommercialization validation testing be used to assess the adequacy of precommercialization environmental testing.** This validation testing should involve testing specific hypotheses related to the accuracy and adequacy of precommercialization testing. This testing should be conducted at spatial scales appropriate for evaluating environmental changes in both agricultural and adjacent, unmanaged ecosystems. This validation testing could be funded and managed through programs such as the USDA Biotechnology Risk Assessment and Risk Management Program, which primarily sponsor university researchers. However, to meet the needs of products being commercialized, these programs will have to be expanded substantially.

The committee recommends that two different types of general ecological monitoring be used to assess unanticipated or long-term, incremental environmental impacts of transgenic crops. One type of monitoring involves use of a network of trained observers to detect unusual changes in the biotic and abiotic components of agricultural and nonagricultural ecosystems. The second involves establishment of a long-term monitoring program that examines the planting patterns of transgenic plants, and uses a subset of species and abiotic parameters as indicators of long-term shifts in an ecosystem.

While validation testing can be put in place using the existing public research infrastructure, it will be much more difficult to institute an effective system of trained observers, and to develop a long-term program for monitoring indicators of ecosystem change. The committee finds that there already are groups of individuals available who could be trained as observers, but it would be inefficient to have these observers assess only the effects of transgenic plants.

The committee recognizes that the ability to monitor impacts of large-scale planting of crops with new traits is hampered by the lack of baseline data and comparative data on environmental impacts of previous agricultural practices. **The committee finds that the United States does not**

have in place a system for environmental monitoring of agricultural and natural ecosystems that would allow for adequate assessment of the status and trends of the nation's biological resources. This problem encompasses much more than concerns about transgenic crops. The committee endorses the development of ecological indicators, as proposed by the NRC (2000a), for both agricultural and nonagricultural environments. Without systematic monitoring data, it will not be possible to separate coincidental anecdote from real ecological trends. The same NRC report recommended development of indicators based on the assumption that monitoring them is more cost effective and accurate than monitoring many individual processes or species. This approach also simplifies communication of findings. Transgenic plants should be one component of such a monitoring system. One essential monitoring requirement will be the spatial distribution of transgenic crops.

The committee recommends that a body independent of APHIS be charged with the development of an indicator-monitoring program. This monitoring program/database should allow participation by agencies, independent scientists, industry, and public-interest groups. The database depository should be available to researchers and the interested public. A scientifically rigorous design modeled after the National Resources Inventory, and with cost sharing among agencies should reduce the burden of costs for such a program. More research is needed to identify organisms and biological processes that are especially sensitive to stresses and perturbations. Finally, there should be an open and deliberative process involving stakeholders for establishing criteria for this environmental monitoring program.

The committee recommends that a process be developed that allows clear regulatory responses to findings from environmental monitoring. Although monitoring may detect some unexpected effects in a time period that allows action to be taken to prevent or ameliorate those effects, in other cases, monitoring may detect such effects so late that environmental damage may be irreversible (e.g., extinction). **Therefore, the committee finds that monitoring cannot substitute for precommercialization regulatory evaluation.**

LOOKING TOWARD THE FUTURE

The significance of biotechnology to environmental risk resides primarily in the fact that a much broader array of phenotypic traits can now potentially be incorporated into plants than was possible two decades ago. The array of traits is likely to increase dramatically in the next two decades. As such, our experience with the few herbicide-tolerant and insect- and disease-resistant varieties that have been commercialized to date provides a very limited basis for predicting questions needed to be asked when plants with very different phenotypic traits are assessed for environmental risks. For example, the production of non-edible and potentially harmful compounds in crops such as cereals and legumes that have traditionally been used for food creates serious regulatory issues. **With few exceptions, the environmental risks that will accompany future novel plants cannot be predicted. Therefore, they should be evaluated on a case-by-case basis.**

In the future many crops can be expected to include multiple transgenes. The current APHIS approach for deregulation does not assess the environmental effects of stacking multiple genes into single-crop varieties. There are at least two levels at which scientists and regulators must look for interactions between such inserted genes with regard to environmental effects. The first level is interactions of genes and gene products that affect the individual plant phenotype. The second is the whole-field or farming systems level. One of the case studies examined by the committee finds early indications that gene stacking can have environmental effects at the farming systems level.

The types of new transgenic crops that are developed, as well as the rate at which they appear, will be affected by the interaction of complex factors including public funding and private financial support for research, the regulatory environment, public acceptance of the foods and other products produced from them, and the resolution of debates over need- versus profit-driven rationales for the development of transgenic crops.

For future novel products of biotechnology, adequate risk analysis for decision support and maintenance of authority will depend on a regulatory culture that reinforces the seriousness with which environmental risks are addressed. Public confidence in biotechnology will require that socioeconomic impacts are evaluated along with environmental risks and that people representing diverse values have an opportunity to participate in judgments about the impact of the technology.

Currently, APHIS environmental assessments focus on the simplest ecological scales, even though the history of environmental impacts associated with conventional breeding points to the importance of large-scale effects. **The committee recommends that in the future APHIS should include any potential impacts of transgenic plants on regional farming practices or systems in its deregulation assessments.**

APHIS has been constrained in its risk analysis and decision-making process by the statutes through which it may regulate transgenic plants. In May 2000, the U.S. Senate and House of Representatives agreed in a conference report on a new Plant Protection Act (PPA). The new regulations that will be used to enforce the PPA have not yet been developed.

The committee recommends that the regulations to enforce the PPA be developed in a manner that will increase the flexibility, transparency, and rigor of APHIS's environmental assessment process.

APHIS jurisdiction has been restricted to the U.S. borders. However, in an era of globalization, environmental effects of transgenic crops on the ecosystems of developing countries will be an important component of risk analysis. As exemplified by the effects of Green Revolution varieties of wheat and rice, novel crop genes often have indirect effects on the environment. These indirect effects can occur because the new crop traits enable changes in other agricultural practices and technologies that impact the environment. They also can indirectly affect vertical integration of agriculture and equality of access to food. Society cannot ignore the fact that people who lack food security often cause major effects on both agricultural and nonagricultural environments, so in a broad context the positive or negative effects of transgenes on human well-being can be seen as an environmental effect.

Environmental concerns raised by some of the first transgenic crops (e.g., gene flow, disruption of the genome, non-target effects) could be ameliorated by expanding our knowledge base in specific areas of molecular biology, ecology, and socioeconomics. Furthermore, such an expanded knowledge base could lead to the production of transgenic plants that would improve the environment. **To increase knowledge in relevant areas the committee recommends substantial increases in public-sector investment in the following research areas: (1) improvement in precommercialization testing methods; (2) improvement in transgenic methods that will minimize risks; (3) research to identify transgenic plant traits that would provide environmental benefits; (4) research to develop transgenic plants with such traits; (5) research to improve the environmental risk characterization processes; and (6) research on the social, economic, and value-based issues affecting environmental impacts of transgenic crops.**

1

Ecological, Genetic, and Social Factors Affecting Environmental Assessment of Transgenic Plants

DEVELOPING A TWENTY-FIRST-CENTURY VIEW OF AGRICULTURE AND THE ENVIRONMENT

At the beginning of the twentieth century U.S. farmers were in general free to use whatever agricultural practices best suited their needs. Although the agricultural extension system (Smith-Lever Act of 1914) was in place at the time, its role was to provide advice, not to enforce regulations. As the twentieth century progressed, the impacts of agricultural practices on human health and the environment became a focus of public attention (e.g., Carson 1962). Regulations and incentive programs were developed for agriculture and now have a major influence on farming, ranging from the choice of tillage practices to the choice of pest control techniques.

Another obvious change in agriculture during the twentieth century was in productivity. Some U.S. citizens see the last 50 years of the twentieth century as a time when hundreds of years of insecurity over food availability came to an end. In their eyes, innovative technologies such as plant breeding, water management, fertilizers, and synthetic pesticides played a heroic role in this drama. Others look back on the same events and see an era when for the first time in history human activity threatened the basic stability of global ecosystems on which all life, including human society, depends. In their eyes, modern agricultural science and technology are inimical to the natural environment.

There also are contrasting perceptions of the agricultural environment itself. In the United States, environmental protection became understood by most citizens to have two priorities: (1) protecting human populations from the toxic effects of pollution and (2) protecting natural areas from human impacts. Both priorities reinforce the view that farms are separate from the natural environment, which must be protected from agriculture. In Europe and some other parts of the world, farms are often seen as an integral part of the natural environment that should be protected (Frewer et al. 1997, Durant et al. 1998). Environmental protection can therefore encompass the protection of a partially domesticated countryside where appropriate farming techniques can foster biodiversity. There certainly are exceptions to this generalization because many U.S. citizens are concerned about the effects of agricultural practices on the flora and fauna of farms, and there are agricultural production regions in Europe that are as thoroughly industrialized as any in the United States (Durant and Gaskell, in press). It is nevertheless important to bear in mind that contrasting cultural values influence the way that different people understand the relationship between agriculture and environment, and this in turn influences their judgment of what constitutes a threat to the environment (Knowles, in press).

The molecular techniques for producing transgenic agricultural crops that came to fruition in the late 1980s arrived on a scene in which advocates for agricultural technology were already prepared to embrace them. Biotechnology was greeted by these advocates as a means of increasing agricultural efficiency, decreasing world hunger, and ameliorating environmental damage caused by previous agricultural technologies. But at the same time, critics of agricultural technology were prepared to view these new techniques with skepticism. Their skepticism was based both on their negative assessment of postwar technologies that contributed to the boom in agricultural productivity and their judgment that the science which could identify the environmental risks of these new technologies was not being adequately supported.

As farming enters the twenty-first century it faces a world in which there is increasing public pressure on governments to more actively protect the environment and conserve biological diversity. Currently, much public concern regarding agriculture and the environment is focused on the potential impacts of transgenic plants. However, it is clear that many of the environmental effects that could result from this specific technology could also occur due to changes in other agricultural technologies and farming practices (see "Environmental Effects of Agroecosystems" below). Concern over the impact of transgenic plants on the environment has led world governments to reassess the standards by which they judge what constitutes a significant negative effect of agriculture on the environment. For example, it is clear that environmental standards being developed for transgenic plant cultivars consider impacts that were rarely even measured when novel conventional crop cultivars were introduced in the 1960s and 1970s (see Chapter 2). Government agencies charged with the regulation of transgenic plants find themselves in the difficult position of enforcing a much higher environmental standard for these plants than the standards currently used to regulate the impacts of other agricultural technologies and practices. If, as we move further into the twenty-first century, stresses on the environment and public concern about those stresses increase, it is likely that new standards developed for transgenic plants will be applied in some fashion to other agricultural technologies and practices. In that sense, decisions now being made with regard to transgenic plants could set a precedent for evaluating all of agriculture. Government agencies and the public must, therefore, keep an eye to the future when working to develop new environmental standards for transgenic plants.

ROLE OF THIS REPORT

Potential controversy over agricultural biotechnology was anticipated by the U.S. government in the early 1980s. Before any agricultural products of genetic engineering had been developed, the federal government began taking steps to develop a regulatory structure that would assure the safety of potential products. In the mid-1980s the U.S. Coordinated Framework for the Regulation of Biotechnology was developed. This framework (OSTP 1986) calls for the Environmental Protection Agency (EPA), U.S. Department of Agriculture (USDA), and Food and Drug Administration (FDA) to work together in assessing the safety of the process and products of genetic engineering. In its current form, the coordinated framework gives the USDA the lead role in assessing the potential effects of nonpesticidal transgenic plants on other plants and animals in both agricultural and nonagricultural environments. The EPA takes the lead role in assessing the health and environmental effects of plants engineered to produce pesticidal substances, and the FDA leads the review of potential health effects of nonpesticidal transgenic plants.

Over the past 15 years the USDA's Animal and Plant Health Inspection Service (APHIS) has developed a system for examining potential environmental effects of transgenic plants. However, there has been concern that an agency with a mandate to promote U.S. agriculture may not be able to objectively assess the safety of new products of agricultural biotechnology. In July 1999, then Secretary of Agriculture, Dan Glickman, publicly expressed concern over this situation, which later resulted in a request for the National Academy of Sciences (NAS) to examine the scientific basis for, and the operation of APHIS regulatory oversight, to ensure that the commercialization of engineered plants is appropriately regulated.

Previous NAS committees have examined a number of issues related to the safety of genetically engineered organisms (NRC 1984, 1987, 2000c), but none specifically examined how the commercial use of all genetically engineered crops could affect agricultural and nonagricultural environments. The task set before this committee specifically included provision of guidance for assessment of the cumulative effects of commercialization of engineered crops on agricultural and nonagricultural environments. Therefore, in this report the committee examines potential effects on the environment that could result from the use of engineered crops on large spatial scales and over many years. In addition to evaluating the potential environmental impacts of single engineered traits in existing agricultural systems, the committee also examines how commercialization of engineered crops with single and multiple traits could actually change farming and thereby impact agricultural and nonagricultural landscapes of the United States. In this report the committee uses the current ecological and risk assessment literature in developing what it finds to be an appropriate framework for assessing the environmental effects of genetically engineered products, and then uses this framework to evaluate APHIS's regulatory process.

The remainder of this introductory chapter presents background information on a set of topics that must be understood before a realistic environmental risk assessment framework can be developed. First, historic evidence is examined of how changes in agricultural technologies and practices have affected surrounding habitats so that readers can gain a sense of the extent of possible interactions. There has been much debate about the potential for genetic modification of crops to cause environmental impacts of a magnitude similar to that caused by the introduction of completely new species. Therefore, the next section of this chapter presents information on the history of environmental effects of conventional crop breeding compared to that of introduced species and also examines the hypothesis that the degree of environmental risk is related to the number of genetic changes introduced into an ecosystem. Next, an in-depth assessment is presented of predictable and unpredictable aspects of both conventional and transgenic processes used to add novel genes to plants. An understanding of the differences and similarities between these methods should give readers a basis for judging how different the side effects of the genetic engineering process can be compared to the side-effects of traditional processes of crop improvement with which we now live. The chapter ends with a brief description of the U.S. Coordinated Framework for the Regulation of Biotechnology and the USDA's role and authority in regulating transgenic crops.

This introductory chapter leads to six detailed chapters. Chapter 2 uses the general principles of ecology and risk analysis to develop a framework that addresses many of the concerns of the public and the scientific community regarding biological risks associated with commercializing genetically engineered plants. Chapters 3 and 4 provide in-depth reviews of how the USDA-APHIS regulates genetically engineered plants, and Chapter 5 assesses how well the USDA-APHIS approach functions and how well it matches the framework described in Chapter 2. Chapter 5 ends with recommendations for specific improvements in the APHIS precommercialization process. One general conclusion from Chapters 1 to 5, and from other published studies is that it will be impossible to assess some types of environmental effects of genetically modified plants based on the small-scale field testing that can be conducted prior to commercialization. Therefore, Chapter 6 examines the prospects and problems associated with developing post-commercialization monitoring programs to detect and measure such effects. The potential environmental impacts of future products of genetic engineering may differ from those of the engineered crops that have recently been commercialized. Therefore, the final chapter of this report (Chapter 7) is devoted to examining potential future products of genetic engineering and to how sets of novel traits in crops could alter the use of land, chemicals, and other resources. This concluding chapter also discusses how the public's view of agriculture and its regulation may evolve in the future.

ENVIRONMENTAL EFFECTS OF AGROECOSYSTEMS ON SURROUNDING ECOSYSTEMS

In recent decades agriculture has been intensified by increases in the use of mechanization, irrigation, high-yielding crops, synthetic fertilizers, and pesticides. This has led to major changes in the structure, function, management, and purposes of agroecosystems (Swift and Anderson 1993, Swift et al. 1995, Matson et al. 1997). Typical structural changes include large reductions in plant, animal, and microbial biodiversity (Swift and Anderson 1993, Lacher et al. 1999) and simplified patterns of abiotic resource stocks and flows (Swift et al. 1995, Matson et al. 1997, Vitousek et al. 1997). Intensified agroecosystems are now predominant in developed countries and in highly capitalized export and commodity production of developing countries. In these latter countries there has also been expansion of less capitalized, smallholder agriculture onto marginal and ecologically sensitive lands, again typically accompanied by reductions in biodiversity and simplification of resource stocks and flows (Holloway 1991, Swift et al. 1995, Perfecto et al. 1996).

These changes have increased the environmental effects of agroecosystems on neighboring ecosystems, relative to those exerted by pre-intensified agroecosystems. These environmental effects are primarily driven by outflows of energy, organisms, and materials from agroecosystems. Materials include resources such as nitrogen (Vitousek et al. 1997) and other substances such as pesticides (Carroll 1990, Soule et al. 1990).

Moreover, changes in agricultural land-use patterns appear to have increased the effects of these outflows on neighboring ecosystems. The most important landscape effect of agricultural development is great reductions in the area of nonagricultural ecosystems (Pimentel et al. 1992, Dale et al. 2000). Today, approximately 50 percent of U.S. land is used for crop and animal production. In agriculture-dominated landscapes, nonagricultural ecosystems have come to function as more or less isolated islands in an ocean of agricultural lands. In Western Europe and North America this habitat loss (and to a lesser degree, habitat fragmentation) has caused reductions in the diversity of birds, small mammals, and insects (Fahrig 1997, Trzcinski et al. 1999). In addition, the habitats that serve as a buffer or interface between agroecosystems and other ecosystems have commonly been degraded or eliminated (Carroll 1990, Risser 1995), as is the case when field borders are extended to the edge of streams or other wetlands. This landscape fragmentation and the degradation of interfaces facilitate the flow and penetration of energy, materials, and organisms into ecosystems adjacent to agroecosystems (Risser 1995).

The specific effects of agroecosystems on surrounding ecosystems have often been considered in isolation from each other (e.g., studies on changes in water quality or wildlife abundance). Now, a growing body of evidence suggests a strong need for a more integrative view of these effects (Carroll 1990, Carpenter and Cottingham 1997, Lacher et al. 1999). Specifically, there is growing evidence (Vitousek et al. 1997, Tilman 1999, Mack et al. 2000) that ecological effects of agroecosystems compound to exert simplifying and destabilizing effects on neighboring ecosystems. These effects are potentially of concern because they appear to weaken or destroy the neighboring ecosystem's capacity for *resilience*—that is, an ecosystem's ability to return to its original ecological structure and function despite disturbance (Carpenter and Cottingham 1997, Ludwig et al. 1997, Tilman 1999, McCann 2000).

For example, agricultural land-use changes and phosphorus enrichment combine to destabilize temperate lake ecosystems by degrading a number of negative feedback mechanisms that collectively provide ecosystem resilience to lakes (Carpenter and Cottingham 1997). In healthy lakes, floating microscopic plants called phytoplankton are generally held in check by limited nutrients and the microscopic predators (zooplankton) that feed on them. Increased inflows of phosphorus can result from greater agricultural use, increased soil erosion, and loss of riparian (i.e., water-edge) vegetation and wetland interfaces adjoining agroecosystems or lakes. The loss of riparian vegetation and wetlands also decreases the inflow of humic substances that typically limit phytoplankton growth by effects on light and temperature in the lakes. Populations of large predatory fishes are often reduced by the biotic and physical changes in lakes, or by overfishing. Reductions in these predators allow zooplankton-feeding fishes to become more abundant. These changes cause reductions in the zooplankton and consequent increases in phytoplankton species, which outcompete larger aquatic plants. The lake food web is further simplified by reductions in crucial habitat provided by large aquatic plants and trees that fall into the lake from riparian areas. The losses of humic substances, adjoining wetlands and riparian forests, large predatory fishes, zooplankton, and large aquatic plants combine to remove a complex of negative feedback factors that act to buffer the lake's response to disturbances and environmental change. As a result, the lake's conditions and attributes become more variable and less desirable to people, as dense algal blooms with associated anoxia and algal toxins become more frequent. The lake also becomes more susceptible to additional disturbances, such as those caused by toxic pollutants or species invasions (Carpenter and Cottingham 1997).

As this example suggests, loss of resilience capacity can place ecosystems in a "poised" or "at-risk" position, increasing the likelihood of further change in ecosystem structure and function (Perry 1995, De Leo and Levin 1997, Ludwig et al. 1997, Gunderson 2000). In addition to affecting the aquatic systems discussed above (Nepsted et al. 1999, Gunderson 2000), intensified agriculture has been shown to reduce resilience of neighboring wetlands, rangelands, and even sections of the Amazonian rainforest (Nepsted et al. 1999, Gunderson 2000).

The committee cannot presently judge whether extensive commercialization of transgenics—and other crops bearing novel traits—will significantly perturb agroecosystems or neighboring ecosystems because of major gaps in our knowledge of these systems. Below, the committee reviews some of the evidence of current agroecosystem effects on neighboring ecosystems as a means of conveying the extent and complexity of potential effects.

Flows of Materials and Organisms from Agroecosystems

Generally, simplified agroecosystems generate greater outflows of materials and organisms than less intensified systems (Swift et al. 1995). These agroecosystems feature both ecosystem processes (e.g., nutrient dynamics, water use) and organism population densities that vary widely in time because they are driven by temporal synchronization of activities of a few dominant plant, animal, and microbial species (Carroll 1990,

Haycock et al. 1993). Peak outflows of materials and organisms are associated with extremes of this pattern of wide variation in activities and abundances. For example, high-yield monocultural cropping systems receive large inputs of mobile nutrients to meet a single period of intense nutrient demand by the developing crop (Robertson 1997). At other times, crop nutrient uptake is modest, and uptake by other plant species is limited by stringent weed management and the absence of plant cover during fallow periods (Swift and Anderson 1993, Robertson 1997). Typically, a significant fraction of total nutrient inputs is not used by the crop, and this residual nutrient pool becomes available for outflow to surrounding systems (Robertson 1997). Outflow commonly occurs by wind or water vectors, and resources also can be carried by organisms, sometimes resulting in highly focused or long-distance transport of materials. For example, in New Mexico, geese feeding on agricultural fields vector large quantities of nitrogen into their roosting areas, which are managed wetlands (Jefferies 2000).

Effects of Outflows of Materials and Organisms on Neighboring Ecosystems

When agroecosystems discharge large flows of resources into surrounding systems, strong ecosystem effects can result. For example, outflows of nitrogen can cause major changes in plant communities, as species that were adapted to low nitrogen are replaced by a less diverse group of species (often mainly composed of exotic invasive species) adapted to higher nitrogen levels (Jefferies and Maron 1997, Vitousek et al. 1997). Plant community changes and resultant changes in landscape-level biodiversity, caused in large part by nitrogen deposition from intensified agriculture, have been particularly dramatic in the Netherlands (Aerts and Berendse 1988), where nitrogen deposition rates are among the highest observed rates globally (Vitousek et al. 1997). High levels of nitrogen deposition have been associated with serious degradation in forest ecosystems, including high rates of tree mortality and problematic soil acidification (Draaijers et al. 1989). Such nitrogen-related reductions in plant diversity appear to reduce ecosystem resilience. In experiments on temperate North American grasslands, nitrogen enrichment caused species loss, and grassland plant communities that had lost species were much less resilient to rainfall and climate variation than more diverse communities that were not nitrogen-enriched (Tilman and Downing 1994, Tilman 1996). Nitrogen deposition also reduces soil quality and fertility in surrounding ecosystems and causes freshwater acidification and coastal zone eutrophication (Jefferies and Maron 1997, Vitousek et al. 1997).

Resource outflows from agroecosystems can also affect population and community dynamics in recipient ecosystems (Polis et al. 1997). One notable response to resource input is a so-called *trophic cascade*, which results in severe reductions in numbers or biomass of organisms that are positioned in a food web below a consumer of some resource that is flowing from an agroecosystem. For example, in northeast Europe, cormorants have increased greatly in both average abundance and variability in abundance due to increased fish populations that have resulted from nutrient flows from agroecosystems into surface waters. After reaching high levels of abundance, these cormorant populations have greatly reduced populations of plankton-consuming fish in other lakes (Jefferies 2000). Also, a "paradox of enrichment" (Polis et al. 1997, Jefferies 2000) has been recognized, in which large inputs of formerly scarce resources cause local extinction of organisms that formerly coexisted in a food web. Again, the net effect of these changes often appears to be in the direction of simplification and reduction in resilience in ecosystems experiencing inflows of resources originating from agroecosystems (McCann 2000).

Nutrient overenrichment from human activities is one of the major stresses affecting coastal ecosystems. There is increasing concern that in many areas of the world an oversupply of nutrients from multiple sources is having pervasive ecological effects on shallow coastal and estuarine areas. These effects include reduced light penetration, loss of aquatic habitat, harmful algal blooms, a decrease in dissolved oxygen (or hypoxia), and impacts on living resources. The largest zone of oxygen-depleted coastal waters in the United States, and the entire western Atlantic Ocean, is found in the northern Gulf of Mexico on the Louisiana-Texas continental shelf. The freshwater discharge and nutrient flux of the Mississippi River system influence this zone (Rabalais et al. 1999). A series of reports that provide an integrated assessment of this hypoxia zone in the Gulf of Mexico document one of the most widespread and substantial outflows of nitrogen from agricultural sources in the United States (Diaz and Solow 1999, Goolsby et al. 1999, Rabalais et al. 1999).

On average, some 1.6 million metric tons of total nitrogen flow from the Mississippi-Atchafalaya River basin annually. Of the various nitrogen sources (atmospheric deposition, groundwater discharge, and soil erosion), agricultural sources of nitrogen such as fertilizers and animal wastes constitute the majority of the annual flux. Nitrogen input from fertilizers into the river basin has increased sevenfold since the 1950s (Goolsby et al. 1999). The impact of these increases in nitrogen outflows (similar patterns are noted with phosphorus) has been linked to the eutrophication of coastal areas and the seasonal hypoxic zone in the Gulf of Mexico (Diaz and Solow 1999). Except in areas of natural upwelling, coastal hypoxia is not a natural condition. The size of the zone fluctuates annually, but in the mid to late 1990s the area averaged 16,000 to 18,000 km^2 (Rabalais et al. 1999).

There are significant ecosystem impacts that result from hypoxia. Hypoxia in the northern Gulf of Mexico has become one of the many factors that at present control or influence the population dynamics of the regions of many pelagic and benthic species. Hypoxia exerts its control at two levels: the primary impact is the loss of bottom and near-bottom habitat through seasonal depletion of oxygen levels. From all indications, few, if any, mobile organisms stay on bottoms that are hypoxic. This forced movement could increase population losses due to predation of species dependent on the seabed for their survival. But when hypoxia dissipates, mobile organisms return. The second major impact is the alteration of energy flows. During hypoxia, significant amounts of the ecosystem's energy are shunted to microbial decomposition, which is the primary biological process that creates and maintains the hypoxia. In the northern Gulf of Mexico the major sources of this energy seem to be organic matter produced in the surface waters and benthic biomass (mostly small worms, snails, and clams; Diaz and Solow 1999). Energy flows between trophic levels (mediated largely by predator/prey interactions) are also altered, as seen in the population movements of fish (Azarovitz et al. 1979) and shrimp species (Gazey et al. 1982, Renaud 1986, Zimmerman et al. 1996).

In addition to outflows of abiotic resources, agricultural activities often support increases in populations of organisms that then enter surrounding ecosystems, where they can exert a wide variety of effects (Carroll 1990). For example, populations of white (or snow) geese in North America have increased greatly since the 1950s, apparently driven by the growth of high-input cropping systems in the United States (Lacher et al. 1999). These large geese populations have altered vegetation on a regional scale in the eastern Canadian arctic by destroying coastal vegetation near their nesting sites (Kerbes et al. 1990, Jefferies 2000). In this case, organism flows from an agroecosystem have had an effect over distances of 3,000 to 5,000 km.

Invasive organisms entering surrounding ecosystems from agroecosystems frequently have properties that increase their colonization success, creating positive feedbacks that amplify their effects on the ecosystems they enter (Mack et al. 2000). For example, certain invasive weeds enter tropical forests from pasture and field crop agroecosystems (Carroll 1990). In addition to being highly competitive with tree seedlings, these species often increase the frequency of fires in forests, entraining a cycle of increasing fires, reductions in forest biodiversity, increased abundance of the invasive species, and more fires. This pattern of invasion driven by a positive feedback cycle appears to be the basis of the success of a variety of grasses derived from agroecosystems that have become major invaders worldwide (D'Antonio and Vitousek 1992).

Landscape-Level Effects of Agriculture

Agricultural land use has always tended to fragment landscapes relative to their pre-agricultural conditions. This pattern of fragmentation has increased the importance of *edge effects* (i.e., effects of one ecosystem on a second that act directly on only the periphery of the second ecosystem).

Populations of many organisms that are resident in nonagricultural lands, including predators (Lacher et al. 1999), can be rapidly depleted by edge effects. For example, populations of migratory forest birds in the United States have been strongly affected by generalist predators (e.g., blue jays, common crows) that are active in peripheral portions of forest fragments in agricultural landscapes (Wilcove et al. 1986). Very few of these fragments are extensive enough to create refuges from predation for these migratory bird populations. These edge-related depletion effects can be particularly significant when they reduce populations of predatory organisms, (e.g., on grazing lands in the western United States; Knowlton et al. 1999). These species are often strong determinants of ecosystem structure and function; in their absence, herbivorous species can increase sharply and quickly deplete vegetation, setting off a wave of ecosystem change.

Fragmentation of nonagricultural lands also affects dispersal and movement of organisms that reside in these lands. In Northwest Europe, red squirrels occur in small populations in forest patches in agricultural landscapes. These small populations are prone to becoming extinct in individual forest patches. After extinction in a particular patch, successful recolonization of the patch by squirrels depends on the distance to other patches and the degree to which patches are interconnected by woody vegetation along field borders. For these reasons, highly isolated patches will usually not be occupied by squirrels (Verboom and van Apeldoorn 1990). Similar effects of fragmentation on the biodiversity of birds, mammals, and insects in nonagricultural lands have been observed (Forman 1997).

Finding 1.1: There is substantial evidence that ecological effects of agroecosystems compound to exert simplifying and destabilizing effects on neighboring ecosystems. These effects are potentially of concern because they appear to weaken or destroy the neighboring ecosystems' capacity for resilience—that is, an ecosystem's ability to maintain a certain state despite disturbance.

Finding 1.2: Potential ecological effects of transgenic crops—and other crops bearing novel traits—may be substantially heightened in the simplified and destabilized ecological milieu that has resulted from agricultural intensification. This prospect is a strong argument for a cautious approach to the release of any crop that bears a novel trait into agroecosystems and the less managed surrounding ecosystems. Equally, it is an argument for a cautious approach to any extensive change in agricultural practices.

Recommendation 1.1: There is a need for a cautious approach when making any extensive change in agricultural practices, including changes in the genetics of crops, because of potential ecological impacts on agricultural and surrounding ecosystems.

ENVIRONMENTAL IMPACTS OF THE DELIBERATE INTRODUCTION OF BIOLOGICAL NOVELTY: FROM GENES TO MINICOMMUNITIES

The practice of modern agriculture introduces biological novelty into the environment for a number of purposes. Conventional plant breeders introduce novel genes into crops to improve yield, taste, agronomic traits, and other characteristics. Pest management specialists often import natural enemies of pest species into the United States from other continents. And horticulturalists are continually seeking new varieties and species of plants that could enhance the aesthetic beauty of urban and rural landscapes.

The biological novelty introduced to agricultural systems by these varied activities can result from very small to very large changes in genetic information relative to preexisting organisms. The general degree of change in genetic information can be measured along two axes: the number of genetic changes and the taxonomic or phylogenetic distance between the source and the recipient of the new genetic information.

Biological novelty within a species' genome can result from a change in a single DNA base pair to more complex changes affecting the constitution of one or more chromosomes. The introduction of new individuals to an ecosystem is another source of biological novelty. Adding an individual of a preexisting species to a given environment introduces biological novelty if that immigrant is genetically different from those already established in that population. The genetic difference can be as small as one novel allele (i.e., one form of a gene) or as large as an individual that represents a different race or subspecies. The introduction of an individual of a new species with preexisting close relatives represents a greater addition in genetic information relative to the prior example. That species may hybridize with those close relatives, and the resulting hybrids represent another type of biological novelty (Abbott 1992). The most substantial addition of genetic information at the individual level would be the introduction or immigration of a species with no close relatives. Of course, an introduced species may arrive bearing other novel organisms in the form of mutualists, parasites, and other symbionts, effectively representing the arrival of a minicommunity.

Within the continuum from minor to major alterations in the genetic information in agroecosystems, the practice of conventional crop breeding has typically been considered to result in minor genetic changes and is not subject to formal regulatory review. On the other hand, the introduction of new species used for biological control, or as new horticultural plants, is considered to involve greater changes in genetic information, and these introductions have been subject to review. Some participants in the debate over genetically engineered crops compare them to the products of conventional breeding because of the small number of genetic changes involved and have concluded that the need for oversight is minimal. Others compare genetically engineered crops to the introduction of new species because the new genetic information often originates from taxonomically distant species. They call for stringent regulation. Both of these perspectives make the assumption that there is a relationship between the nature of biological novelty (number of genetic changes in the first example and phylogenetic distance in the second) and the risk of negative environmental impacts.

The remainder of this section examines the literature on the history of environmental impacts of conventional crop breeding and of species introductions to determine if the extent of genetic change or other general factors can be used as predictors of risk. While the available studies do not permit a precise quantitative comparison, they are informative and suggest that (1) changes at any level of genetic information can have profound environmental consequences, (2) the consequences of biotic novelty depend strongly on the specific environment into which they are released, (3) the significance of the consequences of biotic novelty depend on societal values, (4) the introduction of any type of biological novelty can have unintended and unpredicted effects on the recipient community and ecosystem, (5) it is not possible to qualitatively differentiate the general environmental risk associated with the release of conventionally bred crop cultivars and the introduction of new species.

1. *Small or large genetic changes can have profound environmental consequences.* The introduction of a single gene for short plant stature in rice (see BOX 1.1) enabled major changes in agricultural practices that resulted in much higher yields and substantial stresses on agricultural and

nonagricultural habitats. The introduction of T male-sterile cytoplasm into U.S. maize crops (a very small genetic change relative to the maize genome) permitted the rapid spread of *Bipolaris maydis*, the disease organism causing Southern corn leaf blight (NRC 1972). A single allelic replacement in rose clover (*Trifolium hirtum*) appears to account for a tremendous increase in its colonizing ability (Jain and Martins 1979), and only a few allelic differences appear to be responsible for weediness in Johnsongrass (*Sorghum halepense*), one of the world's worst weeds (Paterson et al. 1995). Honeybees with European ancestors in the New World radically expanded their niche in the tropics after human-mediated, but unintended, introduction of a handful of genetic alleles from an African strain (essentially hybridization between subspecies; Camazine and Morse 1988). Natural hybridization between the introduced New World cordgrass *Spartina alterniflora* and the native *S. maritima* in Britain gave rise to *S. anglica*, an invasive species that rapidly altered the coastal ecosystems of the British Isles (Gray et al. 1991, Thompson 1991). Finally, some of North America's worst exotic invasive plants have close indigenous relatives (purple loosestrife, *Lythrum salicaria*) and others do not (garlic mustard, *Alliaria petiolata*; kudzu, *Pueraria montana*). Clearly, there is potential for small and large changes in genetic information to result in environmental impacts.

In addition to attempts at explaining risk based on the number of introduced genes, studies have examined whether there is a relationship between the taxonomic or phylogenetic distance between the gene donor and recipient and the level of environmental risk. If level of risk increases with the phylogenetic distance between organisms exchanging genes, we might expect to find that documented cases of natural interkingdom horizontal gene transfer (the nonsexual transfer of the genetic material from one organism into the genome of another) correlating with a dramtic ecological change in the recipient organism. That does not appear to be the case (Palmer et al. 2000).

2. *The consequences of biotic novelty depend strongly on the specific environment, including the genomic environment, physical environment, and biotic environment.* "Experience with exotics shows overwhelmingly that an organism's effect and ecological role can change in new environments" (U.S. Congress, Office of Technology Assessment 1993). And these effects and roles of a newly introduced genotype can change with time. Differences in physical environment can make a big difference. *Abutilon theophrasti*, native to Asia, is an important weed in much of the temperate world. In California it is only a weed where available moisture from rain or irrigation will allow its growth (Holt and Boose 2000).

Responses to the physical environment may vary with relatively small genetic differences. In Florida the South American tree *Schinus terebinthi-*

folius is a noxious invader (U.S. Congress, Office of Technology Assessment 1993). But the closely related *Schinus molle* is much more invasive in California's natural ecosystems than *S. terebinthifolius* (Hickman 1993). The differences in invasiveness between the two species in the two locations appear to depend on differences in their germination and growth responses in the different physical environments (Nilsen and Muller 1980a, 1980b). While more subtle, the biotic environment may play the largest role in determining the impact of biological novelty. An obvious example is that many crops initially benefit by being grown in an appropriate abiotic environment that is far from their place of origin. But that benefit eventually erodes as local pests adapt to the new species or as the original pests arrive via unintended human-mediated dispersal (e.g., Strong 1974, Strong et al. 1977, Andow and Imura 1994).

Interactions among genes can be viewed as an internal environmental component. For example, no matter how many alleles a mammal possesses for increased melanin production, the predicted phenotype based on those loci will not be expressed if that individual is homozygous at another locus for alleles that cause albinism (Mange and Mange 1990). If a gene for pest resistance is added to a plant that lacks genes for reproduction without human intervention, it is unlikely to become weedy.

3. *The significance of the consequences of biotic novelty depend on societal values, whether that novelty represents new genotypes or new species.* Ornamental mutants of native American wildflowers are the darlings of those who buy them to grow in their backyards and the bane of restorationists who find them in their seed mix (Montalvo et al., in press). While Africanized honeybees create concern in the New World when they kill humans and livestock, prudent management of the bees has resulted in better honey yields in Brazil compared to the European genotypes of the same species (e.g., Camazine and Morse 1988). Monterey pine is an endangered species in its native range in California, but in Australia it is both valued as an important commercial forest tree and despised as an invader of native eucalypt forests (Burdon and Chilvers 1994). In many cases the majority of the public can agree qualitatively if not quantitatively on the desirability of a specific plant in a specific place, but it is the variation among the public in such value judgments that has made much of environmental regulation so contentious.

4. *Introduction of biological novelty can have unintended and unpredicted effects on the recipient community and ecosystem.* In the past few decades the planting of new varieties of maize in Mexico that require the tools of modern industrial agriculture has rendered such fields inhospitable to several native taxa, most notably teosinte, the progenitor of maize (Sánchez and Ruiz Corral 1997, Wilkes 1997). This outcome would be considered desirable relative to the goal of maximizing yields. Also, natu-

ral hybridization with cultivated rice has been implicated in the near extinction of the endemic Taiwanese wild rice, *O. rufipogon* ssp. *formosana* (Kiang et al. 1979). During the twentieth century, the cultivation of domesticated rice steadily increased on the island of Taiwan. During the same time, collections of *O. rufipogon* ssp. *formosana* showed a progressive shift toward characters of the cultivated species and a coincidental decrease in fertility of seed and pollen. By 1979 naturally occurring populations of this subspecies were on the verge of extinction. Similar substantial effects can occur for successfully established exotic species. The spectacular roosts of overwintering and migrating monarch butterflies along California's coast are on eucalyptus, introduced from Australia in the last 200 years (prior butterfly roosting sites in California are unknown; Malcolm and Zalucki 1993). Another example is the introduction and establishment of European grassland species to California, which have altered natural fire regimes of its inland valleys, resulting in the conversion of native shrublands to European grasslands (Minnich 1998).

5. *There is no strict dichotomy between environmental risk associated with releases of conventionally bred crop cultivars and introduction of new species.* The majority of introduced cultivars and species do not result in long-term environmental establishment. Most registered crop varieties fail to become popular with farmers, and thus do not persist in agroecosystems. The half-life of accepted varieties is on the order of about five years. In the same way, on the order of 10% of intentionally introduced species persist after introduction (Williamson 1993). From this fraction of species that persist, very roughly 10% become an obvious problem in agricultural or non-agricultural environments (Williamson 1993). In sum then, only approximately 1% of species introductions are problematic.

The small fraction of exotics that do cause environmental effects can be tremendously disruptive and have received considerable recent attention (e.g., Pimentel et al. 2000). It is not clear what fraction of introduced crop and horticultural species that naturalize also become pests or otherwise create hardship. Although the fraction is low, it is not zero. Many plants intentionally introduced for agricultural purposes have become noxious weeds (e.g., see review by Williams 1980). A good example is bermudagrass, *Cynodon dactylon*, which is frequently introduced as a forage or as turfgrass but has also become one of the world's worst weeds (Holm et al. 1977). Additionally, the natural hybridization of crops with wild relatives has led to the evolution of both new agricultural weeds and noxious invaders of natural ecosystems (see examples in Ellstrand et al. 1999, Ellstrand and Schierenbeck 2000). Hybridization with crops has also increased the risk of extinction of some native plants (as in the case of *O. rufipogon* ssp. *formosana* discussed above; see also examples in Ellstrand et al. 1999).

But are there any cases where the introduction of a new agricultural *genotype* to a previously extant species has resulted in economic or environmental change? The most dramatic recorded example is the development of semidwarf varieties of wheat and rice (see BOX 1.1). Other examples also illustrate that the introduction of a new crop genotype can have environmental impacts. As noted above, the introduction of modern maize genotypes in Mexico that have replaced traditional maize varieties require agronomic techniques that displace teosinte, leading to increased risk of its extinction. The problems created by new genotypes also can be agronomic, as illustrated in the case of T cytoplasm in maize mentioned above.

The introduction of cytoplasmic male sterile (CMS) genotypes of sugar beet had an indirect consequence that has been economically disruptive to Europe's sugar industry. CMS genotypes are used to create hybrid varieties in this wind-pollinated, outcrossed species (Ford-Lloyd 1995). Natural pollination from wild sea beets growing near the seed production fields in both southern France and near the Adriatic Coast in Italy contaminates the seed with hybrids (Boudry et al. 1994, Mücher et al. 2000). The hybrids are annuals that bolt, resulting in a woody root that yields very little sugar and damages both harvesting and processing machinery. Before they bolt, the weeds are essentially identical to the crop and are therefore difficult to control (Longden 1993). If they are not removed before they set seed, their seed contaminates the soil, frustrating attempts at growing sugar beet for several years. The cumulative cost of three decades of weed beets is at least on the level of hundreds of millions of dollars. Presently there are some sugar beet fields in Europe that produce more weed beets than sugar beets (Mücher et al. 2000).

In summary, the vast majority of crop genotype and species introductions do not persist in the environment and cause environmental damage. A minority causes hardship to humans, agronomic ecosystems, and/or natural systems. The examples reviewed here indicate that general information on the amount or origin of genetic material cannot, on its own, be used to predict the risk associated with a new crop genotype or introduced species. In this regard the committee's findings support those of other scientists who have examined this problem of predicting risk and concluded that risk assessment cannot depend on general characteristics such as the amount of new genetic information introduced but must focus on the ecology of the specific introduced organism (or both the donor and recipient in the case of transgenic organisms) and the characteristics of the accessible environment into which the organism will be released (e.g., NRC 1987, Tiedje et al. 1989, Scientists Working Group on Biosafety 1998). This does not mean that each introduction requires an equally intensive assessment. A number of scientific panels have developed outlines and

BOX 1.1
The Green Revolution

The goal of the Green Revolution was to increase crop yields to feed existing and growing populations in developing countries (Hanson et al. 1982, Conway 1998). Conventional plant-breeding technologies provided the crucial crop cultivars that allowed this revolution in agricultural production to begin. Prior to 1950, tropical cultivars of rice and wheat were about 1.5 to 2.0 times as tall as present-day cultivars. When high rates of fertilizer were applied to these tall cultivars, yields did not rise substantially and the plants tended to lodge (fall over). Lodging caused decreased yield, and made harvest difficult. A few genes were found in temperate zone cultivars of rice and wheat that resulted in what were called "semidwarf" plants (Hanson et al. 1982, Dalrymple 1986). When high rates of fertilizer were applied to these plant types, yield increased dramatically. When these semidwarf genes were transferred to tropical rice and wheat, they too showed very positive responses to fertilization. Thus, these semidwarf genotypes enabled a technology package to be assembled that led to revolutionary increases in yield in many developing countries.

Most semidwarf tropical rice cultivars derive their short stature from a single gene, sd-1 (Dalrymple 1986), and therefore all of these cultivars presumably carried many of the genes that were tightly linked to the sd-1 gene in the donor plant. Some of the early semidwarf cultivars were more heavily attacked by pathogens than traditional cultivars, so genes for disease resistance were transferred from traditional cultivars to what became known as high-yielding varieties or HYVs. With the financial backing of the international community and governments of some developing countries, rates of farmer adoption of these semidwarf HYVs were sometimes very rapid. In an extreme case, adoption of HYVs in the Philippines went from 0 to 50% in five years (Evenson and David 1993).

The semidwarf genes were typically transferred to tropical and subtropical cultivars of rice and wheat along with other non-linked genes that decreased time to harvest (Khush 1995). One of the impacts of these short-season HYVs was the potential, in some areas, for growing two or three crops of rice on the same land in a single year (Greenland 1997). In other areas farmers began to grow one crop of rice and one of wheat in the same year (Hanson et al. 1982). To extract high yields from these fast-growing HYVs, more fertilizer and water were needed. In many areas pest problems appeared to increase, possibly due to the continual availability of the crop in large monocultures (Greenland 1997). These pest problems were often responded to with application of pesticides (Gallagher et al. 1994, Matteson 2000).

As discussed in the text, runoff of fertilizer into wetlands can have major environmental impacts. The extent of these impacts in developing countries has not been measured carefully over time. In many tropical rice ecosystems nitrogen and phosphorus are typically not expected to cause eutrophication because of absorption of these nutrients to soil particles or loss through volatilization (Greenland 1997), but impacts on wetlands have been found (Ghosh and Bhat 1998). Additionally, a study conducted in the Philippines indicated that nitrate nitrogen had leached through the root zone toward the groundwater table. The researchers concluded that over time this leaching could contaminate groundwater (IRRI 1995). In temperate zone areas of Japan, which have the highest rates of fertilizer use in rice, impacts on wetlands have often been reported (Mishima 2001).

Although some short-season rice cultivars can be produced with less water than long-season cultivars, it is difficult to raise yields without more water (Greenland 1997), and when an additional crop is grown each year, water resources can be stretched thin. In production areas not adjacent to river deltas, additional water was generally acquired by increased extraction of subsurface water (Bandara 1977). This has resulted in lowering of the water table in many areas (Bandara 1977, Greenland 1997). For example, in the Punjab area of India the groundwater table has been receding at a rate of 20 cm per year, and it has been proposed that rice cultivation in this area be limited (IRRI 1995). In some arid areas, salinization of water in the rice root zone has become more problematic (Greenland 1997).

Increased use of pesticides has sometimes been successful in controlling insects, but often insecticide use has reduced natural enemies of pest species and triggered worse outbreaks. This resulted in a classical "pesticide treadmill" (Gallagher et al. 1994, Matteson 2000). A few studies have documented the impacts of increased pesticide use on rice agricultural ecosystems (Gallagher et al. 1994, Heong and Schoenly 1998, Matteson 2000) and the effects of pesticide runoff on nonagricultural systems (Nohara and Iwakuma 1996).

As part of the research agenda accompanying the Green Revolution, long-term yield experiments were initiated on research stations to determine if double and triple cropping of rice was sustainable. A 1979 analysis of yields from a triple-cropping experiment begun in 1963 at the International Rice Research Institute showed a pattern of decreased yield (Ponnamperuma 1979). Yield continued to decrease at an annual rate of 1.4 to 2.0% through 1991, when yields were 38 to 58% lower than in the baseline year of 1968 (Cassman et al. 1995). Changes in agricultural practices were able to reverse this yield decline, but the specific causes for yield decline and its reversal are not well understood (Dobermann et al. 2000). A more broad-based analysis of 30 long-term experiments in a diverse set of rice-growing regions found statistically significant yield declines in eight experiments (Dawe et al. 2000). Similar declines have not been seen in farmers' fields, and it is thought that the yield decline may only result when the production system is pushed to nearly 100% of the yield potential and soils rarely go through a drying period (Dawe et al. 2000). There are a number of lessons from these soil fertility studies. One is that these long-term trends could not have been revealed by precommercialization testing (see Chapter 6). A second lesson is that effects on the environment are typically complex, and it is often difficult to determine a specific cause of an observed event. A third lesson is that the monitored environments should closely reflect the diverse environmental conditions in commercial fields.

Because the Green Revolution involved many technical changes, it is sometimes difficult to establish a direct cause-and-effect relationship between the few genes that started the Green Revolution, and resulting environmental changes. However, the Green Revolution clearly demonstrates that commercialization of cultivars with relatively simple genetic changes can have major effects on farming practices that ultimately result in environmental change.

flowcharts for guiding the direction of risk assessments (e.g., Tiedje et al. 1989, Scientists Working Group on Biosafety 1998). With information on the ecology of the organism and the environment in hand, it is possible to quickly rule out some introductions as too risky and to determine that others are unlikely to pose any risks. More importantly, such information can be used to flag cases where only further testing will provide sufficiently accurate information on potential risks. Chapter 2 discusses how a regulatory system can be developed to efficiently utilize this approach. From the above assessment of impacts of adding biological novelty to ecosystems the committee has determined the following:

Finding 1.3: A comparison of environmental impacts from adding biological novelty to ecosystems indicates that:

a) small and large genetic changes have had substantial environmental consequences;

b) the consequences of biological novelty depend strongly on the specific environment, including the genomic, physical, and biological environments into which they are introduced;

c) the significance of the consequences of biological novelty depend on societal values;

d) introduction of biological novelty can have unintended and unpredicted effects on the recipient community and ecosystem;

e) a priori there is no strict dichotomy between the possibility of environmental hazard being associated with releases of cultivated plants with novel traits and the introduction of nonindigenous plant species. However, the highly domesticated characteristics of some cultivated plants decrease the potential of certain hazards.

COMPARING AND CONTRASTING CONVENTIONAL AND TRANSGENIC APPROACHES TO CROP IMPROVEMENT

In debates regarding the need for formal assessment of environmental impacts of transgenic plants, concerns about risks that result from the process of plant transformation and risks associated with the intended products of transgenically-based crop improvement are often not clearly differentiated. Crops produced by transgenic approaches could have risk potential because a novel gene in the transgenic crops makes an intended product with insecticidal properties that affect non-target organisms. In contrast, a transgenic crop could have unanticipated environmental effects if during the transgenesis process some of the normal plant genes were unintentionally disrupted, leading to a change in the products of those genes. In the following section, risks from the genetic processes involved in traditional and modern conventional crop improvement are compared and contrasted with those used in transgenic-based approaches. Other chapters of this report discuss the risks associated with intended products of conventional and transgenic methods.

Traditional and Conventional Processes of Crop Improvement

Humans have been using genetic modification to improve crop plants for thousands of years. Their first experiments consisted of identifying various ecotypes or genetic variants and comparing their ability to produce a good crop when cultivated. Selections (genotypes) that performed well were saved and propagated, while inferior ones did not survive or were simply discarded. These early farmers discovered that crop yields could be increased by modifying the environment. Consequently, they developed methods of irrigating and administering soil fertilizers. Over millennia they learned to modify an important relationship that influences the productivity of all crop species—the interaction between the genotype of the plant and the environment in which it is grown (Simmonds and Smartt 1999).

The types of traits (see BOX 1.2) these early farmers searched for were similar to what is considered important today: higher yield; uniformity in germination, growth and development, flowering time, and shattering (seed dispersal); disease and drought tolerance; resistance to insects and pathogens; and improved functional characteristics of the products used for food (i.e., sugar, starch, protein, oil content, etc.). They had to rely on the genetic diversity that existed in a given ecotype as a consequence of naturally occurring mutations. No doubt there was some degree of allelic diversity among the plants grown, but early farmers had no easy way to identify it nor to deliberately select for the responsible loci. Consequently, their ability to genetically improve crops was limited to the existing genotypic variability and their capacity to discern the consequent phenotypes and select for them. As knowledge of breeding techniques developed, it became possible to improve traits in crop varieties through genetic recombination, which can bring together favorable mutations, genes, or blocks of genes and thereby create more useful varieties (Simmonds and Smartt 1999).

As discussed earlier, the genetic basis for an altered crop trait can simply involve a single gene or can be complex, involving multiple, potentially interacting genes. Simply inherited traits are generally easy for plant breeders to manipulate because they usually show dominant or recessive phenotypes and produce simple ratios of progeny with and without the trait in segregating crosses. Simply inherited traits are often controlled by key enzymes in biochemical pathways. A good example is

BOX 1.2
Traits and Characters

In the conventional genetic sense, a **character** is a feature, such as flower color, while a **trait** is a particular form of a character, such as red in the case of a rose. For this example, the rose would have a red **phenotype** or appearance, and this color is determined by the nature of the genes or alleles (forms of a gene at a particular chromosomal locus) that determine the red trait. The set of genes determining the red color is the **genotype** of this trait. Informally, the term trait is often broadly used for many of these terms, and this leads to confusion. Genes give rise to traits, but they are not the traits themselves.

The alternate states of a character are the **traits** associated with that character. For example, an individual plant that is resistant to glyphosate (the herbicide Roundup®) has the *trait* or *phenotype* of glyphosate-resistance. This trait is conferred on an individual by the gene for 5-enolpyruvyl shikimate-3-phosphate synthase (*ESPS*), which is of bacterial origin. The *genotype* of that individual refers to genes that it has at the location in the genome where the *ESPS* gene is found. The *character*, resistance to glyphosate, is the set of two traits, glyphosate-resistance and glyphosate-susceptibility.

A **character** can be narrowly or broadly defined. For example, *insect resistance* is a plant character that includes resistance in any plant part (roots, stems, leaves, flowers, etc.) to any insect species (caterpillars, beetles, etc.). This character could also be narrowly defined as the resistance to stem-boring Lepidoptera (moths and butterflies), which restricts the range of insect species under consideration. The insect resistance character could encompass traits such as leaf and stem toughness, production of toxic chemicals and growth inhibitors, and modifications in leaf and stem morphology, etc. We also could further narrow the scope of this character to toxin-based resistance in maize, which would focus it primarily on traits involving DIMBOA (a toxin produced naturally in corn) and various transgenic traits based on the production of Cry toxins from *Bacillus thuringiensis*. We also could define a character as Cry toxin-based resistance in maize to stem borers. Alternatively we could define a character based on the expression of a protein, so the character could be Cry toxin production in corn. Even this character could be defined

starch synthesis, which is partially controlled by the enzyme ADP-glucose pyrophosphorylase. In maize this enzyme is composed of two protein subunits encoded by genes that have been given the names *Shrunken2* and *Brittle2* (Hannah 1997). If either of these genes is defective, the result is a smaller seed with reduced starch content. Traits with more complex genetic control, often called quantitative traits, are controlled by many genes that interact with the environment (genotype by environment interaction) and with each other (epistasis) in producing a plant trait (phenotype). Thus, knowledge of each gene is necessary but not sufficient to predict the final plant traits in a specific environment (Lewontin 2000).

more narrowly: transgenic maize with Cry toxins effective against stem borers includes four different proteins, Cry1Ab, Cry1Ac, Cry1F, and Cry9C (see also BOX 2.1), each of which could be defined as a separate character, because they are different proteins and they affect stem borers in different ways. Indeed, even this character can be defined more narrowly. Resistance to late generation stem borers in the various Cry1Ab transformation events is variable. Thus, the character can be defined as full-season Cry1Ab resistance to stem borers in maize, which would include the transformation events *Bt*-11 and Mon 810; another character, partial late season resistance, would include the Event 176 transgene. A similar range of definitions exists for many other genetically engineered characters, including herbicide tolerance, delayed ripening, modified oil content, and virus resistance.

A clear definition of character **facilitates scientific analysis and communication** (see Chapter 2). For a genetic engineer, a specific transgene may be of interest, so a useful definition of character could be an immediate product of the transgene, such as RNA or protein. Use of this narrow definition could facilitate detection and development of products based on the transgene. For a farmer using the transgenic crop, resistance to stem-boring Lepidoptera may be a broad and sufficient definition of a useful character. Additional detail may be irrelevant to a farmer, but knowing which pest is controlled would be quite important.

From the perspective of a risk analyst, a character should be defined in a manner that enables establishment of scientifically rigorous comparisons (that is reference scenarios see Chapter 2). For example, using the broad definition—resistance to stem-boring Lepidoptera in maize—could result in concluding that environmental risks associated with *Bt*-maize are the same as risks associated with DIMBOA (that is, a conclusion of substantial equivalence, see Chapter 4), when the comparison itself may be indefensible scientifically. Instead, defining two characters more narrowly, *Bt* toxin expression, and DIMBOA expression in maize, would lead to a scientifically more rigorous environmental assessment. Thus the definition of character in a risk assessment should include scientific confirmation that the traits included in the definition are similar enough to allow scientifically valid comparison.

Yield is an example of a complex trait. It is easy to imagine that factors such as photosynthesis, water-use efficiency, disease resistance, flower development, and others all contribute to the accumulation of biomass and its deposition in seeds.

For sexually reproducing species, simply inherited or single gene traits are usually incorporated into existing elite cultivars by the process of backcrossing. Backcrossing is conducted by crossing the elite cultivar with a donor (source of the single gene) and is followed by repeated crossing (backcrossing) to the elite cultivar with the goal of recovering all of the genome of the elite cultivar plus the gene of interest. Backcrossing is most useful when the phenotype of the trait can be unambiguously scored by visual, biochemical, or molecular methods. Complex or quantitative traits are typically integrated into cultivars by pedigree breeding. Pedigree breeding is initiated by crossing two lines or cultivars together that complement one another for the traits of interest. This initial cross is typically followed by a series of self-pollinating generations, with each generation evaluated in multiple environments for the traits of interest (Hallauer 1990).

Existence of genetic variation is critical for the genetic improvement of any crop. The ultimate source of genetic variation for both simple and complex traits is mutations. Mutations result from changes in DNA sequence, resulting from errors in replication and recombination, insertion of retrotransposons and unstable genetic elements (transposons)*, gene silencing, and environmental mutagens such as radiation. In recent years the molecular bases for some of the mutations used in crop development and improvement have been characterized. For example, high sugar content in sweet corn and peas results from mutations in genes encoding enzymes required for starch synthesis (Hannah 1997). Sticky rice, which is favored in East Asia, is also caused by a mutation affecting starch synthesis (Isshiki et al. 1998). One example of flower color variation in petunias results from a phenomenon called gene silencing, which occurred as a consequence of a duplicated gene sequence (Quideng and Jorgensen 1998). A large number of mutations, such as one influencing flower color in morning glories (Clegg and Durbin 2000) and another causing parthenocarpic fruit formation in apples (Yao et al. 2001), result from insertion of retrotransposons in genes. Retrotransposons and transposable genetic elements have had a dramatic effect on the size and evolution of plant genomes (Kalendar et al. 2000). A trait for male sterility in maize that was used for many years to produce hybrid corn resulted from a bizarre natural mutation in which part of the DNA in the mitochondrial genome was rearranged to create a novel protein-coding sequence using a

---

*Transposons are small, mobile DNA sequences that can replicate and insert copies at random sites in chromosomes. Such movements often cause mutations. Transposons have nearly identical sequences at each end, oppositely oriented (inverted) repeats, and code for the enzyme transposase, which catalyzes their insertion. Bacteria have two types of transposons: simple transposons that have only the genes needed for insertion, and complex transposons that contain genes in addition to those needed for insertion. Eukaryotes contain two classes of mobile genetic elements. The first are like bacterial transposons in that DNA sequences move directly. The second class (retrotransposons) move by producing RNA that is transcribed, by reverse transcriptase, into DNA, which is then inserted at a new site.

ribosomal RNA gene (Dewey et al. 1986). Normally, ribosomal RNA genes are not translated into protein.

While the nature of gene mutations can be simple or complex, it is not necessary for the plant breeder to understand their bases in order to use them in a breeding program. As long as the mutation creates a stable, heritable phenotype, it can be incorporated into a crop plant by conventional breeding methods. The challenge of improving a trait in a crop plant through breeding depends on whether the trait is simple or complex and the ease with which the phenotype can be measured. Many simple traits can be monitored visually or by measuring biochemical products, such as sugars, proteins, or lipids. For complex traits the end product, such as height, yield, or resistance to stress, is measured.

With sexually compatible species, a useful trait is transferred from one genetic background into a desirable so-called elite genotype by making a cross. In this process, not only are genes encoding the trait or traits of interest transmitted, but so are thousands of other genes, some of which can lead to inferior or undesirable traits. However, most of these inferior genes can be eliminated by making a series of recurrent backcrosses (usually six or more) to the elite parent in the case of simple traits or by selection of the best phenotypes in the case of complex traits. In this way, genes encoding the valuable trait are recovered in the nearly homogeneous elite genetic background. However, the process of backcrossing or recurrent selection cannot avoid introducing a large number of genes that are tightly linked with those encoding the trait or traits of interest. Even after 20 backcrosses there is a theoretical expectation that genes within 9 centimorgans (units of crossing over) of the selected gene will be carried in the new plant line (Naveira and Barbadilla 1992). This 9-centimorgan region could include over 100 genes.

For an empirical example, genes for insect and pathogen resistance are often transferred to cultivated tomatoes from related species. Young and Tanksley (1989) examined eight commercial tomato cultivars that contained the *Tm-2* gene for resistance to tobacco mosaic virus. This resistance gene originated from the wild species *Lycopersicon peruvianum* and had been transferred to a variety of commercial tomato types (e.g., cherry tomatoes, large greenhouse tomatoes) by backcrossing for 5 to 20 generations. Using DNA markers specific to *L. peruvianum*, Young and Tanksley (1989) demonstrated that the introgressed DNA from *L. peruvianum* in the final cultivars encompasses a chromosomal region ranging from 4 centimorgans to over 51 centimorgans. In the extreme case, the cultivar Craigella-*Tm-2* contained an entire arm of chromosome 9 from *L. peruvianum*.

The importance of these linked genes depends on how genetically different the original parents are that were used in the cross. These extra genes, so-called genetic drag, may or may not have important phenotypic consequences for crop production. The plant breeder must, in the end, compare the overall performance of the backcross progeny with the original parent to determine if there has been an improvement in the crop. Sometimes there is no improvement, and it is generally difficult to predict how a given gene or set of genes will affect the overall performance of the elite genotype. Consequently, plant breeders maintain a continuous cycle of creating and testing various genetic combinations for those that improve phenotypic performance.

In some cases, plant breeders introgress completely new chromosomes into a crop from another species or genus, thereby introducing hundreds or thousands of new genes/genetic loci. For example, there are over 200 lines of wheat to which a chromosome of another species has been introduced (Islam and Shepherd 1991). Embryo rescue, which makes use of tissue culture techniques to propagate incompletely developed embryos, permits the great genetic distances between these species to be bridged when the interspecifically produced progeny would not otherwise survive. For example, oats and corn can be crossed and partial hybrids recovered by embryo rescue. A hybrid zygote is formed, but the corn chromosomes are most often eliminated in the early divisions of the embryo; however, corn chromosomes are retained along with the haploid set of oat chromosomes in about a third of the progeny and the resulting plants are fertile (Riera-Lizarazu et al. 1996). Such crosses between exotic plants from germplasm collections or wide hybridization, such as between corn and oats, introduce genes not usually found in the same nucleus, even though these genes arise in the same or related genus. These techniques provide the opportunity to introduce novel and sometimes useful genetic traits. Most breeding programs make relatively little use of such potentially valuable genetic materials. However, as described below, the development of gene transformation technology greatly increases the ability of the plant breeder to utilize diverse genetic resources to a much greater extent than was previously possible.

Traditional plant breeding is an effective way to improve the genotype of elite varieties, but it is largely done in ignorance of the genes involved. The process of making sexual crosses, often followed by backcrossing to a preferred parental line, provides a mechanism to shuffle and then fix gene combinations that can subsequently be evaluated for trait performance. Backcrossing is particularly important following radiation or chemically induced mutagenesis, as a number of defective genes can be simultaneously created and carried into the progeny. Most of these mutations are not obvious to the breeder. Even without mutagenesis, most plants carry a number of inferior alleles of genes that reduce performance when incorporated into an elite genetic background. These mutant genes can be stable or unstable, and they can be responsible for traits that are disadvantageous or even dangerous.

There are many examples of breeding projects that have resulted in phenotypes that were not expected based on knowledge of the parental genotypes. In one cross between the cultivated potato, *Solanum tuberosum*, and a wild potato, *Solanum brevidens*, the hybrid offspring produced a novel steroidal alkaloid, demissine, which was not produced by either parent. A retrospective analysis revealed a potential biochemical explanation for this occurrence (Laurila et al. 1996), but it was certainly not predicted. In another case, potatoes bred for insect resistance were found to be toxic because of high levels of glycoalkaloids and had to be withdrawn from commercial production (Zitnak and Johnston 1970; see also Hellenas et al. 1995). A gene used to create corn hybrids was found several years later to make the plants susceptible to a toxin produced by the fungus *Bipolaris madis*, and this led to widespread destruction of the corn crop by Southern corn leaf blight in 1970 (Dewey et al. 1986; see also "Environmental Impacts of the Deliberate Introduction of Biological Novelty" above). The backcrossing and selection procedures help eliminate undesirable genetic material that is inadvertently introduced when crosses are made between genetically distinct parental lines. However, it is not possible to know the nature of the majority of genes in the genome, so the outcome of a typical genetic cross cannot be known.

Transgenic Techniques for Crop Improvement

The development of techniques that allow the isolation, sequencing, and transformation of genes into plants provides a novel mechanism for creating genetic diversity. Not only is it possible to isolate a gene (or genes) that control a valuable trait in a sexually compatible species and introduce it without simultaneously carrying along thousands of other genes (the so-called genetic drag), it is also possible to introduce genes from species that are not sexually compatible with the recipient (horizontal gene transfer). Thus, completely novel traits can be introduced into grain and fruit crops that will improve their yield, their resistance to biotic and abiotic stresses, their biochemical composition, their shelf life, and other traits.

Currently, there are two common transgenesis procedures by which purified genes are routinely introduced into plant cells (Birch 1997). One makes use of the Ti-plasmid of *Agrobacterium tumefaciens* to transfer the gene as part of this plasmid's DNA (referred to as T-DNA). The Ti-plasmid is a natural vector for transforming genes into plant cells, and a great deal is known about how this occurs (Sheng and Citovsky 1996). Normally, the T-DNA contains genes encoding enzymes that cause the transformed cell to become tumorigenic; however, these tumor-causing genes are removed from the T-DNA vectors used to genetically engineer crop plants.

The Ti-plasmid of *A. tumefaciens* is capable of transforming most dicot plant cells. However, in order to transmit a gene to the progeny, it must be introduced into a totipotent cell (i.e., one that can be regenerated into a whole plant). It is common, therefore, to cocultivate plant tissues, such as very young embryos, which contain totipotent cells with *A. tumefaciens* containing genetically engineered T-DNAs (Birch 1997). In the case of *Arabidopsis*, a model flowering plant used for basic research, developing flower buds can simply be soaked in liquid containing the bacteria in order for germ cells to become transformed with the T-DNA. Typically, the proportion of totipotent cells that become transformed is small. Therefore, a gene encoding a selectable marker, such as a gene for antibiotic or herbicide resistance, is incorporated into the Ti-plasmid to simplify the detection of transformed cells. This allows screening among the embryogenic transformants for those carrying the transgene because only the transformed cells will grow in media containing the antibiotic or herbicide.

The second method for plant cell transformation involves using a physical agent (metal particle or fiber) to pierce the cell wall and carry the cloned DNA into the nucleus. The first application of this technique involved coating tungsten particles with plasmid containing a cloned gene and using the discharge of a 22-caliber cartridge to accelerate the metal particles into the target tissue (Sanford et al. 1993). Thus, it is known as ballistic transformation, and the device used to transfer the DNA into the tissue is called a gene gun. Subsequently, it was found that a similar result is possible using silicon fibers, or whiskers, or electroporation (Asano et al. 1991). Physical insertion does not require any special proteins to transfer the DNA into the nucleus, as occurs with T-DNA-mediated transformation, so the technique is not restricted to species that can be transformed by *A. tumefaciens*. Ballistic transformation was the first technique used to transform monocots, such as cereals, as they appeared to be resistant to *A. tumefaciens* infection. However, in recent years, procedures for transforming cereals with *A. tumefaceins* have been developed, and it is now possible to transform many agronomically important cereals with T-DNA (Chan et al. 1992, Ishida et al. 1996). It is commonly thought that T-DNA-mediated transformation results in fewer, and simpler, transgene insertions than ballistic transformation; however, ballistic transformation can also result in single-gene insertions (Kohli et al. 1998).

As with T-DNA transformation, a gene introduced by physical insertion must be in a totipotent cell in order to end up in the germ line. For this reason, tissues capable of generating somatic embryos, such the cotyledons of young embryos, are also routinely used for ballistic transformation. As only a few of the transformed cells develop into a plant, a selectable marker gene is also valuable in this case to identify cells carrying the transgene.

Ideally, one would target an engineered gene to a specific chromosomal locus, thereby replacing an existing allele (called homologous recombination). While this is possible with certain fungi and mammals (Doetschaman et al. 1987, Hynes 1996), it is not yet a routine procedure for crops. It is possible that directing gene replacement in higher plants is thwarted by the large amount of repetitive DNA contained in their nuclear genomes, or the enzymes required for homologous recombination might not be present in somatic cells. There is only one report of homologous recombination directing gene replacement in a flowering plant (Kempin et al. 1977), and this was in *Arabidopsis*, which contains only a small amount of repetitive DNA. In this case a very large number of transformants had to be examined before a gene replacement was detected.

While it is not possible to target a specific site for inserting a gene into a plant, it is possible to determine the exact location of the gene after it has been inserted. The mechanisms that determine the site of transgene insertion into the DNA of plant cells by either T-DNA or ballistic/physical transformation are not understood. It is widely believed that the transgene is targeted to regions of transcriptionally active chromatin, but this has not been widely investigated (Birch 1997). Since transformation is often done with cells from somatic embryos, it is possible they are expressing genes from a variety of regions throughout their genome. In the case of T-DNA transformation, it is reasonably well documented that insertions take place throughout most regions of the genome (Azpironz-Leehan and Feldmann 1997).

It is common for two or more foreign DNA molecules to combine end to end in the plant cell prior to insertion into the nuclear genome. Many transformed plants contain multiple copies of a transgene, which occur in tandem or inverted repeats. In some cases, fragments of host DNA can become interspersed between the duplicated copies of the transgene, creating complex loci (Kohli et al. 1998, Pawlowski and Somers 1998). Although the mechanisms that create complex transgenic loci are incompletely understood, some evidence suggests they are related to plant cell wound responses and DNA repair mechanisms (Sonti et al. 1995).

Multiple transgene insertions and complex insertions at single loci are highly correlated with transgene silencing (see BOX 1.3), which can lead to inactivation of the transgene and possibly other genes in the genome with sequence homology to regions of the transgenic DNA (Birch

BOX 1.3
Types and Consequences of Transgene Silencing

Two types of transgene silencing have been described: transcriptional and post-transcriptional (Chandler and Vaucheret 2001). Transcriptional gene silencing is often associated with the presence of duplicated DNA sequences in the genome corresponding to the transgene. However, it can also occur through transgene insertion into nonexpressed regions of chromosomes. DNA duplication-directed silencing can occur as a consequence of multiple or complex insertions of the transgene, or it can occur due to other genes in the genome with as little as 100 base pairs of DNA identical with that of the transgene. The consequence can be silencing of the transgene, the endogenous gene, or both. Posttranscriptional gene silencing is associated with sequence homology in gene-coding regions and appears to involve the formation of double-stranded (ds) RNA molecules (Mourrain et al. 2000). It can result either from antisense transcription of a gene, RNA-dependent RNA polymerase copying of an mRNA, or transcription of an inverted repeat sequence. These dsRNAs appear to trigger an antiviral response, resulting in ribonuclease-mediated degradation of homologous RNAs into small 25 base pair dsRNA fragments. These small dsRNA fragments spread throughout the plant via plasmodesmata and phloem tissue, leading to systemic destruction of transcripts containing identical sequences.

1997). Consequently, in order to avoid transgene silencing it is important to screen transformation events to identify those consisting of simple insertions at single loci within the genome.

One way to obtain single gene insertions is by replacing one gene with another. While it is not yet possible to obtain gene replacements by homologous recombination in higher plants, it is possible to target transgene insertion to preset sites by using the P1 bacteriophage *Cre/lox* or the FLP/FRT recombination system of the 2-i plasmid of *Saccharomyces cerevisiae* (Gates and Cox 1988, Hoess and Abremski 1990). These technologies require transformation of one plant with a transgene flanked by *lox* or FRT DNA sequences and a second plant transformed with the gene encoding the *Cre* or FLP recombinase, respectively. When a sexual cross is made between such plants, the enzymatic activity of the recombinase will splice out multimeric transgene insertions, leaving behind single copies of the transgene DNA flanked by *lox* or FRT sequences (Srivastava et al. 1999). Although these genetically engineered recombination systems require additional steps for creating a transgenic trait, they are useful for removing selectable marker genes that are required to propagate a transgene in a bacterial and plant host cell. Transgenes can also be targeted to *lox* or FRT insertion sites, which allows well-characterized transgenic lines to be reused for multiple types of transgene insertions, including introduction of stacked (multiple) transgenes.

Transgene silencing is a phenomenon that can occur with either T-DNA or biolistic/physical insertion of transgenes into the genome, and it may not become detectable until several generations beyond the initial transformation event (see BOX 1.3). Commonly, the goal of inserting foreign genes into plants is to obtain expression of that gene. In these cases gene silencing is a problem. However, there are cases when gene silencing can be used as a valuable tool to block expression of one or more of the existing genes in the crop. For example, this is a way to inactivate a key enzyme in an undesirable metabolic pathway or a means to eliminate expression of genes responsible for producing food allergens.

Because the phenomenon of gene silencing became apparent in the early days of plant genetic engineering, simple types of transgene insertions were often selected from hundreds of transformation events before advancing a few transgenic events for trait evaluation. Typically, these events were characterized by restriction enzyme mapping to determine the number and complexity of transgene insertions, and the nucleotide sequences flanking the transgene may have been sequenced to ensure the integrity of the transgene itself. Because of improvements in and cost reduction of DNA-sequencing technologies in recent years, it is now cost effective, and prudent, to sequence the entire transgene and its site of insertion, unless the objective is silencing a host gene.

In some crops the ability to transform and regenerate plants is restricted to a few genotypes. Typically, these are not the elite lines that have been bred for high agronomic performance. Consequently, a genetically engineered trait is usually transferred by sexual crossing followed by six or more generations of backcrossing into a more useful genetic background. This requirement has many of the same advantages and disadvantages of mutation breeding with conventional crops, as there is a certain amount of genetic drag associated with any nonelite genetic background. However, the process of backcrossing provides the plant breeder with an opportunity to select among a large number of transformation events for those with the best phenotypic outcome and agronomic performance. Through this screening mechanism, complex transgene loci with the potential for unwanted gene silencing or poor expression and heritability are eliminated. Should the genetically engineered trait prove to be unstable or lead to an unpredictable phenotype or a plant with poor agronomic or horticultural performance, it can be culled before being advanced to evaluation for commercial production. As with traditional plant breeding, newly created genotypes must be evaluated in a variety of environmental situations before it is possible to predict their favorable or unfavorable consequences. Thus, there is extensive selection and field testing before a transgenic event becomes propagated to a sufficient degree to allow commercialization of a transgenic crop.

In summary, the above discussion indicates that the genetic processes used in conventional and transgenic approaches to crop improvement each have associated risks. In both processes there are two basic steps. One involves adding novel genetic material to the crop, and the second involves filtering out any of the added genetic material that is undesirable. In conventional approaches there are typically more unwanted genes that must be eliminated from the hybridized lines. But in both approaches there is always a possibility that some undesirable traits will remain in the final product. Furthermore, in both cases there is a possibility that the desired genes themselves will interact with existing genes in the parent line and create unexpected effects.

While transformation of genes into crop plants is a new technique for creating genetic diversity, it fundamentally accomplishes many of the same goals as genetic recombination and reassortment, or making wide genetic crosses between geographically isolated or related species. The key difference is the way in which the genetic variation is created. Once inserted into the genome, transgenes are subject to the same types of genetic regulatory mechanisms as conventional genes. As we are learning from the characterization of plant genomes (Kaul et al. 2000), the nuclear DNA is naturally in a constant state of flux as a consequence of single nucleotide changes and DNA rearrangements and insertions. Even though it is uncommon, horizontal gene transfers are not unprecedented. Even the human genome appears to contain a surprisingly large number of insertions of bacterial DNA without ill effect (Lander et al. 2001, Venter et al. 2001). Thus, the fact that transgenes can originate from another species is not novel. What makes the transgenic approach particularly new is the potential to incorporate novel traits into plants. Current and future transgenic traits are discussed elsewhere in this report.

The line between conventional crop breeding and the creation of transgenic crops has never been perfectly clear, but the distinction between the two approaches is likely to blur even further. The genetic engineering process, per se, presents no new categories of risk compared to conventional breeding, although this technology could introduce specific traits or combinations of traits that pose unique risks. However, crop varieties developed solely by conventional breeding could express traits that would need to be regulated as stringently as those developed by transgenic approaches.

Finding 1.4: Genetic improvement of crops typically involves the addition of genetic variation to existing cultivars, followed by screening for individuals that have only desirable traits.

Finding 1.5: Both transgenic and conventional approaches (i.e., hybridization, mutagenesis) for adding genetic variation to crops can cause changes in the plant genome that result in unintended effects on crop traits.

Finding 1.6: The transgenic process presents no new categories of risk compared to conventional methods of crop improvement, but specific traits introduced by both approaches can pose unique risks.

Finding 1.7: Screening of all crops with added genetic variation must be conducted over a number of years and locations because undesirable economic and ecological traits may only be produced under specific environmental conditions.

OVERVIEW OF CURRENT U.S. REGULATORY FRAMEWORK FOR TRANSGENIC ORGANISMS

In the early 1980s the U.S. government determined that there was a need to regulate transgenic organisms, including engineered plants, to assure health and environmental safety. However, it was decided that such regulation could be accomplished by using existing federal statutes instead of developing new ones specifically designed for transgenic organisms. The U.S. Coordinated Framework for the Regulation of Biotechnology (OSTP 1986) was developed to offer guidance for using existing federal statutes and the expertise of existing regulatory agencies in a manner that would ensure health and environmental safety while maintaining sufficient regulatory flexibility to avoid impeding the growth of the biotechnology industry. What follows is an abbreviated description of the coordinated framework, so that readers can understand the regulatory context of the USDA-APHIS regulation of transgenic organisms. The federal government, in lieu of creating a new law for transgenic organisms, asked that the USDA adapt its interpretation of the Federal Plant Pest Act (FPPA) and the Federal Plant Quarantine Act (FPQA) to ensure environmental safety while not impeding growth of the industry.

Two basic principles were intended to guide regulatory policy. First, agencies should seek to adopt consistent definitions of those transgenic organisms subject to review to the extent permitted by their respective statutory authorities. Second, agencies should use scientific reviews of comparable rigor. (Both products and research are regulated under this framework, but the review here will concentrate on "products," which are defined as transgenic organisms that are released as commercial products into the environment).

The 1986 coordinated framework was constructed to be a flexible policy that would evolve as needed. In 1992 the policy's scope was clarified. While each agency's statute clearly specifies the scope of oversight that each agency can exercise, the diversity of statutes meant that regulatory oversight could vary in degree from agency to agency and statute to statute. This could create a situation where under one statute a transgenic organism is inadequately regulated while under another it is overregulated. To address this possibility, the Office of Science and Technology Policy (1992) issued two principles to restrict the scope of oversight. First, federal agencies were mandated to exercise oversight of transgenic organisms only when there is evidence of "unreasonable" risk. An unreasonable risk is one in which the value of the reduction in risk obtained by oversight is greater than the cost of oversight. The evidence of risk must include information about the transgenic organism, the target environment, and the type of application. The scope policy is vague as to the appropriate valuation system that should be used in making regulatory judgments. Second, federal agencies were required to focus on the characteristics and risks of the biotechnology product, not the process by which it is created.

Since 1992 there have been a few more changes in the policy, including shifts in the responsibilities assumed by each agency. For the purpose of this discussion, one significant shift has been the assumption of responsibility by EPA for a subgroup of transgenic plants that were designated as pesticidal transgenic plants, including insect-resistant crops, such as *Bt* corn, and virus-resistant plants.

In response to the coordinated framework, the Biotechnology, Biologics, and Environmental Protection unit (BBEP) of the USDA-APHIS published regulations for the release of transgenic organisms in 1987 (APHIS 1987), under the statutory authority of the FPPA and the FPQA. While the coordinated framework explicitly provided that federal agencies should focus on the characteristics and risks of the biotechnology product, not the process by which it is created, APHIS explicitly used the process of genetic engineering to trigger its oversight. However, in the 1987 Final Rule (APHIS 1987), APHIS argues that it is not treating transgenic organisms differently than so-called established plant pests or naturally occurring organisms, which may be plant pests. For example, APHIS regulates the movement and release of geographically separated populations of known plant pests because a new geographic population may have genetic characteristics absent in the recipient geographic population, which could increase the plant pest risk in the recipient population. In all cases a permit must be obtained from APHIS prior to importation and interstate movement of potential plant pests. For certain transgenic organisms, APHIS has determined that the release of these organisms into the environment is tantamount to the introduction of a new organism.

The original 1987 APHIS regulations were later to exempt *Escherichia coli* strain K-12, sterile strains of *Saccharomyces cerevisiae*, asporogenic strains of *Bacillus subtilis*, and *Arabidopsis thaliana* from permitting requirements for interstate movement under certain specified conditions (APHIS 1988, 1990). Modifications with broader impact were made in 1993. These provided a more streamlined procedure called "notification" for enabling the field release of transgenic plants not expected to pose serious risks. They also provided a more formal process for applicants to petition the agency for nonregulated status of an engineered plant based on genetic and ecological knowledge of that plant (APHIS 1993). A detailed analysis of APHIS oversight of genetically engineered plants is the focus of Chapters 3, 4, and 5. Chapter 2 provides a more conceptual overview of risk analysis and its application to transgenic plants.

2

Scientific Assumptions and Premises Underpinning the Regulation and Oversight of Environmental Risks of Transgenic Plants

This chapter concentrates on the theoretical and empirical underpinnings of the regulation of environmental risks of transgenic plants and the use of risk analysis to evaluate and manage those risks. The first section summarizes the committee's approach to risk analysis. Two important roles are identified that risk analysis of transgenic crops must fulfill; it must support the decision-making process of regulatory agencies, and it must legitimize the regulatory process, creating public confidence that human well-being and the environment are protected from unacceptable risks.

The second section examines in detail the scientific and logical bases for regulation. In developing an approach that can be applied to both transgenic and conventional crops, the committee endorses the findings of three previous National Research Council (NRC) reports. Transgenic crops do not pose unique categories or kinds of environmental hazards. The entire set of existing transgenic crops is not so different in kind that they pose environmental hazards unlike those caused by other human activities, including conventional crops and other agricultural activities. The committee finds, however, that specific types of transgenic and conventional crops can pose unique environmental hazards. Also, the committee finds that there are good arguments for regulating all transgenic crops. To be effective, such a regulatory system must have an efficient and accurate method for rapidly evaluating all transgenic plants to separate those that require additional regulatory oversight from those that do not.

The third section examines several technical issues related to the underlying risk assessment models. The concentration is on these because they provide a basis for risk assessment of transgenic crops and will frame discussion of the case studies in subsequent chapters. Evaluation of the risks of transgenic crops requires specifying a social and environmental context for the assessment. Depending on the choice of context, different risk comparisons may become relevant. Finally, several formalizations are suggested that could help clarify this dependence on context and enable a regulatory agency to develop formal procedures to learn from its experience to improve the regulatory system.

RISK

Risk is both an intuitively easy and a technically difficult concept to understand. On the one hand, people take risks all the time in daily life, and whether explicitly or not, people are constantly balancing these perceived risks against their needs and desires. Everyone knows it is risky to drive a car, have radon leak into the basement, ice skate, play blackjack, or swim in a lake. But the personal and idiosyncratic ways people have of dealing with risk in their own lives have only tenuous connections to how society as a whole should deal with risk. Personal perceptions of risk may not reflect reality. One person might not care about the risks of eating a fish bone, but another may care so much that she will not eat any fish. Because each of us has our own way of dealing with risk, how do we agree as a society?

Volumes have been written on the technical aspects of risk. In the catalog of the National Academy Press alone, there are 184 titles related to risk. Two that are particularly relevant to this report are a 1983 publication, *Risk Assessment in the Federal Government: Managing the Process*, and a 1996 report, *Understanding Risk: Informing Decisions in a Democratic Society*. The 1983 report outlines a general approach to characterizing hazards, modeling exposure pathways, and quantifying the probability of injury. The 1996 approach argues that any attempt to assess risk involves a series of interpretive judgments and framing assumptions and suggests that democracy is best served when those affected by regulatory decision making can be as fully involved in making those judgments and assumptions as is practicably possible.

This chapter addresses risk from both perspectives. The 1983 report's approach is followed in using science to illuminate technical understanding of the environmental risks of transgenic crops. By following this approach, however, the committee has made several implicit interpretive judgments and framing assumptions, which the 1996 report suggests. Several of these are acknowledged as implicit judgments and assumptions, and by doing so, an alternative perspective can be developed for evaluating risk analysis of transgenic crops. Before developing these particular ideas further, some of the broader assumptions made here to understand risk should be mentioned.

Risk is interpreted primarily as a combination of the probability of occurrence of some hazard and the harm corresponding to that hazard. This is both a highly technical and somewhat vague interpretation, involving the related ideas of hazard, occurrence, and some combining process. A hazard has the potential to produce harm, injury, or some other undesirable consequence. In saying a slippery road is hazardous, it is not meant that any harm or a car accident has occurred. It simply means that the road conditions could potentially cause an accident. In this case the hazard is a car accident, which may or may not happen. Likewise, if a transgenic crop has a hazard, it does not mean that any harm has or will occur. Hazard identification is one of the most subjective and potentially contentious elements of risk analysis. While this report is limited to a consideration of environmental risks, there is some ambiguity in deciding what is and is not an environmental hazard. Does this include or exclude the potential for adverse impacts on human health that are mediated by the environment (not directly by food consumption)? Does it include or exclude the potential for adverse impacts on farming practices and profitability? Is a nonspecific effect on habitats or ecosystems an identifiable hazard? Is an effect on an ecological process an environmental hazard? The characterization of hazards in this chapter reflects an answer to each of these questions.

The occurrence of a hazard is a probability that the hazard would occur. This typically depends on many factors. The probability that an accident will occur on a slippery road will depend on how many cars travel the road, how fast they are going, how much they accelerate and decelerate, the skill of the drivers, and other factors. Clearly, these probabilities will be highly conditional on the environment and other details about the situation, and therefore they will be variable both spatially and temporally. Likewise, the probability that any hazard associated with a transgenic crop would occur is likely to vary spatially and temporally. In much of the literature on risk, the probability of occurrence is called an *exposure probability*, referring to the probability that people are exposed to the hazard. In this report the term exposure is frequently used as a shorthand notation for the probability that a hazard would occur.

The combining of hazard and occurrence probabilities to characterize risk can be contentious, ranging from simple mathematical formulations to complex deliberative processes involving many people. The committee leaves this process deliberately unspecified, so that the range of formulations can be used as needed in the report. In its simplest form, risk can be understood as a weighted probability in which the probability of occurrence is weighted by the magnitude of the hazard. If you are a driver on the slippery road, your risk is a combination of the probability that you will have an accident and the severity of the accident. Presumably (hopefully) the probability of an accident rapidly decreases with the severity of the potential accident, and your risk, which combines across all possible accidents, will be small. If you were a road engineer, you would also average over all combinations of drivers to estimate the total risk for society on that stretch of road. The process of combining hazard and occurrence can become more complicated if some drivers are considered more important than others, so that the risks themselves are weighted by some social criterion reflecting this importance. This can become even more complicated if the weightings are determined by some social deliberative process. Risk characterization is another one of the most subjective and potentially contentious elements of risk analysis.

Risks can be reduced by people's actions, and risk management is an important concept used in this report. By managing risks, people can reduce them to such an extent that they are considered insignificant. For example, one can reduce his or her risk of an accident on a slippery road by slowing down or simply by staying home during inclement weather. Similarly, planting and harvesting transgenic crops in certain ways can reduce the risks associated with them. These methods are discussed in more detail later in this report.

There are a number of aspects of risk that are not explicitly addressed in this report. The committee does not discuss methods for valuation of hazardous events in either economic or other terms. Consequently the committee does not provide a basis for comparative rankings of risks nor a basis for comparing expectations of cost or loss with expectations of benefit associated with commercialization of transgenic crops. Moreover, by emphasizing the possibility and probability of unwanted outcomes, the approach taken in this chapter excludes an interpretation of risk that lays stress on the novelty or unfamiliarity of actions taken (such as fear of the unknown) except insofar as a novelty or unfamiliarity complicates the estimation of risk. The committee's concept of risk also does not imply that some hazards are ones for which agents could be held accountable, while others would be considered works of nature. Finally, the distribution of risks among different groups of people is not emphasized. While we are all concerned with the distributional aspects of risks and benefits, and they appear to influence the debates about the utility of transgenic crops, the focus in this report is on issues where this concern has less influence. Specifically, the committee's focus here is on the roles that risk analysis is expected to fulfill in the discussion of transgenic crops and some of the implications thereof.

Roles of Risk Analysis

Risk analysis has come to play at least two somewhat distinct roles in public discourse. It is often conducted by using scientific information to support decision making in a particular case (referred to as "decision support"). This decision support role presumes that the decision maker (an individual, a group, or an organization) is well defined and has the legitimate and uncontested authority to make a decision. Within this class of risk problems, there are circumstances where decision making consists of selecting from among several well-defined options, such as whether to allow the commercialization of a particular transgenic crop. In these situations, risk analysis often consists of anticipating unwanted outcomes associated with each option and measuring their probability should that option be taken. Much of the present regulatory structure for transgenic organisms in the United States takes this approach to risk analysis.

In circumstances where the decision options are not clearly defined, the role of risk analysis is considerably more ambiguous because its role in helping to inform a decision is not clearly defined. Under such conditions, risk analysis could be used to help inform a decision maker of the general scope of a risk situation and outline potential opportunities for decision making. Risk analysis could be conducted to provide a general survey of hazards and the types of risk management decisions that might confront a decision maker. For example, the World Health Organization (WHO) recently conducted a scientific consultation to determine if indirect human health risks were mediated through the environment from the use of transgenic organisms. This consultation focused on hazard identification because if such hazards could be established, the WHO could use its authority on human health concerns (granted by the United Nations) to justify additional risk analysis.

In addition to its role in decision support, risk analysis can serve to reinforce and legitimize the authority of a particular decision maker to exercise control over a given situation (referred to as "creating legitimacy"). There are a variety of circumstances where the predictive power and presumed value neutrality of scientific risk analysis are implicitly assumed to provide the "best" basis for decision making. A decision maker's ability to produce scientific risk analysis in such circumstances can play an important role in creating legitimacy to set policy or determine a course of events. For example, member-states of the World Trade Organization (WTO) can exercise their legal authority to restrict trade in a good only when they can produce a scientific risk analysis documenting a need to restrict importation of that good. A case where the risk analysis was insufficient in restricting importation was observed with Australia being forced by the WTO to accept frozen Canadian salmon (or pay a steep fine) in spite of serious risk of transmission of fish disease, because Australia was unable to produce a sufficiently quantitative risk assessment (Victor 2000).

Risk analysis can create legitimacy in many extralegal contexts as well. Highly contentious, politically charged issues can be decided on the basis of raw power, whether through political power or public influence, with little regard to the facts. Scientific risk analysis, with its philosophical commitment to value neutrality, can be used to legitimize a decision-making process. In some cases, people may agree to support the outcome of such a risk analysis-based decision process even if it contradicts their own desired outcome (NRC 1996). Thus, risk analysis can be a powerful force to garner legitimacy in a decision process from diverse public interests.

There are many points of tension between the use of risk analysis to create legitimacy and its use as a decision support tool. A risk analysis that is expected to help create legitimacy for decision making must meet different burdens of proof than a risk analysis conducted as a decision support tool. When used in decision support, the authoritative agent is already determined and therefore is free to accept, reject, or modify any of the assumptions or parameters that may have been established in developing any component of the analysis. For example, a decision maker could explore the consequences of weighting the effects of a pesticide on farmers and farm workers less than (greater than) the effects on the general public. It is, in part, the flexibility to adjust assumptions and parameters that makes risk analysis particularly useful in a decision support role. This flexibility allows the decision maker to explore the full range of possibilities before coming to a decision. In contrast, when risk analysis is used to create legitimacy, the potential decision-making agent may find that the choice of assumptions and parameters has a profound effect on the willingness of affected parties to recognize the agent's authority to make a final decision. When it is crucial to gain the acceptance of interested and affected parties, the flexibility to adjust assumptions and parameters may be dramatically reduced or sacrificed altogether. Affected parties know that adjustment of such parameters can influence results in ways that favor one interest over another and may insist on fixing assumptions and parameters in a manner that substantially limits the usefulness of risk analysis techniques. Indeed, the role of risk analysis to create legitimacy becomes particularly perilous when the decision options under consideration are not well defined.

The role of risk analysis in governmental regulatory agencies has become increasingly ambiguous. The legal decision-making authority of such agencies is generally established by specific legislation or administrative findings. Traditionally, officials in these agencies have viewed risk analyses in the role of decision support for the exercise of this legislatively mandated authority. However, their freedom to adjust assumptions and analytical parameters in risk analyses may be constrained by the language of the authorizing legislation or by successful court challenges from affected parties. This suggests that to some degree risk analysis has already become a component in establishing a government agency's legal authority to make determinations. A more serious source of ambiguity flows from the extralegal burdens that regulatory decisions are increasingly expected to meet. The regulatory process is often described as the basis for public confidence in the safety, reliability, and fairness of the regulation of transgenic organisms (Carr and Levidow 2000, Ervin et al. 2000). This confidence may be vested in the integrity and judgment of regulatory officials themselves, but many commentators have expressed the view that it is or should be based on the risk analyses alone. This way of describing public confidence implies that risk analyses must be capable of being seen to embody principles of objectivity, fairness, and rigor by the general public. This is a burden of proof that goes considerably beyond the ability to withstand a legal challenge, and it takes risk analysis well beyond its traditional role in decision support (Funtowicz and Ravetz 1990, Von Winterfeldt 1992).

If risk analysis is expected to play any significant role in establishing the public's confidence in genetically modified crops, it becomes crucial to examine the procedures and methods for conducting a risk analysis with two ends in mind.

> **Finding 2.1: Risk analysis of transgenic crops must fulfill two distinct roles: (1) technical support for regulatory decision making and (2) establishment and maintenance of the legitimacy of regulatory oversight.**

Terminology of Risk Analysis

Although there have been many attempts to standardize the terminology for describing the stages of risk analysis, these attempts seldom gain widespread acceptance. Such stages are only a heuristic device intended to facilitate a collaborative approach to problem solving with respect to risk. Thus, there is no one best set of terminology, and this report shifts among the various sets as the problem merits or demands. However, it is impossible to communicate the committee's ideas clearly without a discussion of the various approaches to describing the stages of risk analysis.

One of the most influential early studies on risk analysis uses the terminology of *risk determination*, *risk evaluation*, and *risk acceptance* to characterize the stages of risk analysis (Rowe 1977). The general idea is that one must have an initial identification of the problem that the risk analysis is expected to address (i.e., risk determination). With a well-defined problem, techniques derived from the natural sciences and engineering are deployed in a process of risk evaluation (measuring and combining the magnitude of the hazard and the likelihood of its occurrence). Finally a decision maker makes a decision (i.e., risk acceptance).

When risk analysis is used to support policy or contentious decisions, the simple three-stage characterization of risk analysis itself becomes controversial. Some have preferred to call the first stage *hazard identification*, implying that a hazard does not become a risk until the probability that it might occur has been measured. Some object to the normative connotations of the word *evaluation* and substitute the term *risk measurement* or *risk assessment* for the middle stage. Many have noted that risks may not be simply "accepted," so the third stage is then relabeled as "risk management." Because risk analysis of transgenic organisms is itself contentious, the committee chose to adopt the terminology *hazard identification, risk assessment*, and *risk management* when discussing risk in its decision support role.

Broadly conceived, the techniques of risk assessment are of five general kinds:

1. *Epidemiological analysis.* Events of interest are observed, and the statistical relations of these events in the sampled populations are analyzed. This epidemiological approach has been very effective in identifying disease risks among populations such as smokers, industrial workers exposed to certain substances, and persons with a specific genotype. The scientific rationale for this method is that empirical correlations provide a basis for predicting effects and may indicate cause. This method could be used to associate risks with particular transgenic plants that are intensively planted in large or specific areas.
2. *Theoretical models.* A theoretical model that mimics or simulates the causal interaction of elements in a complex system is used to identify likely sources of system failure. This approach is widely used to study the risk of failure in engineering contexts such as instrumentation and control design. It has also been applied in biology to develop strategies for ecosystem management. The scientific rationale for this method is that it provides the logical consequences of a set of scientific assumptions about risks. It has played a critical role in analysis of the risk of evolution of resistance to transgenic insecticidal plants (Alstad and Andow 1995, Roush 1997, Gould 1998), and could play a role in evaluating community-level non-target effects of transgenic plants (Andow 1994).
3. *Experimental studies.* Controlled experiments are conducted to iden-

tify cause-and-effect relationships. Variations on this approach (often combined with statistical analysis) are used in product testing, including clinical trials for drugs and therapeutic devices. The scientific rationale for this method is that it establishes the cause or causes of a risk. Laboratory experiments can be used to establish a potential hazard (Hilbeck et al. 2000), and field experiments can be used to evaluate potential risks (e.g., Orr and Landis 1997). With field experiments, however, it is necessary to guard against false negatives (type II statistical errors), which would lead to concluding erroneously that there is no risk when, in fact, the experimental design is incapable of detecting any risk except a very large one (Marvier 2001).

4. *Expert judgments.* A group of experts use personal knowledge of a given system to estimate likely system performance under untested conditions. This is a widely used approach in risk analysis and has become the basis for risk analysis for some invasive species in the United States (Orr et al. 1993). The group is chosen to represent a range in necessary expertise with due consideration to potential conflicts of interest. The scientific rationale for this method is that the consensus of the group of experts represents a synthesis of the best-available scientific knowledge on the risk (Jasonoff 1986). The Environmental Protection Agency (EPA) uses this method frequently to aid in risk assessment of transgenic plants in its SAP (Scientific Advisory Panel) process, and the U.S. Department of Agriculture's Animal and Plant Health Inspection Service Biotechnology, Biologics, and Environmental Protection unit (USDA-APHIS-BBEP) has used it to evaluate the initial notification system for transgenic plants (USDA-APHIS 1993).

5. *Expert regulatory judgment.* Regulatory personnel use personal knowledge of a given system to estimate likely system performance under untested conditions. This is the most widely used approach in risk analysis. The scientific rationale for this method is that regulatory personnel have ready access to confidential business information and understand both current scientific knowledge and the process of risk analysis, so their judgments can be rendered with minimal delays and sufficient scientific accuracy. This is the method used by USDA-APHIS-BBEP to evaluate most of the complex, difficult-to-measure potential risks associated with transgenic plants.

The first three of these approaches are generally accepted as scientifically rigorous methods of analysis. Expert judgment, whether by external experts or by regulatory judgments are less rigorous, but often acceptable. A consensus of multiple external experts is likely to be more rigorous than the expert regulatory judgments because disagreements among external experts are likely to lead to more robust risk assessments (Jasanoff 1986).

Finding 2.2: At least five standards of evidence can be used in a risk assessment for decision support. The scientifically rigorous methods include epidemiological, theoretical modeling, and experimental methods. A consensus of multiple external experts is likely to be more rigorous than the expert regulatory judgments.

Finding 2.3: More rigorous methods used in decision support are likely to help risk analysis fulfill its other social role of establishing and maintaining regulatory legitimacy.

Chapter 5 evaluates circumstances in which these various methods should be used to supplement or displace expert regulatory judgment. In addition, for some questions the committee suggests specific types of data that should be collected to support epidemiological analysis, the simulation models that should be developed, the experimental studies that need to be conducted, and when expert panels should be convened. Implementation of any of these approaches (alone or in combination) requires a process to clarify the role that each general technique will play in measuring risk.

Risks may spark public outrage, and a fourth stage of *risk communication* has been suggested. But in the decision support role of risk analysis, communicating with the public may actually just be a tool of risk management, so the committee treats risk communication as a subset of risk management when discussing risk analysis in its decision support role. However, in its role of creating legitimacy, the greatest source of disagreement over risk often concerns the early stages of conceptualizing and identifying risk. Because some models of risk communication can become a way to exclude affected parties from this crucial part of the process, the 1996 NRC report considers risk communication integral to all stages of risk analysis. While inclusion in all stages blurs the category of risk communication, it must be integrated with all stages of risk analysis, and affected parties must be brought into early discussions.

When this is done, the three-staged framework of risk analysis also becomes less useful. For example, hazard identification is no longer an exercise in listing potential hazards but becomes a deliberative process from which potential hazards are characterized. Moreover, even when science and engineering methods are used to measure the probability that these potential hazards will materialize into bona fide risks, the risk problem is substantially sharpened and redefined by deliberative processes, so it seems appropriate to include *this* part of risk analysis under the heading of risk characterization as well. Clearly, the stages of risk analysis do not, in fact, represent temporally, analytically, or even conceptually distinct and discrete elements. These considerations led the 1996 NRC report to describe risk characterization as:

> the outcome of an *analytic-deliberative process*. Its success depends critically on systematic analysis that is appropriate to the problem, responds to the needs of the interested and affected parties, and treats uncertainties of importance of the decision problem in a comprehensible way. Success also depends on deliberations that formulate the decision problem, guide analysis to improve the decision participants' understanding, seek the meaning of analytic findings and uncertainties, and improve the ability of interested and affected parties to participate effectively in the risk decision process. (158)

In the role of creating legitimacy, deliberation is central in risk characterization. It is important at every step in the process of making risk decisions, and must be incorporated into each stage of risk analysis. Therefore, when discussing risk in its role of creating legitimacy, this committee chooses to adopt the terminology *risk characterization*, which seamlessly integrates processes involved in hazard identification, risk assessment, and risk management. Hence all these processes are better thought of as integrating subcomponents of risk characterization rather than steps in a risk analysis process.

Risk Analysis as Decision Support in the Regulation of Transgenic Plants

Since the use of transgenic crop plants was first discussed, there has been confusion over the basis for distinguishing between the potential environmental risks associated with these plants and the risks associated with conventionally produced plants. As the power of conventional plant breeding has increased over the past 50 years, there has been a corresponding increase in the kinds and number of traits that can be bred into a commercial variety. As reviewed in Chapter 1, these changes and concomitant changes in agricultural production systems have created opportunities for novel interactions between agriculture and the surrounding organisms, habitats, and ecosystems. Transgenic crops are the latest development in this trend. Consequently, it is important for regulatory agencies to pay heed to the possibility that transgenic crops will be involved in novel ecological interactions resulting in novel environmental risks.

A series of scientific studies and reports have consistently found that the risks associated with a crop variety are independent of the means by which the crop was created. A 1987 NRC report concluded that the environmental risks associated with transgenic organisms are "the same in kind" as those of the unmodified organisms or organisms modified by other means. A 1989 NRC report clarified this statement to argue that this meant any hazard that could be identified was similar enough to other known environmental hazards that an existing risk assessment methodology could be found to assess the risk. In other words, a new kind of risk would require a new risk assessment methodology. Below the committee reviews the kinds of hazards that can be associated with transgenic crop plants. None of these kinds of hazards is unique to transgenic plants, and hence no new kinds of hazards are identified. However, with the long-term trend toward increased capacity to introduce complex novel traits into plants, the associated potential hazards and risks, while not different in kind, may nonetheless be novel. For example, widespread planting of *Bt* corn generated concerns over a new potential non-target hazard—that toxic *Bt* corn pollen could reduce monarch butterfly populations.

The 1989 NRC study also found that transgenic plants should pose risks no different from those modified by classical genetic methods for similar traits. Traits that are unfamiliar (i.e., traits for which there has been little or no prior experience) in a specific plant will require careful evaluation in small-scale field tests. A 2000 NRC (2000c) report amplifies these points. It specifically notes that the magnitude of risk varies on a product-by-product basis and does not depend on the genetic modification process. However, the NRC (1989) report points out that information about the process used to produce a genetically modified organism is important in understanding the characteristics of the end product (NRC 1989).

Finding 2.4: For purposes of decision support, risks must be assessed according to the organism, trait, and environment.

Finding 2.5: For purposes of decision support, the process of production should not enter into the risk assessment.

Finding 2.6: The transgenic process presents no new categories of risk compared to conventional methods of crop improvement, but specific traits introduced by *either* of the approaches can pose unique risks.

Risk Analysis for Creating Legitimacy

The way Americans think about the relationship between agriculture and environment has evolved over time. For example, there was a time when a farmer's decision to shift from pasture to crop production would have been thought to be a purely personal economic decision, raising no public concern over environmental issues. Now people recognize that pastures provide habitat for native flora and fauna and, when properly managed, support more complex agricultural ecosystems than do many modern monocultures (see Chapter 1). As such, a crop genetically modified for a trait such as drought tolerance, which might encourage producers to convert pasture to crops in dryland regions, might today be thought to have environmental implications, whereas in the past it would not have been perceived as a problem. There has been a shift in perspective and values that has led many Americans to take an interest in ecological impacts. This shift is not based only on science (see Chapter 1), yet it clearly affects the expectations Americans have for their regulatory agencies.

As such, it will be important for regulatory agencies to maintain some degree of sensitivity to shifts in cultural values and implicit understandings that frame the public's expectations. The novelty of recombinant DNA techniques in the public mind, not to mention the degree of discussion and debate that transgenic crops have received in the mass media, means that it may be important for regulatory agencies to be especially sensitive to transgenic crops and their possible impact for the foreseeable future. Maintaining public confidence in the regulatory system almost certainly requires a focus on transgenic crops in addition to any need based solely on scientific grounds.

Finding 2.7: Risk analysis of transgenic crops has played and is likely to continue to play an important role in maintaining the legitimacy of regulatory decision making concerning environmental and food safety in the United States.

The Precautionary Principle

The Precautionary Principle has been used in multilateral, international agreements to help legitimize a regulatory process based on scientific risk analysis. At its core the precautionary principle is based on the well-accepted folk wisdom that haste makes waste and one should look before one leaps. This folk wisdom, however, allows each individual to decide what is haste and how long and carefully to look. Analogous to the problem of risk, much of the controversy in applying the precautionary principle to societal issues relates to setting socially agreed upon standards for sufficient precaution. In one widely used form, the precautionary principle states that the absence of scientific knowledge about a risk should not impede actions to reduce that risk to society. In other words, even in the absence of clear scientific evidence that a risk is likely to occur, actions should be taken to reduce that potential risk. It is because we do this all the time in our daily lives that the precautionary principle makes such intuitive sense. If I think it might rain but do not know for sure, I often will bring an umbrella. But for the same reasons that risk itself is complicated, and our intuitive understanding of risk has only tenuous connections to scientific risk analysis, the precautionary principle is complicated and our intuitive understanding of it incomplete.

As discussed more thoroughly in Chapter 7, there is no one precautionary principle, and the precautionary principles range from minor procedural changes in risk analysis to major shifts in the burden of proof. The principle is featured prominently in two international agreements about the risks of transgenic organisms—the Biosafety Protocol of the Convention on Biodiversity (CBD 2000) and Directive 90/220 of the European Union (EU), modified in March 2001 (European Commission 2001). In both cases the precautionary principle is specified ambiguously, so its meaning will evolve and become defined by its use. A close reading of the U.S. coordinated framework for regulating transgenic organisms shows that the framework neither excludes the use of a precautionary principle nor endorses one. Thus, the fate of the precautionary principle will be determined in future applications and negotiations among interested parties. As clearly stated in the biosafety protocol and the modified Directive 90/220, science will continue to be the basis of risk assessment.

SCIENTIFIC ASSUMPTIONS UNDERPINNING REGULATION OF TRANSGENIC CROPS

The Categories of Hazards

The initial step in risk analysis and one of the most critical for framing the entire analysis is hazard identification. A hazard is a potential adverse effect (potential harm) from a proposed activity (e.g., release of a transgenic crop plant). Risk is the combination of a hazard and the likelihood that the hazard occurs. This implies that many hazards are not risks, and the existence of a hazard does not imply significant danger from the activity because the hazard might not happen. The set or scope of identified hazards determines the technical capacity and level of deliberation needed to conduct the analysis and therefore is one of the most critical steps.

Four categories of hazards from the release of transgenic crop plants can be identified: (1) hazards associated with the movement of the transgene itself with subsequent expression in a different organism or species, (2) hazards associated directly or indirectly with the transgenic plant as a whole, (3) non-target hazards associated with the transgene product outside the plant, and (4) resistance evolution in the targeted pest populations. A fifth category of hazard, discussed in the other chapters, is that of indirect effects on human health that are mediated by the environment.

Most previous reports have not chosen to classify this as an environmental hazard.

In addition to these categories, the EU recognizes effects on genetic diversity as a separate category of environmental hazard in its modified directive 90/220 (European Commission 2001). The committee has not recognized this category as an environmental hazard because it is the effects of altered genetic diversity, such as increased extinction rate, a compromised genetic resource, inbreeding depression, or increased vulnerability to environmental stresses, that are the actual environmental hazards. These effects are addressed under the effects of movement of transgenes. The EU recognizes this category of potential hazard as a precautionary measure, because the effects of movement of the transgenes are uncertain and are presently incompletely characterized. By recognizing the more easily measured, intermediate effects on genetic diversity as a potential hazard, the EU risk analysis will address and manage all of the effects the committee lists under movement of transgenes without having to assess them specifically.

Hazards Associated with Movement of the Genes

The movement of transgenes does not, in itself, constitute a hazard but can serve as an opportunity for unintentional spread of transgenes in the environment (Nickson 2001). If a specific hazard is associated with that spread, the movement of transgenes constitutes the "exposure" component of a risk. One of three mechanisms—seed dispersal, horizontal transfer, or pollen dispersal—may move transgenes beyond the point of intentional release into environments and organisms other than those intended.

Seed dispersal can be accomplished by (1) unintentional seed spillage during processes that either bring the seed to the field to be planted or involve taking harvested seed from a transgenic crop from the field to market (e.g., in the United Kingdom, some roadside feral oilseed rape populations [*Brassica napus*] appear to be constantly replenished by seed spilling from vehicles on their way to a major oilseed crushing plant; Crawley and Brown 1995) or (2) dispersal of seeds directly from the transgenic crops into the surrounding environment. The hazard usually associated with seed dispersal is the evolution of increased weediness of the transgenic crop itself. (See "Hazards Associated with the Whole Plant" below.)

Horizontal transfer is the nonsexual transfer of genetic material from one organism into the genome of another. For example, it is becoming clear that plants appear to occasionally acquire genes from other kingdoms of organisms, such as fungi (e.g., Adams et al. 1998). The specific mechanisms for such transfer are poorly understood. The rate of such acquisition is extremely low relative to within-species gene transfer but is surprisingly high over evolutionary timescales. For example, flowering plants appear to have acquired a certain mitochondrial gene by hundreds of independent horizontal transfer events over the past 10 million years (Palmer et al. 2000). In essence, horizontal transfer is natural genetic engineering. Specific hazards associated with horizontal transfer are rarely posed, and its significance is largely discussed as a source of unanticipated effects. Presently there are no data to suggest that the extremely low rate of natural horizontal transfer should change for transgenic organisms. As Rissler and Mellon (1996) state, "Although currently no evidence exists that genes move horizontally from plants to any other organism, the phenomenon is so recently described that it is too early to conclude that transfers from plants never occur...New discoveries in this area could change the assessment and significance of risk."

Pollen dispersal provides an opportunity for the sexual transfer of transgenes to relatives of the crop, including other varieties of that crop, related crops, and wild relatives. The specific vectors for pollen dispersal vary with the crop. Wind and insects are the most frequent agents that carry pollen from plant to plant. Almost all crops are expected to disperse some pollen. Although many crops are self-fertilizing (Frankel and Galun 1977), even crops with very high selfing rates are capable of mating with plants many meters away (e.g., Wagner and Allard 1991). Furthermore, although certain crops are typically harvested before flowering (e.g., sugar beets), occasional plants flower prematurely (e.g., "bolters"; Longden 1993) or are missed by harvesting equipment. Very few crops are apparently 100% male sterile (e.g., certain potato clones, certain ornamental varieties).

One hazard associated with pollen dispersal that has received considerable attention is that of sexual transfer of crop alleles to wild relatives, resulting in the evolution of increased weediness (e.g., Goodman and Newell 1985, Ellstrand 1988, Snow and Moran-Palma 1997). When wild relatives are nearby, natural hybridization with them is apparently a common feature of most cultivated plants (Ellstrand et al. 1999). The precedent has been set with traditional crops. Spontaneous hybridization between crops and their wild relatives has already led to the evolution of difficult weeds, such as weed beets in Europe (Boudry et al. 1993, Mücher et al. 2000) and weed rye in California (Suneson et al. 1969, Sun and Corke 1992). One could imagine that certain crop transgenes might also contribute to the evolution of increased weediness.

Another hazard associated with pollen dispersal is that locally common species can overwhelm those that are locally rare with their pollen, increasing the risk of extinction by hybridization (see Chapter 1). Extinction may occur in one of two not necessarily exclusive ways (Ellstrand and Elam 1993). The rare population may produce so many hybrids that it becomes genetically absorbed into the common species (genetic assimilation). Additionally, hybrids may have reduced fitness (outbreeding depression), and therefore the rarer of the species may be unable to maintain itself. The problem of extinction by hybridization has long been recognized as a conservation problem for animals (e.g., Rhymer and Simberloff 1996) but in plants has only recently received attention (e.g., Levin et al. 1996, Huxel 1999, Wolf et al. 2001). Nonetheless, theoretical models have demonstrated that extinction by hybridization can proceed rapidly, resulting in local extinction of a population in as few as a handful of generations (which, for some plants, could be less than a decade). For example, spontaneous hybridization between crops and their wild relatives has been implicated in increased extinction risk to wild species, ranging from the disappearance of wild coconuts (Harries 1995) to the contamination of California's wild walnut populations with genes from the cultivated species (Skinner and Pavlik 1994). There is no reason to imagine that this hazard should be any different for transgenic crops.

Hazards associated with the movement of transgenes within a crop, from one variety to another, has rarely been discussed. However, crop-to-crop hybridization may lead to the unanticipated natural "stacking" of transgenes, as in the case of the evolution of triple herbicide resistance in oilseed rape in Canada (Hall et al. 2000). Likewise, it is possible that crops transformed to produce pharmaceutical or other industrial compounds might mate with plantations grown for human consumption, with the unanticipated result of novel chemicals in the human food supply (Ellstrand 2001). While experience with traditional crops appears to offer no precedents for the latter hazards, it seems that if hybridization occurs so readily between crops and their wild relatives, it should occur even more easily between adjacent crops of the same species.

Hazards Associated with the Whole Plant

The transgenic plant itself may become an environmental hazard because the traits it receives may improve its fitness and ecological performance. Many crop plants may pose little hazard, insofar as they are unable to survive without human assistance. Frequently, traits that make them useful to humans also reduce their ability to establish feral populations in either agroecosystems or nonagricultural habitats. For example, lack of seed shattering and seed dormancy greatly reduces the ability of an annual crop to persist without human intervention. Without major changes in its phenotype, corn is unlikely to survive for multiple generations outside agricultural fields no matter what transgene is added to it (see discussion of epistatic effect in Chapter 1).

However, it is generally the case that most crops have weedy and/or wild populations in close association with cultivated forms in some part of their global distribution (De Wet and Harlan 1975). For example, sugar beets establish wild populations in the United Kingdom (Longden 1993). Depending on the location, certain crops (e.g., tomatoes) evolve to a wild-type phenotype very quickly and could become viable wild populations by the F_2 generation. The existence of these populations demonstrates that transgenes that confer adaptation to significant limiting factors can create significant whole-plant hazards, particularly if the ecological effects of transgenic crops are evaluated on global basis. The frequency of feral crop populations also reveals the difficulty of distinguishing gene flow and whole-plant hazards. Gene flow between feral crop populations and transgenic crops may create weeds that bear adaptations derived from the feral plants—such as seed dormancy—that suffice to produce new invasive-plant hazards in an agroecosystem or beyond.

From a U.S. perspective, whole-plant hazards are limited by the fact that many crops are highly unlikely to establish self-reproducing populations in the United States. Potatoes can establish wild populations in South America, but they are not known to establish wild populations in this country. Wheat is established in wild populations in the Middle East but not here. Many tropical plants cannot overwinter in most of the United States, and certain polyploid crops cannot produce viable seed. The factors that limit establishment of many crop species can be subtle and are not well understood.

For example, an annual crop could produce large quantities of viable seed with good seed dormancy characteristics. If, in addition, seedlings establish at high rates and produce viable F_1 and F_2 populations but germinate at the wrong time relative to weed control measures or herbivory, no viable population may be established. Thus, the mere presence of a transgene should not be taken as prima facie evidence that the weediness of a crop has been altered. Many crops are unlikely to be made more weedy by the addition of a single trait (Keeler 1989).

Some crops are capable of establishing wild populations in the United States. These particularly include crops that are only slightly modified from wild progenitors and that are adapted to U.S. conditions. This class includes many forage grasses, poplars, alfalfa, sunflowers, wild rice, and many horticultural species. Some domesticated species are important weeds of natural plant communities, including birds foot trefoil (*Lotus corniculatis*) and Bermuda grass (*Cynodon dactylon*). In these crop species the addition of a single transgene that improves some ecological characteristic could increase the weediness or invasiveness of the species, and these risks merit evaluation.

Non-target Hazards

Non-target organisms are any species that is not the direct target of the transgenic crop. To date, the vast majority of the published studies that examine non-target hazards have focused on *Bt* crops. For example, *Bt* corn is presently targeted to control the key pests, European corn borer (*Ostrinia nubilalis*) and southwestern corn borer (*Diatraea grandiosella*) (Ostlie et al. 1997) and *Bt* rice is targeted against striped stem borer (*Chilo suppressalis*) and yellow stem borer (*Scirpophaga incertulas*) (Cohen et al. 1996). Any other species affected by *Bt* corn or *Bt* rice is a non-target species; consequently, the list of potential non-target species is very long. These organisms can be grouped conveniently into five categories that are not mutually exclusive: (1) beneficial species, including natural enemies of pests (lacewings, ladybird beetles, parasitic wasps, and microbial parasites) and pollinators (bees, flies, beetles, butterflies and moths, birds and bats); (2) non-target pests; (3) soil organisms, which usually are difficult to study and identify to species; (4) species of conservation concern, including endangered species and popular, charismatic species (monarch butterfly); and (5) biodiversity, which may include the species richness in an area.

Transgenic pesticidal crops present a challenge because in some respects they are similar to chemical pesticides that are routinely regulated and in other respects they are similar to conventional breeding and other agricultural technologies that are not regulated at all. Scientific progress has been an awkward accommodation between these perspectives that has yet to reach scientific consensus. This issue is examined in detail later in this chapter.

Hazards can be difficult to demonstrate scientifically because it can be tricky to reproduce experimentally the route by which organisms are exposed. The present status of knowledge indicates the following possible hazards:

Beneficial Species. In well-controlled laboratory studies, *Bt* toxin similar to that in *Bt* corn increased mortality in green lacewing larvae (a predator of insect pests; Hilbeck et al. 1998a, 1998b). Both direct consumption of the *Bt* toxin and indirect consumption by eating caterpillars that had themselves consumed *Bt* toxin resulted in higher lacewing mortality. Field evaluations of this effect have been inconclusive (e.g., Orr and Landis 1997), although none of the published studies have documented an effect. Herbicide-tolerant crops might cause indirect reductions on beneficial species that rely on food resources associated with the weeds killed by the herbicides.

Non-target Pests. Most transgenic crops have effects on non-target pests. Although these effects are expected to sometimes be positive and sometimes negative, studies documenting only reductions in non-target pest populations have been published. For example, with *Bt* rice, it will be very important to confirm that it does not increase attack by rice planthoppers, the most important insect pests of rice (K. Sogawa, personal communication, Chinese National Rice Research Institute, 2000).

Soil Organisms. No effects on soil organisms have been documented in either the laboratory or the field. Surprisingly, however, *Bt* toxin leaks out of corn roots to detectable levels in the soil and may persist adsorbed to soil particles longer than nine months (Saxena and Stotzky 2000). The consequences of this unusual exposure route are not fully understood, although a number of *Bt* toxins have been found to be toxic to bacteria-eating soil nematodes (see BOX 4.2). Because there are so many species of soil organisms and their ecology is poorly understood, it is difficult to conduct comprehensive tests of many soil organisms. A different approach to non-target effects that might be considered would be to evaluate the effects of transgenic plants on targeted soil function, such as soil respiration, organic decomposition rates, and emissions of greenhouse and ozone-depleting gases. This has been done using nonisogenic comparisons (Donagan et al. 1995, 1996), but such evaluations probably would require the use of isogenic controls to the transgenic crops so that the effect of the transgene could be clearly evaluated (see discussion below on the need for isogenic varieties). Changes in these soil functions would indicate that some non-target species were being affected in ways that could have significant effects on the environment. Research is needed to evaluate the utility of such an approach.

Species of Conservation Concern. The effect of *Bt* corn on monarch butterflies (Losey et al. 1999, Jesse and Obrycki 2000) has captured widespread attention in part because it is a species dear to many people. Such charismatic non-target species gain great importance because of their symbolic significance to humans. The scientific data on this issue are still emerging (see BOX 2.1). Because some *Bt* corn events have higher toxin concentrations in pollen than others, the detrimental effect of some types of *Bt* pollen on monarchs is less likely for other types of *Bt* corn. In the upper Midwest, herbicide-tolerant soybeans might cause indirect reductions of monarch populations because their milkweed host plants are killed by the herbicides.

BOX 2.1
Effect of *Bt* Corn on Monarch Butterflies

Several types of transgenic *Bt* corn that are toxic to Lepidoptera have been commercialized in the United States. All have been deregulated by APHIS, but none have received permanent registration from EPA at the time of this writing.

Transformation Event	APHIS Deregulation Decision	Initial EPA Conditional Registration	Toxin	Trade Name and Company
Event 176	5/17/1995	1995, expired 4/1/2001	Cry1Ab, truncated, activated	Maximizer, Syngenta
Mon 80100	8/22/95	5/29/1996, voluntarily cancelled 5/8/1998	Cry1Ab, truncated	YieldGard, Not commercialized, Monsanto
Bt-11	1/18/1996	1996	Cry1Ab, truncated	YieldGard, Syngenta
Mon 810 Mon 809	3/15/1996	7/16/1996, Mon 810	Cry1Ab, truncated	YieldGard, Monsanto
DBT 418	3/28/1997	1997, expired 4/1/2001	Cry1Ac, truncated, activated	Bt-Xtra, DeKalb
Mon 802	5/27/1997	Not registered	Cry1Ab, truncated	Not commercialized, Monsanto
CBH-351	5/8/1998	5/1998, voluntarily cancelled 10/12/2000	Cry9c, not truncated, activated	StarLink, Aventis
TC 1507	6/14/2001	7/2001	Cry1Fa2, truncated, activated	Herculex, Dow and Pioneer

APHIS regulatory authority for evaluating risks to monarch butterflies and other non-target insects is indirect (see Chapter 3). These non-target species may feed on agricultural weeds, so reducing their populations could indirectly harm crop plants, causing an "indirect plant pest" risk. This differs substantially from EPA regulatory authority, which is directly aimed at protecting non-target species from the effects of a pesticide. In other words, it is possible for APHIS to find that monarch populations are harmed but that there is no increased plant pest risk and therefore no need to manage the risk. This could occur if milkweed abundance were not regulated by monarch density and, even if it were, milkweed might not be competitive enough against other plants to alter plant community composition. Hence, even if monarchs were to disappear, there might be no change in plant pest risk. This argument, however, has not been used by APHIS in its risk assessments of *Bt* crops.

APHIS evaluations of non-target effects on *Bt* corn have concentrated on the potential effects on beneficial organisms, such as insect natural enemies and pollinators (honeybees) and threatened or endangered species (see Chapter 4). APHIS did not request information on broader non-target effects, but most petitioners submitted data on potential effects on earthworms, other moths and butterflies, *Daphnia*, and other species. APHIS concluded that *Bt* corn could detrimentally affect only a specific group of Lepidoptera, but such effects would cause no additional plant pest risk. Threatened and endangered species were considered not at risk because none of these species feed on corn plants. But the potential toxic effects of *Bt* corn pollen were not assessed, and effects on monarchs were not considered explicitly in any of the APHIS risk assessments for deregulation up through 1998.

EPA granted a conditional registration to all of these *Bt* crops some time after the APHIS deregulation decision. The main reason for the conditional registration rather than a permanent registration was to enable the registrants to develop and implement scientifically based resistance management strategies prior to permanent registration. The effect of the conditional registration, however, was to maintain all of the *Bt* crops under active regulatory oversight, so that if any unexpected potential risks were identified, EPA could act to manage those risks. This has given EPA both the authority and the practical ability to make additional risk assessments of *Bt* corn long after APHIS had decided that it presented no plant pest risk.

At the time of its conditional registration of Event 176 (EPA 1995), EPA noted that corn pollen containing Cry1Ab toxin can drift out of cornfields, but the agency considered the amount of such pollen that would drift onto food plants of susceptible endangered species to be very small. Therefore, EPA did not expect that any endangered or threatened species would be adversely affected by *Bt* pollen-containing Cry1Ab toxin. Significantly, both the EPA and the APHIS evaluations concentrated on only endangered and threatened species. Monarch butterflies are neither endangered nor threatened, so the risk to this species was not considered in risk analyses conducted by either EPA or APHIS.

By the end of 1998 APHIS had deregulated five transformation events of *Bt* corn, finding that none of them created additional plant pest risk, and EPA had granted four of them conditional registration.

In May 1999, *Nature* published a brief communication by Losey et al (1999) that showed that *Bt* corn pollen was a potential hazard to monarch butterflies. The authors applied large amounts of *Bt* pollen from *Bt* corn to milkweed leaves and showed that larval mortality was higher, development was slower, and size was smaller in monarch larvae fed the leaves with *Bt* pollen compared to leaves with non-*Bt* pollen or no pollen. The authors suggested that *Bt* pollen could dust the leaves of milkweed and monarchs could eat it and suffer reduced fitness. The fact that the *Bt* toxins were harmful to monarchs was not the significant element of the paper. This effect could be anticipated because the toxins were known to be harmful to a specific group of Lepidoptera, of which monarchs were likely to be a member. The significance of the paper was to suggest that the movement of pollen onto other plants could create hazards for the insects feeding on those plants. Previous risk assessments had not given this exposure pathway thorough consideration.

This paper stimulated a number of responses during 1999. EPA initiated an internal evaluation of its *Bt* corn risk assessments to determine what, if anything, it should do in

continues

BOX 2.1 continued

response to the information. USDA directed the corn insects research group in Ames, Iowa, to conduct research to estimate the risks to monarchs, increasing the laboratory's budget. At this time, several agricultural biotechnology industries were cooperating informally in a *Bt* working group. This group initiated a series of studies by selecting and funding several public sector scientists to conduct studies to estimate the risks to monarchs. All of this research was planned and implemented rapidly during the early part of the growing season. During the fall of 1999, EPA issued a data call-in for evaluating the risks of *Bt* corn to monarchs. This is an infrequently used formal mechanism to require sufficient data from registrants and other interested parties so that EPA can make a science-based risk assessment.

A major assumption of the 1999 research was that monarchs do not occur in sufficient numbers in cornfields to merit investigation. The 1999 assumption meant that the researchers focused on estimating the risk that monarchs are killed near *Bt* corn. These unpublished research results from 1999 were used by the Agricultural Biotechnology Stewardship Technology Committee (ABSTC) to construct an initial evaluation of the risks of *Bt* corn to monarchs. The essence of the argument was that so few monarch caterpillars encounter *Bt* pollen near cornfields that monarch populations are not at risk (EPA 2000b). Careful observations at a cornfield in Iowa, coupled with results from two other fields, suggested that *Bt* corn pollen did not move in quantities sufficient to affect monarch larvae much more than a few meters from a cornfield. Spatial and temporal variations in pollen movement and retention on milkweed leaves were not characterized.

The industry group reorganized formally into the ABSTC in January 2000 to provide a mechanism for cooperation among companies to address common interests related to risks of agricultural biotechnology without risking lawsuits on collusion or other monopoly practices (members are Monsanto, Syngenta, Aventis, Dow, and DuPont). Prior to the 2000 field season, the ABSTC and the USDA organized to fund jointly research for the 2000 field season. After considerable negotiation, USDA agreed to match the money contributed by ABSTC (about $200,000) under the condition that a multi-stakeholder advisory panel established by ABSTC would advise the USDA on how the money should be allocated. This advisory panel was comprised of Eric Sachs (Monsanto), Margaret Mellon (Union of Concerned Scientists), J. Mark Scriber (Michigan State University), Adrianna Hewings (USDA), Eldon Ortman (Purdue University), and Richard Hellmich (USDA Agricultural Research Service). One of the major features of the proposed research for 2000 was to challenge the assumption that monarchs did not occur in significant numbers in cornfields. The research also focused on obtaining more accurate toxicity data for the various *Bt* corn events under both laboratory and field conditions, and further estimation of pollen deposition rates in the field.

Early in the summer of 2000, Jesse and Obrycki (2000) published their paper on *Bt* corn and monarchs in *Oecologia*. This paper confirmed the finding of Losey et al. (1999)—pollen from *Bt* corn was toxic to monarchs. Jesse and Obrycki (2000) collected milkweed leaves in and around cornfields and tested the toxicity of these leaves on monarchs in laboratory trials. The new finding was that the level of natural deposition of *Bt* pollen on milkweed leaves in cornfields was sufficient to kill monarch larvae. The authors also reported that pollen deposition was too low to kill monarchs at distances greater than 5m from a field edge. In addition, preshadowing results to come from the ABSTC/USDA cooperative research, the authors reported variable results from Mon 810 and *Bt*-11 *Bt* corn varieties.

Also during early summer 2000, Wraight et al. (2000) published a study showing that *Bt* corn pollen from Mon 810 had little effect on *Papilio polyxenes*, black swallowtail butterfly caterpillars. In laboratory bioassays, they found that Event 176 pollen, but not

Mon 810 pollen caused significant caterpillar mortality. These results indicate that the different events have different pollen toxicity to swallowtail caterpillars.

The results of the ABSTC/USDA funded research was published in September of 2001 in the Proceedings of the National Academy of Sciences (Hellmich et al. 2001, Oberhauser et al. 2001, Pleasants et al. 2001, Sears et al. 2001, Stanley-Horn et al. 2001). A number of significant results were found. Oberhauser et al. (2001) found that monarchs do use cornfields as oviposition sites. In fact, throughout the northern corn belt, more monarchs emerge from cornfields than any other habitat. Cornfields are not the most productive habitat per area, but because they occupy such an extensive area in the monarch's breeding range, they produce more monarchs. Moreover, monarchs use these cornfields when corn is shedding pollen in the northern part of their breeding range. Thus monarchs are more likely to be at risk inside *Bt* corn than near it. Hellmich et al. (2001) showed that Event 176 pollen is indeed quite toxic to monarch larvae, but Mon 810 pollen and *Bt*-11 pollen were not acutely toxic at average pollen densities that were observed in the field (Pleasants et al. 2001). Finally, Stanley-Horn et al. (2001) found that in the field, monarchs protected from natural enemies and exposed to pollen naturally deposited from Event 176 *Bt* corn had reduced survival and slower growth than monarchs exposed to non-*Bt* corn pollen, while pollen from *Bt*-11 and Mon 810 had no detectable effects. Thus, the researchers concluded (Sears et al. 2001) that the risk to the monarch is negligible for the two types of *Bt* pollen that comprised over 90% of the *Bt* corn area in the United States in 2001 (Sears et al. 2001).

In an independently developed study comparing field survival of monarchs and black swallowtails exposed to naturally deposited pollen, Zangerl et al. (2001) found that Event 176 pollen significantly reduced growth of black swallowtail, but had no detectable effect on monarchs. Mon 810 pollen had no significant effect on either species in the field. This lack of a significant effect of Event 176 pollen on monarchs contrasts with the significant effect observed by Stanley-Horn et al. (2001). Zangerl et al. (2001) did not protect monarch or swallowtail larvae from natural enemies, while Stanley-Horn et al. (2001) did. This raises the speculative possibility that in monarchs, in contrast to black swallowtails, density-dependent mortality from natural enemies compensates for any effects caused by Event 176 pollen.

Although the recent articles certainly establish a lack of significant risk of acute toxicity to monarchs, there are a number of questions that should be addressed in follow-up research. While the recent work was able to test for large sublethal effects in early larval development, it would be useful to determine if there are any sublethal effects of *Bt* corn pollen on monarch caterpillars that result in reduced female fecundity, reduced probability of overwintering survival, or reduced male mating ability. It would also be useful to determine if plant parts other than pollen land on milkweed leaves in cornfields and are eaten by monarch larvae (see Hellmich et al. 2001) because other floral parts have a high concentration of Cry1Ab toxin even in Mon810 and *Bt*-11 varieties.

Event 176 had a sizable share of the *Bt* corn market when it was first introduced. By the 2001 field season, Ciba Seeds (Novartis) did not even make an effort to sell it. The outcome for monarchs would have been substantially different had Event 176 become the dominant transgenic *Bt* event in corn given that pollen from Event 176 is over an order of magnitude more toxic to monarchs than that of Mon 810 or *Bt*-11. There is no reason to expect that a *Bt* corn using the 35S promoter must necessarily produce a better corn variety than Event 176, which uses other promoters. Indeed, had the toxin content of grain been a significant market force during 1996–1999 and had Event 176 been used in some of the best corn germplasm in the corn belt, it might have become the most popular variety of *Bt* corn grown because unlike Mon 810 and *Bt*-11, event 176 does not produce the *Bt* protein in it kernels.

Biodiversity is elusive because it embraces many dimensions, including the number of species, their relative abundances, the way they interact, and whether they are indigenous or exotic. From a conservation perspective, preservation of native species is a priority. Relatively little is known about the potential effects of transgenic crops on biodiversity from this perspective. However, herbicide-tolerant crops might reduce biodiversity by eliminating weeds that harbor many species. Such a reduction in biodiversity might then reduce the population densities of birds that rely on this biodiversity for food. Although this topic has been discussed specifically for genetically engineered herbicide-tolerant crops, it is, in fact, a potential impact of all herbicide-tolerant crops, regardless of their origins.

The state of knowledge about non-target effects of transgenic plants is improving slowly, but controversy surrounds each published study. The biggest gap in the research is in establishing scientifically rigorous assessment protocols that take into account the unusual exposure routes of transgene products. Consequently, considerable scientific work remains to be done before evaluation of non-target effects is standardized.

Hazard of Resistance Evolution

Resistance evolution can occur in pests that are targeted for control by or associated with a transgenic crop. This is a potential environmental hazard because if the pest becomes resistant to control, alternative, more environmentally damaging controls may be used. In addition, new control tactics may be rushed into use before their environmental risks are completely assessed. Insects, weeds, and microbial pathogens all have the potential to overcome most control tactics used against them (Barrett 1983, Georghiou 1986, Georghiou and Lagunes 1988, Green et al. 1990, NRC 2000c). Insect resistance to *Bt* crops is considered inevitable, and efforts are being made by the EPA to manage resistance evolution to these transgenic crops. Virus-resistant transgenic crops have not been used extensively, but many viruses have evolved resistance to conventional virus-resistant crops (Fraser 1990). Fungal and bacterial resistance is not yet commercially available in transgenic crops, but both groups of organisms have evolved resistance to conventional crop resistance, sometimes within five years (Delp 1988). Evolution of herbicide-tolerant weeds is an indirect environmental risk. Herbicide-tolerant transgenic crops are designed such that specific herbicides can be used to control weeds, usually after the crop has emerged. These postemergence weed controls might allow herbicides to be used only as needed, reducing herbicide applications to crops. In some crops these postemergence herbicides might replace ones that are more damaging to the environment. As weeds evolve resistance, though, these potential environmental benefits could be lost.

Finding 2.8: There are potential direct environmental hazards associated with the environmental release of transgenic crops. The kinds of direct hazards (or categories of hazards) are those associated with the movement of the transgenes, escape of the whole plant, non-target effects, and resistance evolution.

Indirect Hazards

Transgenic crops can have indirect environmental hazards, especially when scaled up for commercial production. Although analysis of these indirect hazards is complex, they may in some cases be as important or more important than direct hazards. Some of these indirect hazards are discussed in other chapters.

Environmental Risks of Transgenic Crops and Conventionally Bred Crops

Despite nearly 20 years of scientific discussion, scientific debate continues about whether the risks associated with transgenic crops are similar to or different from those associated with conventional crops. The source of scientific disagreement is that in some ways transgenic crops are just like conventionally bred crops, but in other ways they are quite unlike conventionally bred crops. Clarifying the ways that they are similar and dissimilar can sharpen the scientific dimensions of this issue. This issue is of more than academic interest. Its clarification can provide a scientific justification for the present U.S. regulatory system.

Conventional crop breeding can be broadly understood to consist of two relatively distinct processes. The first is a set of activities to create genetic novelty in the plant population. In the vast majority of cases this is done by crossing several elite varieties together to create a mixed population (see Chapter 1). The second set consists of activities that select out of that genetic novelty genotypes that are useful new crop varieties. These methods include a number of techniques, including recurrent selection, mass selection, marker-assisted selection, and "selfing selection." Crop breeding must be understood as the integration of these two sets of processes.

Transgenic methodology is a new process for introducing genetic novelty into a crop plant. It is not a process of selection. Transgenic methods as they are applied to crop variety production are only a part of the crop-breeding process. Indeed, it is because transgenic methods provide new ways of introducing genetic novelty into a crop that they are not similar to conventional plant breeding. As pointed out in Chapter 1, new techniques such as wide hybridization, embryo rescue, and functional genomics that are categorized under the umbrella of conventional breeding are also novel approaches for introducing new qualities into crops.

Transgenic crops bear many similarities to conventionally bred crop varieties because they are both selected by the same conventional methods; crop varieties therefore bear some similarities. It is this confounding of parts of the breeding process with the whole breeding process that has created some of the scientific confusion.

At a deeper level, however, the general comparison of environmental risk of transgenic and conventional crop varieties probably cannot be resolved scientifically at the present. The unresolved scientific challenge is to determine if the range of genetically engineered crop varieties has a similar degree of environmental risk as the range of conventionally produced crop varieties. Theoretically, the most quantitatively robust estimate of environmental risk for transgenic crops would compare the probability of environmental damage associated with transgenic crops to the probability of environmental damage for all crops. This would allow the risk from commercialization of transgenic crops to be expressed as a conditional probability given the presence of transgenes. Such comparison is impossible. There has never been a systematic attempt to measure the probability that a randomly selected agricultural crop will cause environmental damage; hence, the baseline data that would be needed to derive a conditional probability estimate simply do not exist. In addition, because the range of possibilities attainable by both conventional and transgenic methods is expanding rapidly, any scientific determination would be provisional and rapidly outdated. As argued in Chapter 1, while our long experience with some conventional crop breeding techniques indicates that for the most part it has been safe for humans and the environment, there are cases where it has led to environmental harm.

Most inquiries into this issue have focused on one of two narrow questions. First, does a commercial transgenic crop variety with a particular trait have similar ecological risk as an isogenic commercial variety with the same functional trait produced by a conventional method? APHIS expands this argument to assert that the risk of the transgene is not greater than the risk of a comparable conventional variety that is neither isogenic nor possessing the same trait (see Chapter 4). There are no published scientific studies that demonstrate the similarity of risk at either of these levels of comparison (see BOX 1.2).

Second, does a commercial transgenic crop variety with a particular trait have similar ecological risk as an isogenic conventional commercial variety without the trait? Crawley et al. (2001) provide some evidence to this point, although it is not clear if the comparisons were isogenic ones. For the several crop-trait-environment combinations that were tested, no difference in weediness was found within the pairs. Crawley et al. (2001)

warn not to generalize from these cases because in none of the cases would the investigators have hypothesized that the added trait could have increased weediness of the crops tested. Moreover, they evaluated only one hazard within the large class of potential hazards associated with transgenic crops. Similarly, claims that the lack of effects from the tens of millions of hectares of transgenic crops that have been planted in the United States during the past three years are nonscientific. There has been no environmental monitoring of these transgenic crops, so any effects that might have occurred could not have been detected. The absence of evidence of an effect is not evidence of absence of an effect.

The Trigger for Risk Analysis

Although it is generally agreed that the risks associated with any transgenic organisms should be evaluated on the basis of the trait, organism, and environment without reference to the process of transformation, there is considerable disagreement about the scientific basis for triggering regulation of transgenic crops. At one extreme, some might argue that the regulatory trigger should also be based on some ex ante risk evaluation based on the trait, organism, and environment, but at present there is no scientific basis on which to build such an ex ante system. In the absence of such a system, the committee argues below that transformation is both a useful and logically justifiable regulatory trigger. Transgenic organisms have potential environmental risks, but the committee expects that most of them will not produce significant actual environmental risks. Consequently, the committee also suggests that for environmental risk regulatory oversight should be designed to winnow the potentially riskier transgenic crops from the less risky ones before a substantial regulatory burden is imposed on the less risky ones.

This report makes a number of comments about the properties of transgenic and conventionally bred crops that seem, at least superficially, to be in conflict. One of the committee's key findings (2.6) is that "the transgenic process presents no new categories of risk compared to conventional methods of crop improvement, but specific traits introduced by *either* of the approaches can pose unique risks." Chapter 1 points out that any addition of genetic novelty can have ecological impacts and that there is no predictive relationship between the number of new genes and the degree of impact. It also finds that there is no relationship between the taxonomic distance between the donor and recipient of a gene and the environmental risk. In addition, another committee finding (2.4) is that regulation should be based on the trait, organism, and environment. Thus, a transgenic stress-tolerant trait in canola and a conventionally produced allelic trait in canola should receive equal regulatory scrutiny. However,

APHIS would not examine the conventionally bred, stress-tolerant canola because of the process used to breed it. Therefore, it might appear that the APHIS trigger is not justifiable.

Before discussing the scientific details of the issue related to transgenic crops, the committee addresses two related issues. First the Office of Science and Technology Policy (OSTP) scope statement (1992) concerning the regulatory trigger is reviewed. The guidance is found to be ambiguous. Second, another regulatory trigger used by APHIS under the same statutory authority that it regulates transgenic crop plants is reviewed. The committee finds that APHIS has adopted a different process-based trigger to meet these other regulatory responsibilities.

OSTP Scope

The entire coordinated framework (OSTP 1986) is premised on regulating transgenic organisms (i.e., using the process of transformation to trigger the entire regulatory apparatus it describes). Hence, from its very inception, use of a process-based trigger to regulate transgenic organisms has existed. The coordinated framework did not fully address how oversight should be exercised by federal agencies under the various statutes, so in 1992 OSTP issued a statement on the scope of oversight so that common principles would govern decisions on how to regulate transgenic organisms.

The Final Statement of Scope (Section III) states that:

> federal agencies shall exercise oversight of planned introductions of biotechnology products into the environment only upon evidence that the risk posed by the introduction is unreasonable. A risk is unreasonable [when] the full value of the reduction in risk obtained by oversight exceeds the full cost of the oversight measure. (6,756)

Any regulatory trigger that adheres to this principle is consistent with the OSTP scope statement. This statement is silent on the use of the transformation process as a trigger for regulatory oversight. While the final statement supersedes all previous discussion on scope, it is nevertheless instructive to examine some of the principles in the 1992 document that the final statement supercedes.

In Section II.A (6,756), three fundamental scope principles are articulated:

> 1. A determination to exercise oversight within the scope of discretion afforded by statute should not turn on the fact that an organism has been modified by a particular process or technique, because such fact is not alone a sufficient indication of risk.
>
> 2. A determination to exercise oversight in the scope of discretion afforded by statute should be based on evidence that the risk presented by

> introduction of an organism in a particular environment used for a particular type of application is unreasonable.
>
> 3. Organisms with new phenotypic trait(s) conferring no greater risk to the target environment than the parental organisms should be subject to a level of oversight no greater than that associated with the unmodified organism.

The first principle is the only one that directly relates to the issue of using a process-based trigger, and it clearly proscribes the use of one if it turns on the fact that an organism is transgenic. The principle does allow the fact that an organism is transgenic to be a part of the basis of a trigger; it simply proscribes its use as the primary basis for a regulatory trigger. More significantly, principle 2 is adopted as the core of the relevant part of the Final Statement on Scope, and principle 3 is mentioned prominently in the implementation of the document, as discussed below. Only principle 1 fails to appear in any form in either the final statement or the implementation. Consequently, principle 1 is fully superseded by the final statement, and a fully process-based trigger is consistent with the 1992 OSTP scope document.

The scope document provides several paragraphs of guidance on implementation (Section IV) of the Final Statement on Scope. Section IV.A (Exercising Discretion Within the Scope of Statutory Authority) clarifies that the final statement governs all federal oversight of transgenic organisms. This section is otherwise silent on the issue of using a process-based trigger. Section IV.B (Evaluating Risks) clearly embraces principle 3 (above) as a core principle (paragraph 4). Section IV.D (Use of Categories of Exclusion/Inclusion) is the only section to mention process-based triggers explicitly. The document notes that:

> several commenters suggested that the [previously proposed categories of] exclusions were inconsistent with a risk-based approach because they were "process-based." . . . Because this Final Statement is to be a guidance document to the agencies, it is not meant to provide the risk rationales for these examples [the exclusions]. Any agency that wishes to use any of these categories in the context of a specific statute would provide a rationale based on risk. (6,758)

Thus, it is clear that the Final Statement on Scope allows an agency to use a trigger that is process-based as long as it provides a rationale based on risk.

Federal Plant Pest Act and APHIS Regulation

Invasiveness, which is the alteration of community or ecosystem structure or function, is the main environmental risk of nonindigenous species. Invasiveness is a function of both the organism phenotype and the environment. As discussed in Chapter 1, only a small minority of nonindigenous species is invasive (see Figure 2.1), but at present we are unable to predict ex ante which species will be invasive and which will not. Consequently, it is impossible to construct a scientifically justified regulatory trigger based on the actual risk of invasiveness.

APHIS asserts its regulatory authority over the importation of nonindigenous species under the Federal Plant Pest Act (FPPA). The vast majority of nonindigenous organisms introduced into the United States arrive inadvertently in commercial shipping or deliberately through the trade of living organisms (aquarium fish, pets, horticultural plants). There is also a small but significant possibility that nonindigenous species are brought into the country by people arriving in airplanes and ships. For these passengers, APHIS uses two triggers for additional oversight. If a passenger is carrying animal or plant products, including fruits, meats, and living plants and animals, APHIS inspectors may confiscate or quarantine such products. If a passenger has visited a farm, APHIS inspectors may confiscate or quarantine clothing. While these triggers have a risk-related basis, they are also process-based triggers. Merely visiting a farm does not imply that a passenger is carrying an invasive nonindigenous species. However, as long as there is information indicating that, on average, passengers that have visited farms have a higher probability of carrying such a species, the process of visiting a farm is a useful risk-based trigger. These processes can be logical, operational ways to identify a subset of potentially risk-inducing activities, that will trigger oversight at U.S. borders.

Finding 2.9: A logical scientific argument can be made without reference to conventional crops that all transgenic crops should enter into regulatory oversight.

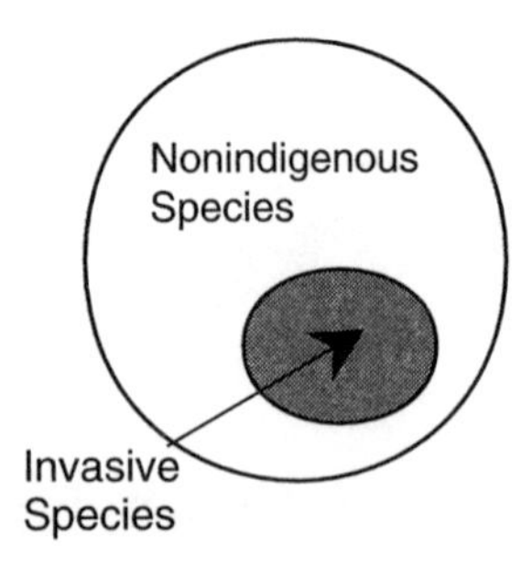

FIGURE 2.1 Venn diagram illustrating that invasive species are a small subset of all possible nonindigenous species.

Justification: Chapter 1 argues that, although most introduced genes will have no significant adverse environmental effects, any change in genetic composition can cause ecological impacts. (This is especially true for traits that are specifically introduced to change the ecological performance of a plant.) Because we cannot judge, ex ante, which genetic changes will have detrimental effects, it is important to take a case-by-case approach to the review of potential impacts. Most genetic changes are expected to have low risk, so they should be reviewed quickly and determined to have low risk. A smaller fraction will need closer scrutiny and can be judged to have potential for undesirable impacts. Therefore, there is a scientific basis to examining all genetically engineered crops, at least briefly, and then to examine some more carefully when concerns are raised during the quick examination.

Finding 2.10: Even if the risks of all conventionally bred crops are considered to be "acceptable," there is still a logical scientific justification for genetically engineered crops to enter into regulatory oversight.

Justification: In the 1980s when the United States began to develop its system for regulatory oversight of genetically engineered crops, an assumption was made that, even though conventionally bred crops were not considered to be completely risk free, the risks associated with the entire class of crops could be considered "acceptable" to society (see Figure 2.2). This assumption that any trait produced in a crop by conventional means was acceptable gave a clear baseline for judging traits added by other means. Anytime it could be shown that a trait added to a crop by genetic engineering was substantially equivalent to a trait added by conventional breeding, the genetically engineered trait could be judged to have acceptable risk. Because some transgenic traits that have and could in the future be added to crops that do not have conventional equivalents, some will need to be assessed. It is not possible for a regulatory agency to judge ex ante whether a transgenic crop is similar to a conventional crop variety and therefore has acceptable risks, so some regulatory evaluation must be done. Moreover, because this assessment should be done on a case-by-case basis, all transgenic crops should be reviewed through regulatory oversight. Again, most genetic changes would be expected to have low environmental risk, so all transgenic crops should be reviewed quickly to identify the smaller fraction that will need closer scrutiny. Unlike the risk assessment called for in the argument accompanying Finding 2.9, the initial determination called for here would be an assessment of whether there is a conventional analog to the transgenic plant. Therefore, there is a logical basis to examining all genetically engineered crops

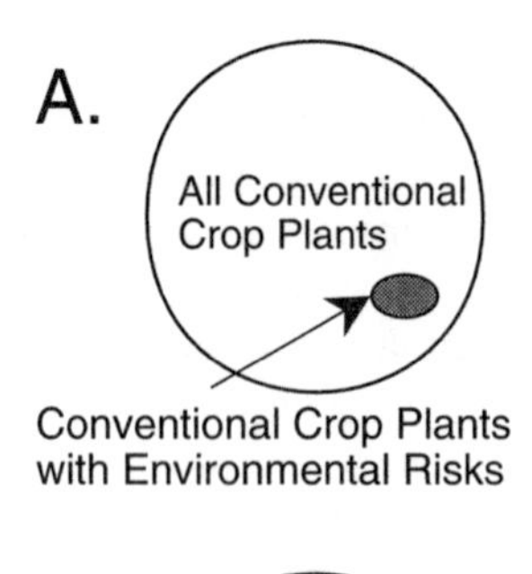

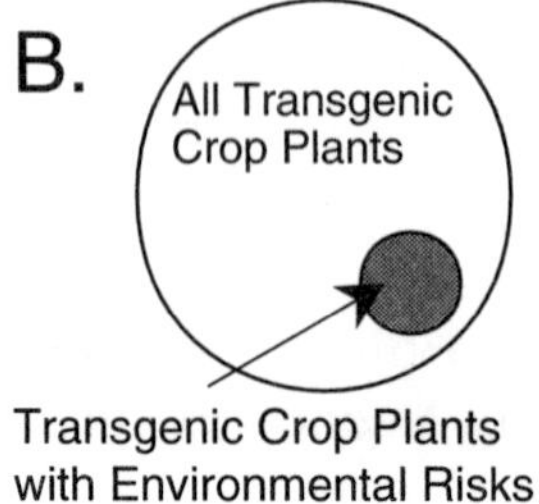

FIGURE 2.2 A. Set of all conventional crop plants and those with unacceptable environmental risk (assumed to be none). B. Similar diagram for transgenic crop plants. Arrow points to the subset of plants with risks.

to determine if they are similar to some conventionally produced variety. If they are dissimilar, a more detailed risk assessment would be called for.

The assumption that all conventional crops have acceptable risk, however, leads to a logical contradiction. If a conventional method, such as mutational breeding, had been used to develop a sulfonyl urea-resistant canola cultivar, then as long as the cultivar produced by genetic engineering could be shown to be similar to the conventional cultivar, the risks of the engineered cultivar would be considered acceptable. If, by chance, no sulfonyl urea-resistant cultivar had been produced by mutational breeding or any other conventional method, the engineered cultivar would not be considered, a priori, to have an acceptable risk but would need greater oversight. In other words, the level of oversight does not depend on the characteristics of the transgenic plant—namely, the trait, plant, and environment—but on the existence of another conventional plant with similar characteristics. This logic also holds if the conventional herbicide resistance was to an environmentally damaging compound, such as atrazine. This could lead to the ludicrous conclusion that a transgenic atrazine-resistant plant would be considered safe for the environment while a transgenic sulfonyl urea-resistant plant would need to undergo considerable regulatory scrutiny.

Finding 2.11: The risks associated with crop cultivars that have been or could be developed through conventional breeding should not be assumed to be acceptable.

Justification: Chapter 1 provides examples of conventionally bred cultivars that have had negative environmental impacts that would have been considered serious enough to question the deregulation of a genetically engineered crop or the introduction of an exotic species. There are also examples of traits such as tolerance to a particular herbicide and drought tolerance in conventional crops that would have been carefully assessed for environmental impacts had they been introduced by genetic engineering. The committee's finding in Chapter 1 that specific traits developed by both conventional and transgenic methods could have unique risks underscores the fact that the assumption that all conventionally bred crops always have "acceptable risks" is not scientifically justified (see Figure 2.3).

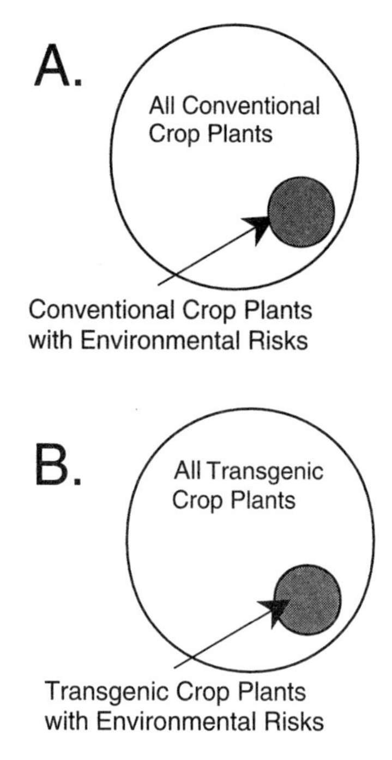

FIGURE 2.3 A. Set of all conventional crop plants with the small subset that have environmental risks, which is the actual situation. B. Similar diagram for all transgenic crop plants. Arrow points to plants with risks.

Finding 2.12: If we reject the assumption that all risks associated with conventionally bred crops are acceptable risks, there is still scientific justification for regulatory oversight of genetically engineered crops. However, substantial equivalence to conventionally bred crops cannot always be used as a reference baseline.

Justification: If we accept the fact that some conventionally bred crops can have significant risks, we cannot automatically conclude that a transgenic crop with a similar trait needs no regulation. For example, just because a sulfonyl urea-tolerant conventionally bred canola exists does not mean that the potential environmental effects of engineered canola with this trait should not be carefully reviewed.

Finding 2.13: The exclusion of all conventionally bred crops from regulatory oversight cannot be justified on a scientific basis. While it would be imprudent to immediately capture all conventionally bred crops under regulatory oversight, there is a need to reexamine the potential environmental impacts of products from traditional and emerging crop improvement technologies that fall under the general category of conventional breeding.

Justification: Because at least some crop varieties produced by conventional breeding can have substantial negative environmental impacts, it is impossible to come to a scientific conclusion that no conventionally bred crops should enter into regulatory oversight. It is nevertheless important to reexamine the safety of conventionally bred cultivars, especially because a number of novel plant-breeding techniques that fall under the broad category of conventional breeding have recently expanded our ability to transfer and accumulate new genetic traits into crop cultivars. The considerable experience and evidence associated with most conventionally produced crop varieties will probably be useful in developing a sound scientific approach to such oversight. Specifically, new conventional varieties of many of the dominant crops in the United States are quite unlikely to cause detrimental environmental effects, so the proportion of risky conventional varieties may be small (see Figure 2.3). As discussed in Chapter 1, there has been an evolution in our concepts of agriculture and the environment over the past 40 years, from a situation where the environmental effects of agriculture were largely ignored to the present recognition that what happens on the farm can affect the health of other ecological systems. In a world with this changed perspective, it is reasonable to reexamine the conventional techniques developed in previous eras. However, a change in regulations that resulted in the environmental assessment of a substantial fraction of conventionally bred crops could have negative effects on conventional plant breeding, so such oversight would need to be developed carefully.

Because transgenic crops are new genotypes introduced into the environment, they fall under the legal and regulatory authority of the Federal Plant Quarantine Act (FPQA) and the FPPA. Clearly, this authority could be extended to regulate any aspect of conventional crop variety production. Because there are no general scientific principles for identifying safe transgenic plants and there is a scientific expectation that most transgenic plants will have little environmental effect, the entire category of transgenic crops has been brought in for regulatory review under the FPQA and FPPA. As outlined above, when this is done, there should be a rapid initial evaluation to separate transgenic crops needing additional regulatory review from those that do not. At first, APHIS had no rapid sieve to separate transgenic crops. Consistent with the scientific and logical rationale discussed above, the agency constructed an early rapid sieve empirically, based on the results of its actual risk assessments. Empirical construction avoids the difficulties and pitfalls of ex ante prediction inherent in many of the scientific arguments about the safety of transgenic crops. However, if it is constructed on an overly broad generalization, the sieve can be faulty. The APHIS notification system is the early rapid sieve. While the committee agrees that the general scientific and logical principles underlying this approach are sound, Chapter 5 evaluates the implementation of these principles.

REFERENCE SCENARIOS—THE COMPARATIVE RISK APPROACH

Reference scenarios or reference values are those situations to which the focal risk is compared to contextualize understanding of the risk situation. This will depend on the risk characterization process. For example, if the context is understanding whether the transgenic crop is likely to increase or decrease risks in the area of introduction, the risks of transgenic crops could be compared to the dominant agricultural practice in the area of adoption or to the best agricultural alternative in the area. Thus, *Bt* cotton is compared typically to conventional cotton grown with high rates of insecticide application, while *Bt* corn is typically compared to conventional corn without any insecticides applied. In a broader context, transgenic crops could be compared to alternative technologies or technological systems for reaching similar production and environmental goals. For example, the risk characterization might evaluate the policy issue of which technologies or technological systems have lower environmental risks and should be encouraged for future development. Under this context, transgenic crops could be compared with sustainable agriculture and organic production methods.

The context could be considered even more widely in a developing country where a transgenic crop might replace local varieties and cause concentration in the seed distribution system so that relatively few people control it. In some of these countries, one instructive reference scenario might be a different agricultural input that changed from being created on farms with relatively little external supplementation to being purchased from a limited number of suppliers. In some parts of the world, soil fertility had been created on farms as animal manure but is now purchased as chemical fertilizers. The kinds of hazards associated with concentration of the seed industry might be paralleled by the hazards that occurred during concentration of the fertilizer industry. A second instructive reference scenario might be a strengthened on-farm seed production system. This would address the policy concern of which systems should be supported for future development. Clearly the reference scenario determines the context for comparison. Thus, there is no one "best" reference scenario because the risk characterizations may need to meet various contexts.

Several key implications of this perspective about reference scenarios for the characterization of risks associated with transgenic crop plants are discussed in the following sections. While some of these implications are relevant to conventionally produced crop varieties, the following discussion concentrates on their application to transgenic crop varieties. The first is that for the evaluation of risks of transgenic crops several reference scenarios can be found, which imply that risk comparisons to conventional breeding are sometime irrelevant, as illustrated by some of the examples described above. Thus, general statements about the relative risk of particular transgenic crops to their conventional counterparts will not be the only basis for risk characterization. Indeed, it is the specifics of the trait, plant, and environment that have the greatest utility in risk characterization, not the specific comparison to conventional crops. The second issue relates to how to characterize the transgenic organism in a way that it can be compared to appropriate reference scenarios. No one method is used worldwide, and two different conceptual approaches are compared here. A third issue concerns a technical component of risk analysis. By introducing some of these more technical methodologies, new questions and issues can be addressed. Specifically, a fault-tree analysis could be helpful for developing a regulatory system that self-corrects itself by learning from its past activities. Finally, the need for closer, more scientifically rigorous experimentation is discussed.

Appropriate Reference Scenarios

Depending on the potential exposure pathway and the potential hazard, various reference scenarios could be chosen. For example, the potential risks from activated Cry toxin in the soil can be evaluated by comparisons to the effects of other synthetic pesticides in the soil, of formulated *Bt* insecticides, or by experimentation using traditional soil toxicology methods. Non-target risks from activated Cry toxin might be evaluated using standard species and methods for non-target testing for insecticides, but these methods may use inappropriate test species and inappropriate exposure methodologies. Typical high-dose toxicology testing is designed to make qualitative assessments of whether a compound has toxicity to an organism, and to measure small quantitative effects. As discussed more fully in Chapter 3, short-term laboratory toxicology testing may be useful in assessing acute effects of a toxin, but when a toxin has more than one mode of action, chronic effects on non-target species fitness may be missed. Furthermore, pesticide toxicology testing was developed for chemicals that had broad effects on the nervous systems of invertebrates and vertebrates. The model organism approach may have been more relevant for these chemicals than for proteinaceous toxins with specific physiological effects on a narrower range of species that are common in transgenic plants.

The potential hazard that antibiotic resistance genes could be transferred from transgenic crops (such as *Bt* corn with the *nptII* gene) to soil bacteria is unique to transgenic crops. However, even this unique hazard can be evaluated using appropriate reference models for comparison. A laboratory model optimized so that the bacteria are readily transformed by naked DNA in the culture environment provides a worst-case scenario (Nielsen et al. 2000). The rate of occurrence in the natural environment is assumed to be less than that in the permissive experimental environment. The experimental transformation rate was low, which suggests that gene transfer via transformation in natural environments is correspondingly rare. Another informative comparison is the number of *nptII* genes in the transgenic crop plant compared to the number in the soil. In several agricultural soils, the *nptII* gene is already present at 10^5/g soil (Smalla et al. 1993), a very high rate of occurrence. This high rate of occurrence may be characteristic of only this antibiotic resistance gene, but this comparison suggests that a transgenic crop may not be a significant source for this resistance gene in nature.

It should be clear from these examples that appropriate reference comparisons depend on the nature of the risk. Conventional plant breeding is not always an appropriate reference comparison and for the cases presented here, it was not used as an appropriate reference. Second, there are unique risks associated with certain transgenic crops. The categories that these risks correspond to, however, have been previously identified, and many methods exist to evaluate them. In this sense, transgenic crops present no new types of risks, although specific risks can be unique to specific transgenic crops.

Finding 2.14: There are typically multiple reference scenarios that are appropriate in assessing the risks of transgenic crops.

Characterizing the Transgenic Organism

There are several approaches to characterizing the transgenic organism for comparing risks. In the United States, APHIS divides the transgenic organism into two parts—the unmodified organism and the transgene and its products. The hazards associated with these parts are identified, and the likelihood that the hazard will occur is assessed using a simplified fault-tree analysis (Lewis 1980, Haimes 1998). One alternative to this two-part model is a whole-organism analysis. And in addition to the fault-tree approach, an event-tree analysis also is considered. These appear at first to be very subtle distinctions, but they lead to a different structure in the risk analysis.

Two-Part Model

In the two-part model the transgenic plant is conceptually divided into (1) the unmodified crop and (2) the transgene and its product. The risk assessment then evaluates the risk of the unmodified crop separate from the risks associated with the transgene and its products. For example, a herbicide-tolerant soybean would be evaluated as an untransformed soybean plant for which no environmental risks have been found and as the appropriate herbicide resistance gene and its transcribed product or products (see other examples in Chapter 4). The theoretical rationale for this approach is that the transgene is a small genetic change that is likely to have only a small phenotypic effect. Therefore, the untransformed soybean is unlikely to be changed very much and provides a good baseline comparison for understanding the risk of a transgenic plant. The transgene itself is considered an incremental change in the soybean, and its risks are assumed to be able to be assessed independent of the soybean.

After these assessments are completed, the whole transgenic organism is assessed to determine if there are any additional hazards that would require assessment. If so, appropriate reference comparisons are developed. Conceptually, this reductionist methodology attempts to identify and assess most of the risks in the initial step. The final step is to pick up any residual effects stemming from any interactions between the trait and the untransformed organism that require additional assessment.

The two-part model makes a procedural commitment to making the untransformed crop plant the central reference scenario for contextualizing the risks of a transgenic crop. As discussed below, to evaluate the effect of the transgene, the most scientifically revealing comparison is the transgenic variety and its near-isogenic non-transformed parent. However, near-isogenic lines are not always available to allow the best scientific comparisons to be made. In addition, the two-part model has not used some alternative reference scenarios, such as a range of cropping systems or alternative approaches to accomplishing similar production or environmental goals as the transgenic crop. Finally, the two-part model assumes that single gene changes have small ecological effects. As discussed in Chapter 1, this is not always the case.

Whole-Organism Model

In a whole-organism model, the reference comparisons to the whole transgenic organism are developed at the beginning of the risk analysis. This opens up the scope of reference comparisons to include some that might not be considered under the first model. This model does not require that all risks are assessed using a whole organism; this is a conceptual model that provides an approach for identifying appropriate hazards and reference scenarios.

For example, under the two-part model, *Bt* corn was assessed by considering first the unmodified corn plant and the Cry toxin. Because corn was considered to pose insignificant environmental risks, it has been dismissed routinely as a risk. Cry toxins were evaluated for their insecticidal properties during the registration of *B. thuringiensis* as an insecticide, and these evaluations were used to assess the risks of the toxin. The appropriateness of this comparison can and has been challenged, in that where APHIS found no hazards in its initial assessments (see Chapter 4), EPA identified the evolution of resistance in the target pests to the toxin as a hazard. In posing the question of whether there are any additional hazards arising from the interaction of the transgene and the plant, relatively few hazards have been identified, perhaps in part because the methodology has not been firmly established. In general, as shown in Chapter 4, APHIS relies on an analysis that is based on comparing the transgenic crop to a conventional crop with a similar phenotype. If a similar phenotype can be identified, APHIS contends that because the conventional crop phenotype has a record of safe use, the transgenic crop plant is expected to be safe too. In the case of *Bt* corn, APHIS compared the transgenic crop to other insect-resistant corn varieties because no conventional *Bt* variety exists. While the validity of such comparisons can be challenged (see Chapter 5), the method is consistent with the two-part model. A whole-organism model would initially focus on identifying the potential hazards associated with the transgenic crop variety. The appropriate reference scenario for hazard identification may include the untransformed crop and conventionally produced crops with comparable phenotypes, which are used in the two-part model, but risk comparisons are not restricted to these two scenarios. For example, the transgenic crop will be expected to replace other agricultural production systems and possibly some nonagricultural native or natural habitat, and it may affect neighboring agricultural and nonagricultural habitats. Hence, the risks associated with a *Bt* corn might be compared to corn sprayed with insecticides in the irrigated regions of northeast Nebraska because it might be expected to replace this production system at that locality. In Minnesota, Iowa, Illinois, and the rest of Nebraska, *Bt* corn might be logically compared to an unsprayed conventional crop. In a similar way, *Bt* poplar might be compared to conventional poplar plantations sprayed or not sprayed with insecticide, but because it might lead to an increase in area planted to plantation poplar, it could also be compared to the risks associated with second-growth forest, which it might replace. In addition, because *Bt* corn might affect neighboring organic farmers through contamination via cross-pollination, and *Bt* poplar might affect the ecological functioning of neighboring forest and savanna ecosystems, these ecosystems might also be appropriate reference scenarios. The theoretical rationale supporting the whole-organism model is that the risks and hazards associated with a transgenic crop plant occur as a result of the trait *in* the crop plant *in* a particular environment. The risks and hazards do not occur from the trait separated from the crop plant separated from the environment.

Comparison of the Two-Part and Whole-Organism Models

In principle, the two approaches should end up with the same characterizations of risk, but in practice they probably will not. The two-part model addresses whole-plant concerns in its second step and should therefore capture all hazards and risks at this stage. The methodology used in the second step of the two-part model, however, must be different from the whole-organism model because if it were the same, the first step would not be necessary. It is in this difference that the models probably diverge. There are potential biases in the two-part model, which may underestimate or overestimate risk. The initial step of dividing the organism and evaluating the risk of the parts may lead to an underestimation of risk when the parts are found to have no significant risk and, vice versa, may lead to an overestimation of risk when the parts are considered risky. This is because these initial findings are likely to influence subsequent investigations used in the second step. If no significant hazards or risks are characterized in the initial step, it may be difficult operationally to justify and sustain a lengthy second step risk analysis. Conversely, an initial finding of risk may unleash a cascade of additional investigations, perhaps leading to an overestimation of risk.

Risk Assessment Models

Risks can be assessed in numerous ways, with various assumptions and biases. Transgenic crops have been assessed in the United States by APHIS utilizing a rigid risk analysis framework that lacks a formal approach for detecting potential for mistakes. Certainly the system is designed to limit mistakes, but mistakes do occur, and their significance is unknown. Two different analytical frameworks are discussed below that could be used to supplement present risk assessment practices—a fault-tree analysis and an event-tree analysis (Lewis 1980, Haimes 1998). Fault-tree analysis can be used to investigate potential risk. For example, one can hypothesize that *Bt* corn adversely affects some non-target organism. A fault tree could lead the analyst to investigate thoroughly all potential causal pathways.

Fault-Tree Analysis

Fault-tree analysis logically evaluates risk by tracing backwards in time or backwards through a suspected causal chain the many different ways that a particular risk could happen (Lewis 1980, Haimes 1998). The analysis is conditioned on a given hazard, that is, the analyst must have a particular hazard in mind before the analysis can be conducted. This is both its strength and its weakness. It is a strength because the analysis focuses on the ways that risks occur and does not waste time evaluating the ways that risks do not occur. By concentrating on the known hazards, the analysis provides a comprehensive and efficient methodology for assessing risk. It is a weakness because unknown or unanticipated hazards cannot be evaluated simply because they have not been identified. Because the hazards associated with complex systems cannot all be unambiguously identified ex ante, this is its most serious weakness.

Figure 2.4 is a simplified fictitious example of a fault-tree analysis of the risk of *Bt* corn to a non-target insect species. The total risk is a weighted "or" gate of the risks in cornfields, near cornfields, and far from cornfields, where the weight is the relative proportion of the insect population in each habitat area (in the example in Figure 2.4 these weights are equal). In an "or" gate the risk would occur if any one of the inputs occurs, and in an "and" gate the risk would occur only if all of the inputs occur. This risk calculation is detailed for risks to the species inside cornfields, which is an "or" gate of the risks associated with each of the three most common transformation events in *Bt* corn. This risk, in turn, is an "and" gate of the probability that monarchs are in *Bt* corn of the specific event and the probability that they will be killed by pollen from the event. This box is further detailed for Event 176, in which mortality is related via an "and"

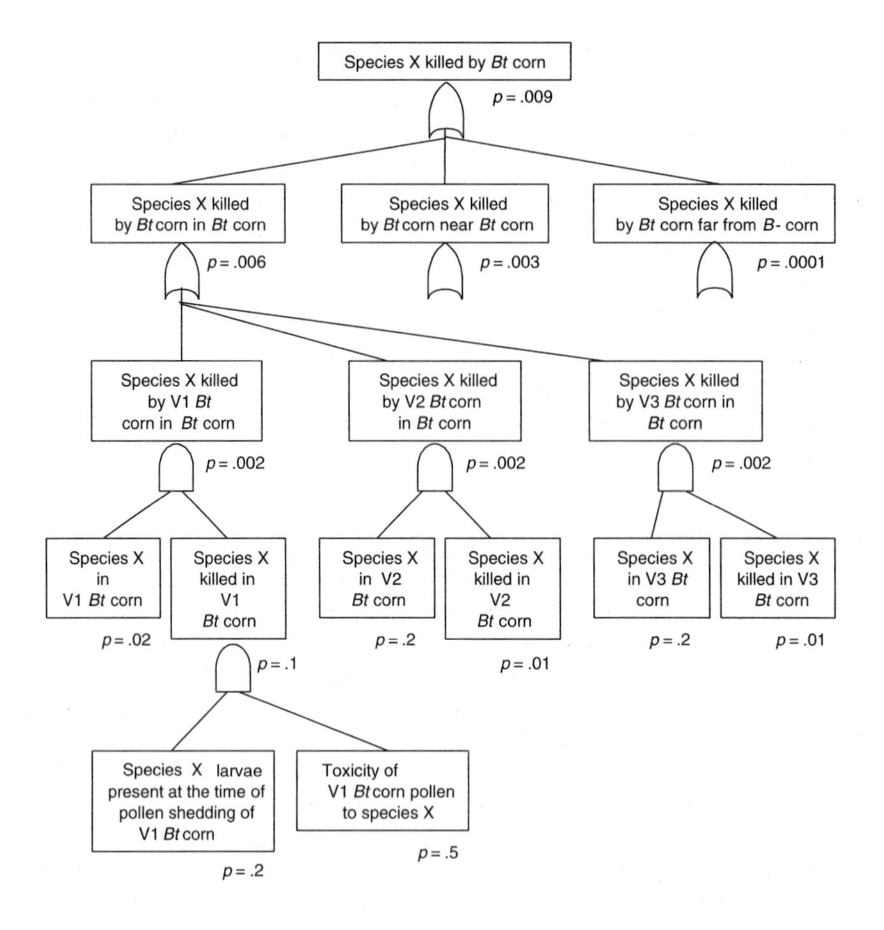

FIGURE 2.4 Simplified fictitious fault-tree analysis of the risk that individuals of species X are killed by the sum of three varieties of *Bt* corn (V1,V2,V3). This fault-tree traces backwards through the causal chain. In this example the logical structure of the model is Boolean algebra, which consists of "gates" that indicate how the faults lower in the tree should be combined to estimate the probability of occurrence of the higher levels in the tree. The upper levels of the tree consist of "or" gates (the pointed arrowheads). An "or" gate signifies that if any of the input statements are true, the output statement is true and therefore the inverse of the product of the inverse of the input probabilities can be used to compute the output probability, assuming the input statements are independent. The lowest level consists of "and" gates (the half-rounded symbols). An "and" gate signifies that all input statements must be true before the output statement is true, and therefore the product of the input probabilities can be used to compute the output probability. *p* values in the figure are entirely fictitious and are provided for illustrative purposes.

gate to the probability that the species is exposed to the *Bt* pollen and the toxicity of Event 176 pollen.

Fault-tree analysis is often used to assess the risk involved in potential failures of the safety system itself. This tradition is well developed in the aeronautics industry, where airplane safety and safety systems are subjected to rigorous fault-tree analyses (Lloyd and Tye 1982). A simplified fault-tree model for the safety system associated with APHIS regulation of transgenic crop plants is illustrated in Figure 2.5. This emphasizes one potential adverse effect that could occur through failure in the present oversight system. Although conducting such an analysis would require too much effort for this committee to complete, the elements of such an analysis are discussed in Chapter 5. Clearly, conducting such an analysis

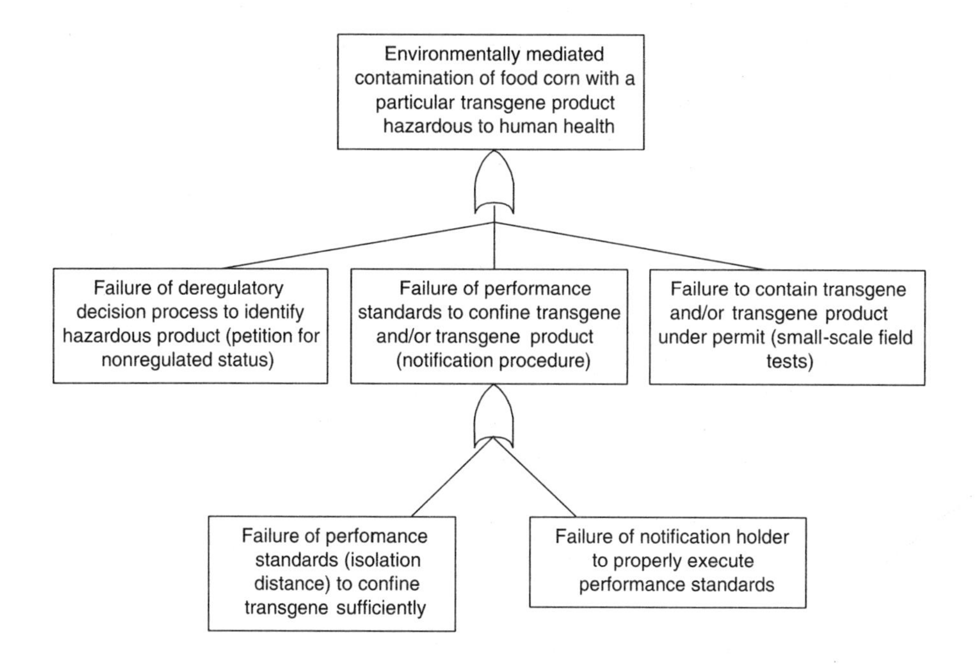

FIGURE 2.5 Simplified fault-tree model of the risk that a commercialized hazardous transgene product enters the human food chain via the environment. The logic is Boolean logic as in Figure 2.4, except that in this tree all the gates are "or" gates. This fault tree focuses on the possibility that the risk occurs because of failures in the safety network. Most industries with safety systems to manage risks analyze the potential failure of the safety system itself. This kind of analysis has led to the development and implementation of "fail-safe" systems, such as in the aeronautics industry.

would provide a regulatory agency with considerable information to improve its regulatory procedures and could contribute to a more formal methodology for a learning system in an agency. The cost of such an analysis is unknown.

Event-Tree Analysis

In contrast to fault-tree analysis, event-tree analysis logically works forward in time or forward through a causal chain to model risk (Lewis 1980, Haimes 1998). Event-tree analysis is not conditioned on the existence of a known hazard. Starting with an initiating event, the next steps in a logical chain of events can be assessed until the risk probability associated with a hazard can be calculated from the probabilities associated with the chain of events. Interestingly, event-tree analysis does not have to be aimed at estimating the risk associated with a particular hazard. Event-tree analysis can be initiated to investigate an entire category of hazards, during which the risk associated with particular hazards is assessed. If the processes in a complex system are understood well enough, event-tree analysis can identify hazards by following the complex sequence of events to their logical conclusions.

For example, an event chain analysis of the potential non-target risks associated with a transgenic crop producing a toxin could begin quite generally, because we know the toxin that would cause any of the non-target risks that could occur (see Figure 2.6). Consequently, an event-tree analysis of non-target effects can be initiated through a toxin fate and transport analysis, tracking where and when a toxin goes until it is degraded into nontoxic forms. If there is enough information, such a fate and transport analysis could reveal when apparently independent pathways converge to allow toxin to concentrate in some part of the ecosystem.

For *Bt* corn, the first step in the event-tree analysis requires understanding what toxin is expressed in the plant, where and when it is expressed, and how that toxin might reach the environment. All *Bt* corn varieties produced a truncated Cry toxin, which is significantly smaller than the protoxin produced by the bacterium. It is hypothesized that truncation may alter the host specificity spectrum, creating novel non-target risks dissimilar from *Bt* insecticides (Jepson et al. 1994, Hilbeck 2001, but see MacIntosh et al. 1990). The toxin is expressed in all of the plant tissues when the plant is actively growing, with variation among the transformation events. The toxin moves in the pollen and remains in the dead plant tissue after senescence. In *Bt* corn, truncated Cry toxin can enter the soil via root exudates (Saxena et al. 1999), exposing soil organisms. The activated toxin readily binds to montmorillic clay particles and

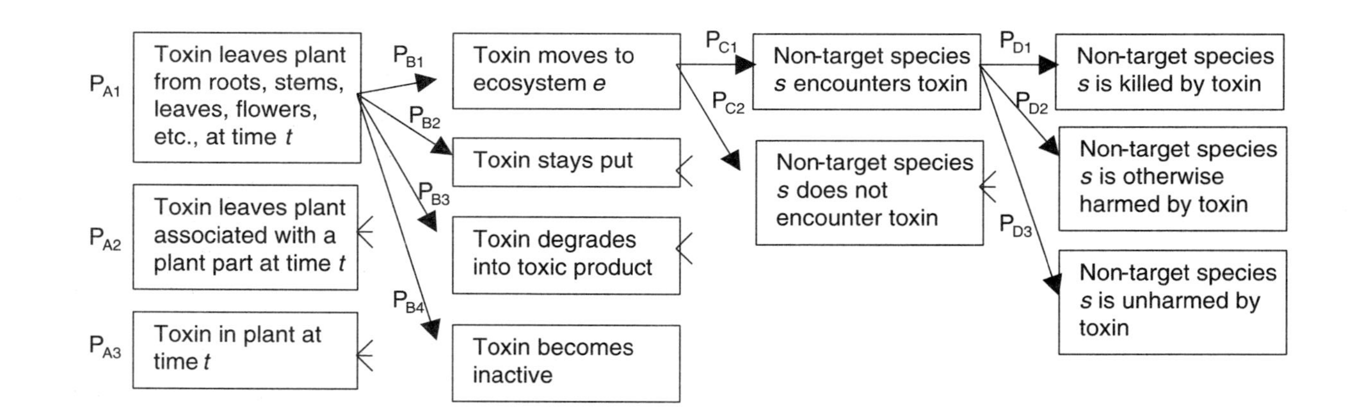

FIGURE 2.6 Simplified event-tree analysis of the non-target risk of a toxin produced in a transgenic plant. Event-tree logic is based on a chain of mutually exclusive events. An event tree begins with an initiating event and follows the consequences of the event through a chain of subsequent events, assigning to each branch a probability of occurrence. The end result is a long list of the possible consequences of the initiating event in which each consequence has associated a certain probability of occurrence. The sum of the probabilities at a level of the tree, $S_i PW_i = 1$ (here W = A, B, C, or D), that is, $P_{A1} + P_{A2} + P_{A3} = 1$. The branches for many of the lower boxes are incomplete. Non-target species *s* could be affected through multiple causal chains, so the overall effect of the toxin on species *s* may be a complicated aggregation of these effects. This event-tree analysis also is initiated by a general event that would be common to many non-target hazards. The specific hazards to particular species are evaluated later in the analysis.

can retain insecticidal activity for more than nine months (Saxena et al. 2000). The untruncated protoxin in *Bt* insecticides does not readily bind to clay and is degraded rapidly in soil (Venkateswerlu and Stotzky 1992). Consequently, Cry toxin from *Bt* corn has several unique exposure pathways not characteristic of either corn or the toxin from bacteria. Each of the potential pathways can then be assessed to determine if there is a potential hazard associated with them; if so, the risk can be characterized by choosing an appropriate reference comparison for that exposure pathway and potential hazard. Event-tree analysis may become particularly useful when risks associated with transgenic crops with many traits are characterized.

Finding 2.15: The two-part model may bias a risk assessment either toward a finding of no significant risk or a finding of significant risk. As presently implemented, the former bias may predominate.

Finding 2.16. Additional research is needed to evaluate the utility of the whole-organism model and a more formal use of fault-tree and event-tree analysis.

Experimental Comparisons

Plant breeders and geneticists make a variety of comparisons using appropriate statistical designs and analyses. New varieties of crops are not generally released for large-scale use until a variety review committee reviews the data. In the case of transgenics a single gene is introduced into cells growing in tissue culture, and the regenerated plant displaying the trait is backcrossed several generations to an elite line. A reasonable genetic comparison for risk assessment is the untransformed elite line. For example, six backcrosses should produce progeny that are genetically over 99% identical to the elite line used as the recurrent parent. This value of 99%, however, is an average or expectation and is not always reached.

The ideal comparison of a transgenic following backcrossing is to its near-isogenic line derived by self-pollination of plants heterozygous (hemizygous) for the transgene in the most advanced backcross generation. This will give plants homozygous for the transgene as well as plants without the transgene; all unlinked regions of the genome will segregate independently of the transgene locus and should not influence comparison of the two homozygous lines derived from the homozygous plants. This comparison is not absolutely perfect because the transgene chromosome likely will carry a genomic region on each side of the transgene from the transgenic parent. However, this comparison is theoretically the best available.

If near isogenic lines are not available the only possible comparisons are among unrelated plant lines. In such comparisons it will always be difficult or impossible to determine if an observed difference between the plant lines was caused by the insertion of the transgene or was simply due to another genetic difference between the two plant lines. Conversely, if no difference is observed, it will always be difficult, or impossible to determine if the transgene had no effect or if other genetic differences masked the effect of the transgene.

The committee recommends that near-isogenic lines with and without the transgene be developed and evaluated as part of the process of risk analysis. In addition, a way should be found to make such near-isogenic lines available for study by public scientists. The availability of such genetic materials would substantially strengthen comparisons, save time and money, and provide much more legitimate comparisons. The potential to mislead the public—positively or negatively—is too great unless the best-available genetic materials are used. Clonally propagated crops can be readily compared to transgenic versions. Self-pollinated crops raise most of the same concerns expressed above for cases where backcrossing is involved. If backcrossing is not the breeding scheme, the process of advancing the materials must be considered to determine the most appropriate comparisons.

Finding 2.17: For certain questions, the availability of near-isogenic lines can allow scientifically legitimate comparisons and facilitate the assessment of risk.

CONCLUSION

This chapter has reviewed general approaches and roles of risk analysis, the scientific, social, and logical bases for regulation of transgenic plants, and some technical issues related to the risk assessment process itself. It explains that risk analysis typically must fill two roles: providing technical information to decision makers, and creating confidence among stakeholders in the risk analysis process. The specific importance of the second role in the risk analysis of transgenic plants is discussed. The use of the most rigorous scientific approaches in conducting risk analyses is critical for both roles of risk analysis.

Because there is no way to evaluate the risks of transgenic plants without empirical examination, there is a scientific reason to examine the characteristics of transgenic plants on a case-by-case basis. Although there has been a general assumption that the products of conventional plant breeding are safe, while products of transgenic processes could pose environmental risks, this chapter and others in the report find no logical or scientific support for this assumption.

Environmental assessment of transgenic crops relies on risks associated with the transgenic variety being compared with multiple appropriate reference scenarios (ways to accomplish the same goal). For example, the risks associated with a transgenic herbicide-tolerant variety could be compared with risks of chemical weed management used on nontransgenic varieties, cultivation practices, crop rotations, and other approaches for weed management. This chapter ends by pointing out that there are a number of general approaches for formally examining a risk analysis to determine where errors could occur. These formal examinations could be helpful in assuring the rigor of the current system used for transgenic plants.

3

APHIS Regulatory Policy for Transgenic Organisms

OVERVIEW

This chapter reviews the regulations and procedures currently used to guide the Animal and Plant Health Inspection Service (APHIS) in overseeing the environmental safety of testing and commercialization of transgenic plants. It also describes the historical development of these regulations and procedures. The broad overview provided here leads into Chapter 4, which presents case studies of APHIS oversight of specific transgenic crops.

Chapter 1 described the U.S. Coordinated Framework for the Regulation of Biotechnology created in 1986 for the regulation of development and commercialization of transgenic organisms in the United States and Chapter 2 described the scope and regulatory triggers that can be used as clarified in 1992 (OSTP 1992). Except for pesticidal plants, USDA-APHIS-BBEP (Biotechnology, Biologics, and Environmental Protection) has primary regulatory authority for environmental assessment of all transgenic plants. This chapter specifically examines the oversight of transgenic plants by APHIS under the Federal Plant Pest Act (FPPA) and the Federal Plant Quarantine Act (FPQA). The final chapter of this report considers how that oversight might change under the Federal Plant Protection Act of 2000.

The scope of APHIS's review includes, under the FPQA, importation of regulated articles and interstate movement of regulated articles and, under the FPPA, environmental release of regulated articles. "Regulated article" is defined below. "Environmental release" is defined as "the use of a regulated article outside the constraints of physical containment that are found in a laboratory, contained greenhouse, or a fermenter or other contained structure" (APHIS 1987). For example, growing regulated articles in a controlled greenhouse would not constitute a release, but growing them in a standard greenhouse would.

When a regulated article is determined to have "nonregulated status," it no longer is subject to APHIS oversight, but the agency has the option of taking action on any plant that it believes to present a plant pest risk as long as they have not determined it to have nonregulated status. In this case the article has been evaluated, and determined with a reasonable degree of certainty not to present risk as a plant pest, and APHIS cannot regulate it under FPPA or FPQA.

In 1987 APHIS published regulations for the release of transgenic organisms into the environment (APHIS 1987). These were modified in 1988 and 1990 to exempt *Escherichia coli* strain K-12, sterile strains of *Saccharomyces cerevisiae*, asporogenic strains of *Bacillus subtilis*, and *Arabidopsis thaliana* from permitting requirements for interstate movement under specified conditions (APHIS 1988, 1990) and again in 1993 and 1997 to provide a notification procedure and a process to petition for nonregulated status (APHIS 1993, 1997a).

The rules and their revisions are promulgated under the statutory authority of the FPPA and FPQA, which grant APHIS broad powers to regulate potential plant pest risks. APHIS must grant a permit for the introduction of any regulated article. Although there are several fine distinctions in its definition (see BOX 3.1), a regulated article is a genetically engineered organism that is or contains genetic material from one of many taxa listed in 7 CFR 340.2 or that meets the definition of a potential plant pest. The definition of "plant pest" is discussed in more detail below. Anyone wishing to "introduce" such an organism must request a permit from APHIS. APHIS has streamlined the permitting process for most transgenic plant species under specified conditions; for these, formal notification of APHIS in advance of a release is often sufficient for obtaining permission to introduce the transgenic plant. In addition, APHIS has developed a process whereby an applicant can petition the agency to determine that a transgenic organism does not pose a plant pest risk and is therefore no longer a regulated article subject to APHIS regulatory oversight.

The coordinated framework explicitly provides that federal agencies should focus on the characteristics of risks posed by a biotechnology product, not on the process by which it is created, but APHIS uses the process of genetic engineering to trigger oversight. As discussed in Chapter 2, APHIS has argued that it was not treating genetically engineered organisms differently from so-called established plant pests or naturally occur-

BOX 3.1
Key Definitions Used by APHIS

Introduce or introduction (APHIS 1997b) "To move into or through the United States, to release into the environment, to move interstate, or any attempt thereat."

Plant (APHIS 1987) "Any living stage or form of any member of the plant kingdom including, but not limited to, eukayotic algae, mosses, club mosses, ferns, angiosperms, gymnosperms, and lichens (which contain algae) including any parts (e.g., pollen, seeds, cells, tubers, stems) thereof, and any cellular components (e.g., plasmids, ribosomes, etc.) thereof."

Plant pest (APHIS 1987) "Any living stage (including active and dormant forms) of insects, mites, nematodes, slugs, snails, protozoa, or other invertebrate animals, bacteria, fungi, other parasitic plants or reproductive parts thereof; viruses; or any organisms similar to or allied with any of the foregoing; or any infectious agents or substances, which can directly or indirectly injure or cause disease or damage in or to any plants or parts thereof, or any processed, manufactured, or other products of plants."

Regulated article (APHIS 1987) "Any organism which has been altered or produced through genetic engineering, if the donor organism, recipient organism, or vector or vector agent belongs to any genus or taxon designated in a list (7 CFR 340.2) of taxa known to have plant pests, and meets the definition of plant pest, or is an unclassified organism and/or an organism whose classification is unknown, or any product which contains such an organism, or any other organism or product altered or produced through genetic engineering which the Director, BBEP (Biotechnology, Biologics, and Environmental Protection division of APHIS), determines is a plant pest or has reason to believe is a plant pest. Excluded are recipient microorganisms which are not plant pests and which have resulted from the addition of genetic material from a donor organism where the material is well characterized and contains only non-coding regulatory regions."

Release into the environment (*Federal Register* 52:22908–9) "The use of a regulated article outside the constraints of physical containment that are found in a laboratory, contained greenhouse, or a fermenter or other contained structure."

Well-characterized and contains only noncoding regulatory regions (APHIS 1987; e.g., operators, promoters, origins of replication, terminators, and ribosome binding regions) "The genetic material added to a microorganism in which the following can be documented about such genetic material: (a) The exact nucleotide base sequence of the regulatory regions and any inserted flanking nucleotides; (b) The regulatory regions and any inserted flanking nucleotides do not code for protein or peptide; and (c) The regulatory region solely controls the activity of other sequences that code for protein or peptide molecules or act as recognition sites for the initiation of nucleic acid or protein synthesis."

ring organisms, which might be plant pests. For example, APHIS regulates the movement and release of geographically separated populations of known plant pests because the source population can have characteristics that are absent in the recipient geographic population, which could increase the plant pest risk in the recipient population. In all cases a permit must be obtained from APHIS before importation and interstate movement of potential plant pests. For some transgenic organisms, APHIS has determined that release into the environment is tantamount to introduction of a new organism; therefore, it requires that a permit be obtained beforehand. In other words, APHIS considers that transgenic organisms are new organisms because they potentially have new ecological characteristics that could make them plant pests. That is consistent with the principle that transgenic organisms might have new characteristics that require oversight and that the genetic engineering process might trigger oversight, even if it is not considered in the evaluation of risks, as discussed in detail in Chapter 2.

In 1993 APHIS proposed a simplification of the regulatory procedure in order to:

- create a simplified notification procedure for transgenic plants that meet eligibility criteria and are tested using performance standards that minimize risks;
- allow under notification corn, cotton, potato, soybean, tobacco, tomato, or any additional plant that BBEP has determined may be safely introduced;
- extend notification procedures to include a number of virus resistance modifications;
- provide a petition process allowing for a determination that certain plants are no longer regulated articles; and
- allow permissions for determination for nonregulated status to be extended to closely related articles (an "extension" process).

APHIS received 84 comments ranging from complete opposition to enthusiastic support for even greater reduction in regulation. According to the *Federal Register* notice (APHIS 1993), the majority of commenters expressed general or qualified support for the suggested changes. Prior to this proposed change, however, APHIS consulted with a scientific subcommittee of the USDA's Agricultural Biotechnology Research Advisory Committee (ABRAC) in a public forum about the scientific basis for the proposed reduction in regulation. APHIS argued that the six specified crops had no weedy relatives, that only cotton had wild populations (in Florida), and that only cotton and potato had wild relatives in the United States. It also argued that the confinement criteria in the performance standards should protect against gene flow to these wild relatives and populations, and the subcommittee generally concurred. (ABRAC held its last meeting in January 1996.)

In 1995 APHIS proposed a further simplification of the regulatory procedure to:

- allow environmental release under the notification system of any plant not considered a noxious weed as long as the plant is not considered a weed in the intended area of release;
- extend the notification procedures to include a number of genetic modifications resulting in virus resistance;
- discontinue the requirements for each relevant state to concur with interstate movement;
- simplify reporting requirements for field trials conducted under notification or permit; and
- allow permissions for determination of nonregulated status to be extended to closely related articles (an "extension" process).

These proposed changes also proved to be controversial. A total of 50 comments were received from industry, universities, state departments of agriculture, science policy organizations, environmental groups, industry organizations, professional societies, consumer organizations, individuals, and a university cooperative extension office. The majority (over 60%) supported the proposed amendments. Those expressing opposition were concerned with the potential for increased risk posed by particular transgenic plants, especially those with wild or weedy relatives. Some opponents were concerned that regulatory oversight would be decreased in the shift from a permitting to a greater emphasis on the notification system and that notification was harder to enforce and therefore compliance might be compromised.

Part of the APHIS response to the expressed public concerns was publication of a revised user's guide (APHIS 1997b), which provided examples of how applicants might meet the performance standards and other information to help applicants design their field trials for specific organisms. Although not a formal regulatory document, the user's guide provides information to applicants on how to comply with the regulations. The guide gives specific examples and describes approaches to compliance on such matters as shipping and maintenance of regulated articles, how to avoid admixtures, devitalizing treatments, elimination of viable vectors, and minimizing dispersal and persistence in the environment. The guide emphasizes that these are examples only, that following the examples exactly does not necessarily ensure acceptability, and that alternative approaches might be equally acceptable. APHIS also argued that field inspections of trials conducted under the notification system (as well as those under permit) achieved a high degree of compliance.

The controversial changes proposed in 1995 that went into effect in 1997 were created by APHIS internally, and the scientific response of APHIS also was internal. This subcommittee recognizes that the proposed changes of 1995 were precedent setting (e.g., extending the notification procedure from six crops to all nonweed plants) and created without external scientific input beyond those respondents to the *Federal Register* notice. It is certainly possible that APHIS may continue to propose and make important policy changes without external scientific input.

Recommendation 3.1: For changes in regulatory policy, APHIS should convene an external scientific advisory group and hold at least one meeting to solicit public scientific input to review proposed changes.

SCOPE AND REGULATORY PROCEDURES USED BY APHIS

APHIS currently regulates transgenic plants under 7 CFR part 340, "Genetically Engineered Organisms and Products: Simplification of Requirements and Procedures for Genetically Engineered Organisms," which was published in 1997.

Anyone introducing (importing, transporting interstate, or releasing into the environment) a regulated article must have authorization through either a notification or a permit and must comply with other restrictions as described in 7 CFR part 340. The regulations provide APHIS with the authority to regulate such introductions for certain transgenic organisms. The regulatory objective of 7 CFR 340 is to allow the evaluation of transgenic organisms with sufficient scrutiny to identify any plant pest risks at an early enough stage to allow remedial action prior to the occurrence of any real damage.

The basis for regulation is the broadly defined plant pest. Part 340.1 provides definitions for the terms used (see BOX 3.1). Plant pest is defined as "any living stage (including active and dormant forms) of insects, mites, nematodes, slugs, snails, protozoa, or other invertebrate animals, bacteria, fungi, other parasitic plants or reproductive parts thereof; viruses, or any organisms similar to or allied with any of the foregoing; or any infectious agents or substances, which can directly or indirectly injure or cause damage in or to any plants or parts thereof, or any processed, manufactured or other products of plants." Under this definition, almost all organisms (and their derivatives) can be potential plant pests, including, for example, herbivorous invertebrate animals (anything that eats a plant can be considered a plant pest). A transgenic organism is considered a regulated article if it is a plant pest or if it or a gene donor or vector used in its construction are plant pests according to a long list of taxa listed in 7 CFR

340.2 (this list can be amended as outlined in part 340.5). In addition, a transgenic organism can be considered a regulated article if APHIS has reason to believe it presents a plant pest risk.

The definition of plant pest is in many ways extremely broad but in other ways surprisingly restricted. The breadth of the definition is provided by the inclusion of indirect injury or disease. Indirect injury occurs when a plant pest has an effect on another species that eventually leads to a detrimental effect on a plant. Thus, any species that interacts ecologically with a species that directly injures a plant (e.g., by feeding on it) can be considered a potential plant pest. Indeed, species that interact with species that interact with direct plant pests can be considered indirect plant pests, and this leads to the potential inclusion of nearly every species. In contrast, the taxonomic list restricts the definition of plant pest in an important way. According to the definitions, no vertebrate can be considered a plant pest, despite considerable evidence to the contrary. (This situation might change under the new Plant Protection Act of 2000; see Chapter 7.) For example, feral pigs and goats have had serious effects on native plants in Hawaii and other oceanic islands (Allen 2000). The exclusion also means that genetically engineered fish cannot be regulated as potential plant pests. There is some uncertainty about the status of algae as plants that could be affected by potential plant pests. Modern classification excludes most algae from the Kingdom Plantae (Campbell et al. 1997). Thus, many algae are not members of Plantae, as required by the definition of plant, but they are listed as a group under the definition. That is an important distinction because shellfish, which consume unicellular algae and are invertebrates, could be regulated under the APHIS rules if unicellular algae were considered members of the Kingdom Plantae.

Note that if a transgenic plant was created from a nonweedy species without the insertion of genes from a plant pest and it was transformed without the intervention of a plant pest (as would be the case with the use of particle bombardment or electroporation), it would not necessarily be considered a regulated article. In such cases, the creators of transgenic plants to be field released apparently have always sent a "courtesy" notification or permit application to APHIS, but the possibility remains that field release of certain transgenic plants could escape APHIS oversight.

Notification System for Introduction of Certain Regulated Articles (7 CFR 340.3)

Most transgenic plants are field tested under a notification system. In recent years "nearly 99% of all field tests, importations, and interstate movements of engineered plants [have been] performed under this system" (OSTP/CEQ 2001). Notification for the movement, importation, and field testing of transgenic plants follows a rigid but streamlined format that allows an expeditious review. For a specified list of plants and characteristics, an applicant may simply notify APHIS of its intent to release a regulated article. Upon receipt of such a notification application, the document is logged into the APHIS database and reviewed by one of the scientific staff for qualification and completeness, and then a recommendation is sent to appropriate state officials for concurrence. The entire process must be completed within 30 days (10 days for interstate movement). Scientific evaluation of a notification application by APHIS personnel is typically completed within a few days. Acknowledgment by APHIS that an article meets the notification requirements means that a permit is not required for field testing within one year of the date of introduction. Renewal can be accomplished by submitting an additional notification to APHIS.

The assessment of notifications is not subject to external scientific review or any other public input. A few transgenic plants are now grown to produce commercial products under notification.

To meet APHIS criteria for notification:

- The plant must not be listed as a noxious weed under the Federal Noxious Weed Act or be considered to be a weed in the area of release into the environment.
- The inserted DNA must be stably integrated into the host genome. According to the user's guide (APHIS 1997bs), this means that the trait is inherited in a Mendelian fashion for at least two generations.
- The function of the inserted DNA is known, and its expression does not result in plant disease. Function is not precisely defined in the user's guide, but the intent of this criterion is that expression does not result in plant disease.
- The inserted DNA does not
  —cause production of an infectious entity;
  —encode substances that are known or likely to be toxic to non-target species known or likely to feed on the plant (According to the user's guide, toxicity to non-target species is restricted only to those non-target species that feed on the plant, not dispersed plant parts, such as seeds, pollen, or plant residue.);
  —encode products intended for pharmaceutical use. (According to the user's guide, "intention" becomes clear for regulatory purposes only when clinical testing of the product is proposed to the Food and Drug Administration. Until such time, a product is not considered to be intended for pharmaceutical use.)

• Any virus-derived sequences must be known noncoding regulatory sequences or otherwise unlikely to facilitate virus virulence and spread in plants.

• The inserted sequence must not be derived from human or animal viral pathogens or other potential human or animal disease-causing agents.

In addition to meeting these criteria, introductions under the notification system must meet specified performance standards designed to ensure such confinement that the transgenic plant or its progeny will not unintentionally persist in the environment.

The standards state general concerns but leave applicants the flexibility to meet them according to their own circumstances:

The transgenic plant must be transported and stored in a way that minimizes the escape of viable plant parts into the environment. According to the APHIS user's guide, this performance standard refers to seeds, cuttings, buds, and other plant parts that would be planted to grow the crop but not the movement of pollen.

For environmental release, inadvertent mixing of the transgenic plant with nonregulated plant material must be avoided. According to the APHIS user's guide, this performance standard refers to seeds, cuttings, buds, and other plant parts that would be planted to grow the crop but not the movement of pollen.

The identity of the transgenic plants and their parts must be maintained; plants and their parts must be contained or devitalized after use. According to the APHIS user's guide, this performance standard does not cover the movement of pollen.

There must be no pathogenic vector associated with the plant.

Field trials must be conducted in a manner that precludes persistence of the plant and its progeny.

Management practices to prevent persistence of the plant or its progeny in the environment must be applied.

APHIS provides detailed guidance for notification applications in its user's guide including sample notification letters and guidance on how to meet performance standards. The notification application consists of the name and identity of the responsible person (applicant) and a description of the regulated article. The description must include the identity of the transformed plant species, the method of gene transfer, and a full description of the inserted sequences, including the functions, encoded proteins, and source donor organisms for each segment. In addition, the notification application must include, for field releases, the geographic details, size, and duration of field trials and, if appropriate, information about intended transport. Finally, the applicant must sign a statement affirming that the article meets the eligibility requirements and that any actions taken will meet the mandated performance standards listed above.

Any applicants whose notification application is denied may apply, without prejudice, for a permit for the same regulated article.

Permits for Introduction of a Regulated Article (CFR 340.4)

Permits are required for the movement, importation, and field testing of transgenic plants that do not qualify for notification (such as pharmaceutical-producing plants) and for plants denied notification. Under the permitting process described in CFR 340.4, APHIS presumes that the regulated article is a potential plant pest and requires anyone who wants to introduce it into the environment to obtain a permit. The permitting process allows APHIS to evaluate the potential plant pest risk and prescribe prevention measures to reduce it (e.g., through confinement procedures, or limiting the spatial, temporal, and numerical scales of the release). At least one regulated article is now grown to produce a commercial product under permit.

It is anticipated that commercial production of pharmaceutical products will occur under the permitting process. The process provides APHIS with the authority to request further information from applicants, but it does not give the agency the authority to require the requested information. Although applicants are not legally required to provide additional requested information, so far all have complied, presumably to maintain their desired cooperative relationship with APHIS.

The application for a permit for release into the environment is more detailed than the notification application. The primary emphasis for field release under permit is information regarding confinement. That is, the confinement imposed should effectively eliminate the potential for significant environmental impact. An application for a permit must provide data so that a decision can be made to ensure that (a) the transgenic plant is adequately characterized; (b) no transgenic plant material will persist in the environment; (c) that unintentional or unanticipated effects, if any, can be restricted to the receiving facility or the confined field site; and (d) in the case of field testing, plants must be managed in such a way that there are no environmental risks after the confined field release is terminated. APHIS must receive field trial results within six months of trial completion. The applicant must allow APHIS and state inspectors access to the trial and must notify the agency of any unusual occurrences.

APHIS provides detailed guidance for permit applications in its user's guide including sample applications. In the permit application the applicant lists the regulated article; the donor organism; vector or vector agent;

date of importation, movement, or release; and port of importation or site of release. Additionally, detailed information is required, as appropriate, on the:

- anticipated or actual expression of the altered genetic material in the regulated article and how it differs from the nonmodified parent organism;
- molecular biology of the system;
- locality where the donor, recipient, and vector were collected and produced;
- experimental design at the release site;
- facilities at the destination;
- measures to ensure confinement; and
- final disposition of the regulated article.

The permit application should be submitted to APHIS at least 120 days prior to intended release into the environment. APHIS conducts an initial review of the dossier (within 30 days) to determine if all necessary information is supplied. If not, the 120-day "clock" stops until the applicant provides the missing information. The permitting process allows APHIS to request monitoring and reporting of the results of small-scale releases.

Petition for Determination of Nonregulated Status (CFR 340.6)

APHIS has a procedure whereby an applicant can request that the agency determine that a particular transgenic plant is not a regulated article—that is, that it does not fall under the definition of a plant pest (BOX 3.1). This procedure is the sole route for commercialization of transgenic plants (e.g., sale of transgenic soybean seed) and the primary but not sole route to commercialization of transgenic plant products (e.g., when the plants are never sold but a product such as an industrial protein extracted from the plant is sold).

Once APHIS decides that a transgenic plant is not a regulated article, it cannot exercise any additional oversight on the plant or its descendants. Those descendants include all plants of the same species that receive the transgene through sexual reproduction. Therefore, separate deregulated lines can be mated with one another via conventional crossbreeding to bring together different transgenes in the same plant, and such plants are not subject to regulatory evaluation (see discussion of methods to create multitransgenic plants in Chapter 4).

Descendants may also include distantly related members of the same crop species. For example, after deregulation, a breeder could use crossing to move a transgene not only from one flax variety to another flax variety, or from one cabbage to another, but also from flax to linseed and from cabbage to broccoli. Although it is difficult to move genes among more distant relatives through crossing, it is not impossible. For example, crosses among different species have been used to create modern cultivars of sugarcane and raspberry (Smartt and Simmonds 1995; see also Chapter 1). Finally, intended or unintended crosses may result in descendants that are wild species, when those species receive the transgenes from introgression from the crop. Also, APHIS cannot require monitoring of any deregulated article.

Finding 3.1: Currently, APHIS deregulation is absolute, completely removing the article from the agency's regulatory authority.

Finding 3.2: Presently, APHIS deregulation includes all progeny and descendants of a deregulated item.

Decisions to deregulate articles are evaluated on a case-by-case basis; the procedural requirements are specified in the *Federal Register* (APHIS 1993). APHIS has made decisions to deregulate dozens of genetically modified crops. The current list of deregulated crops (and notification and permit applications) can be obtained at a website maintained by Information Systems for Biotechnology (*www.nbiap.vt.edu/*).

Anyone can petition, with supporting evidence, that a specified transgenic plant be determined a nonregulated article. Essentially, the applicant must show that the regulated article is free from any risk under 7 CFR 340. Petitions for determination of nonregulated status are comprehensive data packages to APHIS for scientific review. Assessment of petitions by APHIS scientists relies on data supplied primarily or exclusively by the applicant, but APHIS may seek additional information (the case study on squash in Chapter 4 is an example). The accumulated information is used to determine whether the regulated article displays no plant pathogenic characteristics; is not more likely to become a weed than its nontransformed parent; is unlikely to increase the weediness of cultivated, feral, or wild-related plants; does not damage processed agricultural commodities; and is unlikely to cause unintended significant harm in other organisms. Because reviews are conducted on a case-by-case basis, the information needed may vary with plant species, the specific type of modification(s), and end use of the transgenic plant. The following is a summary of the required data and information in the *Federal Register* (APHIS 1987).

1. a description of the biology and taxonomic identification of the nonmodified recipient plant;

2. relevant experimental data and publications;
3. a detailed description of the genotypic differences between the regulated article and the nonmodified recipient organism;
4. a detailed description of the regulated article's phenotype, especially regarding known and potential differences from the nonmodified recipient organism that would indicate whether or not the regulated article poses a greater plant pest risk; and
5. field test reports for trials involving the regulated article conducted under permit or notification.

To provide guidance for petitions, the APHIS user's guide has a sample application for determination of nonregulatory status representing a suggested format for submission. APHIS has inserted in the left-hand margin of the sample application comments and issues that may need to be addressed by the applicant. The guidance detailed under each heading is directly quoted below from the user's guide:

I. Rationale for Development of the Product

II. Relevant Biology of the Plant

Description of the biology of the nonmodified recipient organism should include taxonomy, genetics, pollination, evidence of reported weediness (e.g., noting whether the crop or sexually compatible species is listed in the relevant publications of the Weed Society of America), discussion of sexual compatibility with wild and weedy free-living relatives in natural crosses or crosses with human intervention. The applicant should provide source of recipient (cultivar name or accession number) and the weed status of its sexually compatible relatives.

The applicant should explicitly identify the lines that are to be considered in the petition and the cultivars from which they are derived. If there are multiple lines, each line must be given a unique identifier that must be listed in the application.

For virus-resistant plants, applicants should provide in an additional section the following information on the nature of the virus that provided the sequences encoding the resistance phenotype:

i) the taxonomic name of the virus including family, genus, and strain designation including any synonyms;
ii) the type of nucleic acid contained in the virus;
iii) whether the infection is systemic or tissue specific;
iv) whether the virus is associated with any satellite or helper viruses;

v) the natural host range of the virus;

vi) how the virus is transmitted;

vii) if transmitted by a vector, the identity of the vector including mode of transmission (e.g., persistent or non-persistent);

viii) whether any synergistic or transcapsidation interactions with other viruses under field situations have been reported in the literature; and

ix) the location and the name of the host plant the virus was originally isolated from.

The above information can be provided in a table format. This information can be supplemented by listing references that report the host range, insect vectors, etc., for the virus.

III. Description of the Transformation System

For Agrobacterium-based transformation protocol, the applicant must indicate how Ti plasmid-based vector was disarmed (i.e., all tumorigenic DNA was removed). Applicants can provide citations that describe the transformation procedure. However, any significant modifications of transformation, strain designation, etc., should be described.

For other methods of transformation, the applicant can describe the sources of various components of the plasmid (or other DNA including possible carrier DNA) and method of transformation by citation. However, any significant modifications of transformation, strain designation, etc., should be described.

The applicant must provide a detailed restriction map of the plasmid that is sufficient to be used in the analysis of Southern data. Description of added restriction sites is helpful in the interpretation of Southern data and should be provided.

Indicate the functions of the gene(s), promoters, leader sequences, enhancers, introns, and any other sequences that are used for gene expression in the plant and a reference describing from where the sequences were obtained. Discussion should include whether the inserted sequences are responsible for disease or injury to plants or other organisms. The nucleotide sequence(s) of the plant-expressed gene(s) should be provided by citation and not submitted in the application. If there has been a significant modification to sequences and the modified sequence has not been published, they should provide the complete sequence highlighting the modifications. If there have been minor modifications to the sequence of the plant-expressed gene, they should be provided. For example, if in a chemically synthesized *Bt* gene amino acid 23 was changed from methionine to alanine, it should be stated without providing the complete sequence.

IV. Donor Genes and Regulatory Sequences

V. Genetic Analysis and Agronomic Performance

In general, it always is prudent to analyze data statistically when such analysis is possible. When unpublished information or an opinion has been supplied by a scientific expert, a letter communicating the information should be included in the petition. If the unpublished information provided is data resulting from scientific research then these data can be provided as a personal communication either in a letter from the researcher or in the text of the petition. In either case the materials and methods, data analysis, and discussion of the data analysis should be provided in detail. Unsupported assertions about the results of the experiment are not acceptable.

Applicants must report any differences noted between transgenic and nontransgenic plants that are not directly attributed to the expected phenotype. Differences observed could include changes in leaf morphology, pollen viability, seed germination rates, changes in overwintering capabilities, insect susceptibilities, disease resistance, yield, agronomic performance, etc. Applicants must also note the types of characteristics that were compared between transgenic and nontransgenic plants and found to be unchanged.

The applicant should describe whether data submitted are from inbred or hybrid plants; if hybrid plants, state which generation.

The applicant should state whether data with respect to plant performance were generated in a greenhouse or field environment, and if from the field, indicate how many sites, states and number of years the data represents.

Seed germination, seed dormancy, seed production, growth rate, and other data relating to the plant's performance will be required when the nature of the gene and the biology of the plant (including sexually-compatible relatives) warrant such data. This type of data will usually not be required for plants that have some of the following attributes: highly domesticated (e.g., corn), exclusively self-pollinating (e.g., soybean), male sterile, and have high seed germination rates (>90%), and whose phenotypes are unlikely to affect performance with respect to weediness or fitness (e.g., delayed ripening or oil seed modification). Phenotypes that might require performance data (depending on the plant) include but are not limited to the following: cold tolerance, salt tolerance and tolerance or resistance to other biotic or abiotic stresses.

Southern analysis should include DNA isolated from nonmodified recipient, all or selected transformed lines, and the vector. Parental plasmid DNA (e.g., PUC 18) not containing intended donor genes may be labeled and hybridized to Southern blots to demonstrate that only the intended sequences have been incorporated in the genome of the transgenic plant. Restriction enzymes to be used might include enzymes that do not cut within the transforming plasmid but will cut the "entire insert" into one fragment from the DNA of the transgenic plant.

In the case of an Agrobacterium-based transformation system, the applicant should determine if genes that reside outside the LB/RB are inserted in the genome of the regulated cultivar. If a complete copy of any of these genes is present, the applicant should determine whether it is expressed in the plant. For direct transformation systems, applicants should determine which sequences are inserted in transgenic plants and whether they are expressed. PCR analysis may be used to prove that only the targeted DNA has been incorporated. Sequencing of the transgene in plant and adjacent sequences is not required. Determination of the number of copies of integrated transgenes is not required, but the number of insertions may be used to support analysis of inheritance data.

If the inserted DNA sequence order is complex, as is often the case for plants engineered via direct transformation systems (e.g., electroporation, polyethylene glycol transformation of protoplasts, or particle bombardment techniques), the applicants should summarize the data by providing the following information for all the genes (whether under the direction of plant or bacterial promoters). Is there a complete copy of the gene present in the regulated article? Is the protein expressed in the plant? If multiple complete copies of a gene are present, applicants do not have to determine if each copy of the gene is expressed. Applicants should provide a table, like the one shown below [the user's guide provides a table with hypothetical data], that summarizes the results and indicates where specific data is to be found.

Mendelian inheritance data and Chi square analysis for at least 2 generations are appropriate to demonstrate whether the transgene is stably inserted and inherited in Mendelian fashion. Such data are generally not necessary for infertile vegetatively propagated crops such as male-sterile potatoes.

RNA—Northern analysis is generally not required except for virus-resistant plants. However, such analysis may be necessary for ribozyme, truncated sense, or antisense constructs, when protein levels cannot be provided.

PROTEINS—Expression levels of gene(s) of interest and marker genes in various tissues, developmental stages of plant, and experimental conditions (induced or noninduced) are required. Assays can be of enzyme activity. Serology, ELISA, and Western blots may also be used. Describing the source of the immunogen is critical for serological analysis.

For virus resistant plants, the amount of viral transgene RNA produced should be determined and compared to the amount of the RNA produced by the viral gene in an infected nontransgenic plant. Applicants should address whether the transgene RNA (or protein) is present in the same tissues as are infected during natural infections. From transgenic plants singly infected with each of the widely prevalent viruses in the U.S. that normally infect the recipient plant (contact APHIS for the list of these viruses) determine the amount of both coat proteins (i.e., from both the transgene and the naturally infecting virus). For comparison, provide the amount of both coat proteins produced in the nonengineered plant in mixed infections of the virus from which the coat protein gene was derived and the same widely prevalent viruses used in the single infection study. Provide description of symptoms of infected plants in all cases.

For all diseases and pathogens surveyed, names of the diseases and the scientific names of the pathogens should be provided. Data from field tests in foreign countries are acceptable. If the data on diseases and pests were obtained in the foreign country, the applicant should submit information about the distribution of those pests, disease or pathogens in the U.S. Disease and pathogen susceptibility on wild type and transgenic plants should be determined preferably from natural infestations. However, if the applicant must use direct inoculations, i.e., with virus resistant transgenic plants, the source and taxonomic classification of the virus should be provided.

Certain plants have minute quantities of known toxicants which may adversely impact nontarget organisms and beneficial insects; e.g., tomatine in tomatoes, cucurbitin in cucurbits (APHIS identified cucurbitin as a known toxicant while they probably meant curcurbitacin. Cucurbitin is an amino acid that to the subcommittee's knowledge is not highly likely to affect nontarget species, whereas cucurbitacins are highly toxic "bitter principles" long known to have effects against both herbivores and predators (which may attack cucurbitacin-sequestering herbivores), gossypol in cotton, etc. If such plants are recipients of transgenes, the applicant should provide information as to whether the level of toxicants is altered. If the plant produces no known toxicant, the applicant should state so and provide the reference to support the claim. Plant toxins can be assessed by the tests and criteria that plant breeders traditionally use in the crop. In some instances, this may be done qualitatively, e.g., taste testing of cucurbits.

VI. Environmental Consequences of Introduction of the Transformed Cultivar

VII. Adverse Consequences of Introduction

Assuming that the levels of known toxicants in the regulated article reported in Section V are in acceptable range; that there were no notable differences reported in Section V between transgenic and nontransgenic plant; and that the gene(s) engineered into the recipient plant have no known reported toxic properties; then, toxicological data on effects of the plant on nontarget organisms and threatened and endangered species will usually not be required.

A separate petition should be submitted for each category/phenotype combination. For example, a petition for coleopteran insect resistant potatoes or PVY resistant potatoes should be submitted separately. However, when a single plant contains more than one phenotype modification, submit only one petition. For example, one petition should be submitted for potatoes that are both PVY and PVX resistant.

Upon receiving a petition for nonregulated status, APHIS assigns a petition number and publishes a notice in the *Federal Register* to solicit comments from the public. Comments submitted within the 60-day comment period become part of the file. Within a 180-day period, the administrator responds in writing to the petitioner, either approving (in whole or in part) the petition or denying it. If the administrator denies a petition, the reasons are provided in writing to the applicant. The unsuccessful applicant may appeal the decision within 10 days of receipt of the written notification of denial.

Requests for Extension of Determination of Nonregulated Status to Additional Regulated Articles

APHIS may extend a determination of nonregulated status to additional regulated articles if it finds that such articles do not pose a potential for plant pest risk. Extension requests are handled somewhat like petitions. Under the extension provisions, the administrator may determine that a regulated article is substantially similar to another nonregulated article and does not pose a plant pest risk. Any person may request consideration under this argument to other regulated articles, based on submitted evidence of similarity to equivalent nonregulated articles.

The finding is based on an evaluation of the similarity of the additional regulated articles to an organism that has already been the subject of a determination of nonregulated status by APHIS and is used as a reference (antecedent) for comparison to the regulated article under consideration.

In its user's guide, APHIS provides examples that illustrate molecular manipulations that create organisms the agency believes are unlikely to pose new risks beyond those that would have already been considered in the initial determination of nonregulated status:

- "Modifications in which the amino acid sequence of any encoded proteins is unchanged with respect to the corresponding sequence in the antecedent organism (i.e., synonymous codon changes)."

- "Production of new transformants of the same cultivar as the antecedent organism, or new cultivars of the same plant species or variety that do not differ significantly from the antecedent cultivar in reproductive fitness, obtained via transformation using the same transforming nucleic acid as was used in producing the antecedent organism."
- "Production of new transformants of the same cultivar as the antecedent organism, using a transformation vector that differs from that used to produce the antecedent organism only: in noncoding regulatory sequences used to control expression of any of the introduced genes; or in other vector DNA sequences that were not incorporated into the recipient plant cells, unless the new regulatory sequences cause the expression of any introduced genes in plant tissues in which the introduced genes were not expressed in the antecedent organism."
- "Production of new transformants of the same cultivar as the antecdent organism, in which the genetic material transferred into the recipient plant to produce the antecedent organism is identical to that in the regulated article in question, but in which said material was introduced into the recipient plant using a different transformation vector or technique."
- "Modifications in which the antecedent organism and the regulated article in question contain different donor genes, but the donor gene used in producing the antecedent organism and the donor gene used in producing the regulated article in question encode enzymes catalyzing the same biochemical reaction (i.e., molecules that have the same substrates and products) or encode other proteins performing the same molecular function (i.e., molecules that bind to the same target molecule in vitro and either inhibit its function via the same mechanism, or cause the same biochemical change in the target molecule)."
- "Modifications in which the antecedent organism and the regulated article in question differ only in [the] marker genes they contain, which were used in their identification or selection, provided that the new marker genes in the organism in question do not raise any new risk issues, i.e., do not encode substances toxic to plants; do not encode substances with pesticidal properties; do not confer resistance to any antibiotic of significant importance for veterinary or human use; and do not confer resistance to any different herbicide than was conferred by the marker genes in the antecedent organism."

To provide guidance for extension requests, the APHIS user's guide provides a sample request, which suggests that certain information be provided by the applicant. Essentially, the request is a streamlined petition that focuses on differences between the regulated article and the antecedent organism. In particular, APHIS focuses first on a "precise description of the genetic modifications in the regulated articles under consideration and detailed comparison of the modifications in those regulated articles with those in the antecedent organism." For organisms not covered in the APHIS examples above, the applicant may address why the particular organisms in question do not raise any issues different from those considered in the determination for the antecedent organism, thereby meriting separate consideration under a new petition.

Also, the requesters should provide information on the phenotypic expression of the genetic modifications in the regulated articles, indicating any expected or unexpected differences in phenotype between the regulated articles and their antecedent organism. Data from at least one field trial should be included, and all data reports from completed field trials with a regulated article should be submitted with the extension request.

CONCLUSION

APHIS regulation of transgenic organisms has evolved over the past two decades. To meet the growing number of field tests an expeditious notification system has evolved, assuring rapid turnover of field test applications. The permitting process, originally used for all field tests, is now used to deal with field testing situations that require substantial scrutiny (such as products intended for pharmaceutical use). Both the notification and permit processes require that transgenic organisms be grown and handled in such a way as to prevent their escape into the ambient environment. Petitions of the determination of nonregulated status (and a similar extension system) represent what has been the primary pathway to commercial plantation of transgenic plants. However, commercial products have also been created from regulated transgenic organisms that have been grown under notification and permit. It is not clear which of these regulatory pathways will become the option of choice for future field-based commercial plantations of transgenic plants.

4

Case Studies of APHIS Assessments

Chapter 3 provided an overview of the regulations and general procedures used by the Animal and Plant Health Inspection Service (APHIS) to review the large number of transgenic plants coming through the research and development pipeline each year. This chapter examines in detail the specific procedures and judgments made by APHIS in its assessment of a set of cases that have moved through the notification, permitting, and deregulation pathways. The specific set of cases examined was chosen in order to cover a broad array of products, procedures, and potential risks. For each case the types of risks considered by APHIS are noted, and the information and processes used by APHIS in making judgments about these risks are assessed. Risks not considered by APHIS also are pointed out. For all of the case studies, the degree of public and external scientist involvement in the decision-making process also is assessed.

First presented is a case study for a transgenic plant that was field tested through the notification process and one that went through the permitting process. Then four types of transgenic plants that have been deregulated by APHIS are examined. Chapter 5 develops a more general assessment of APHIS oversight and makes recommendations for specific changes.

NOTIFICATION PROCESS CASE STUDY

Notification for Salt- and Drought-Tolerant Bermudagrass

Bermudagrass (*Cynodon dactylon* and a few related species) is an important grass of lawns and pastures in the United States and elsewhere

(Simpson and Ogorzaly 1995, Taliaferro 1995). Consequently, strategies have been sought to improve the performance of this plant under a variety of environmental stresses (Cisar et al. 2000).

Cynodon dactylon is also considered one of the world's worst weeds (Holm et al. 1977), especially in the tropics and subtropics but also in warmer parts of temperate zones. It is an especially important weed of sugarcane, cotton, and corn. It is a troublesome weed in some parts of the United States; for example, in the West it has been described as "posing a serious threat to crop production and turf management" (Ball et al. 2000). The species is wind pollinated and reproduces by seed but more frequently by vegetative spread of plant parts (stolons and rhizomes).

Since 1999 APHIS has received and acknowledged the eligibility of five notifications from Rutgers University for New Jersey field tests of salt- and drought-tolerant bermudagrass. One of the notifications (99-308-10n) is discussed here.

The notification application names the transformed organism as a bermudagrass hybrid, *Cynodon dactylon* × *C. transvaalensis*. The application also describes the mode of transformation (in this case, particle bombardment). The added genes also are detailed. The gene inserted to confer possible drought and salt tolerance was betaine aldehyde dehydrogenase from *Atriplex hortensis*, a plant species that shows considerable drought and salt tolerance. The promoter for that gene was maize ubiquitin. The terminator was nopaline synthase polyadenylation sequence from *Agrobacterium tumefaciens*. The plants also are transgenic for a selectable marker, hygromycin B phosphotransferase from the bacterium *Streptomyces hygroscopicus* with a rice actin promoter and a 35s polyadenylation sequence from cauliflower mosaic virus (CaMV) for a terminator. The specific planned introduction was a field test at Rutgers University Horticulture Farm II. The applicant certified that the regulated article "will be introduced in accordance with the eligibility criteria and performance standards set forth in 7 CFR 340.3."

Because the primary concern for the notification process is containment, adherence to performance standards is important. And because the transformed organism is closely related to an important weed and contains transgenes that might confer an advantage to that weed, that containment is not just an academic exercise. Indeed, the applicant took containment very seriously when reporting his containment procedures to APHIS in a letter dated March 26, 2001. The transformed organism is the variety "TifEagle" (Hanna and Elsner 1999), used primarily as a turfgrass for putting greens in golf courses. The variety is a triploid bermudagrass that is both male and female sterile. Thus, dispersal by pollen and seed does not occur. The field test involved a comparison of the performance of the transgenic grass to nontransgenic strains in an area considerably less than an acre. Twice weekly mowing maintained the grass at a height of 5/32 inches. The borders of the field test were maintained by application of an herbicide treatment every two weeks. The applicant also states that TifEagle cannot withstand central New Jersey's cold winters but that the plots would be monitored to see if the transgenic plants had developed winterhardiness. The applicant does not describe other methods for preventing accidental spread by fragments of the grass that may attach to equipment or shoes. But the ultradwarf, dense-growing nature of this particular variety probably makes such fragmentation extremely unlikely—especially compared to the easily broken, rambling runners of the wild type.

Environmental Risks Considered by APHIS

APHIS does not conduct environmental assessments on notifications, which are assumed to be safe based on meeting the notification criteria and based on using plant-specific performance standards that minimize any chance of plant or gene escape beyond the confines of the field plot.

Involvement of Potential Participant Groups

There is no public or external scientific involvement for this or any other plant that goes through the notification process.

THE PERMITTING PROCESS

Permitting of Maize-Expressing Proteins with Pharmaceutical Applications

Background

This case is a permit application (00-073-01r, dated March 8, 2000) in which the applicant (ProdiGene) requested permission to grow maize transformed with one or more transgene-expressing proteins with pharmaceutical properties (with the date of intended release 60 days later). The specific phenotype is listed as "antibody production in seed." A description of the transgenic plant was not available because it is confidential business information (CBI). The purpose of the permit was to grow the transformed maize for seed increase and genetic improvement. The test plots, totaling no more than 2 acres, were to be grown in Nebraska.

Environmental Risks Considered by APHIS

Although APHIS decided that this application did not require an environmental assessment, the permit application itself provides information that addresses a number of environmental issues (see Chapter 3). The following information was gleaned from the permit application and from APHIS's letter giving notice of its review to the state of Nebraska.

Environmental Impact Related to Donor. Maize was transformed with four genes, one or more of which encode a protein or proteins (one or more antibodies) with pharmaceutical properties. The identities of the genes, their enhancers, their products, and their sources were marked CBI and therefore are not available to the public. The plants were also transformed with the selectable marker, the gene coding for maize-optimized phosphinothricin acetyl transferase (*moPAT*) derived from the bacterium *Streptomyces viridochromogenes*, a nonpathogenic soil bacterium. The selectable marker results in tolerance to the herbicide glufosinate. That promotor and terminator for that gene were from CaMV. APHIS concluded that none of these genes contained any inherent plant pest characteristics. APHIS also concluded that the promoter and terminator from CaMV cannot cause plant disease by themselves or in conjunction with any of the genes introduced into the maize plants.

Environmental Impact Related to the Vector and Vector Agent. Transformation was facilitated with *Agrobacterium tumefaciens*. This is a well-characterized transformation system resulting in stably integrated and inherited transgenes. APHIS found no inherent environmental impact.

Quarantine of Organism and Final Disposition. The focus of the permitting process is to ensure that the transgene escape does not occur by the dispersal of seed, pollen, or plants and that transgenic plants do not persist after the experiment ends. The applicants stated that an isolation distance of 1,320 feet would be used to minimize transgene flow by pollen (this is double the 660-foot isolation distance recommended in the APHIS 1997 user's guide). The applicants stated that all seed would be harvested and that the plants would be plowed into the soil. The field would then be monitored for volunteers, which would be destroyed by hand or with an appropriate herbicide. APHIS concluded that these measures were adequate to confine the transgene and the transgenic organism. Supplemental permit conditions included monitoring by the appropriate officials, the reporting of field data to APHIS within six months of the termination of the field test, monitoring the field for one year after the test for volunteers, and notification of any changes in protocols.

An issue taken at face value is the isolation distance requirement for corn, a highly outcrossing, wind-pollinated species. Although corn has no weedy or wild relatives in North America with which it can freely cross, isolation between the transgenic corn and other corn crops remains an issue. Isolation distances are set to prevent some minimum level of contamination but were not set up to provide for zero levels of contamination. And zero levels are what would be needed to absolutely prevent escape of the transgene into the environment. For example, the APHIS-recommended isolation distance of 660 feet for corn is presumably derived from that required by the U.S. Department of Agriculture (USDA 1994a) for producing foundation seed (used for seed increase); the maximum proportion of contamination is 0.1%. There is no reason to assume that absolute isolation should be attained at twice that distance. It is likely there would be some very low level of contamination of any corn grown at or near the 1,320-foot isolation distance from the test plots. If adjacent corn were grown for a purpose such that its seeds were not replanted, there would be no permanent escape into the environment. However, as outlined in Chapter 2, some consider that the risk of these genes entering the food supply should be considered an environmental risk.

If contaminated corn were grown such that its seeds were to be replanted, it is possible that the transgenes (for antibody production and glufosinate tolerance) could end up in the genetic stocks. One possible example is adjacent plots of other experimental corn varieties grown for seed increase prior to commercial use. Another example could be nearby plantations of open-pollinated corn grown by a farmer who keeps the seed from year to year. Either of these scenarios could result in perpetuation of the transgenes indefinitely unless they were lost from the breeding stock by random drift. Whether the transgenes for antibody production in seed have an impact in a different corn crop depends on a variety of factors, including the specific pharmaceutical compound created through expression of the transgene; the levels at which that compound is created and stored in tissues that might be consumed by humans, farm animals, or non-target organisms; and the level at which contamination has occurred, and the threshold effects of that compound if used for animal or human food. In this case, because identity of the pharmaceutical compound is not given in the application because it is CBI, it is not possible to judge that impact.

Whether the transgene for glufosinate production has an impact in corn-breeding stocks contaminated by pollen would depend on whether the contaminated stocks would find themselves under selection by that herbicide. The environmental impact of glufosinate tolerance in corn is discussed at length in the *Bt* corn case study below.

Involvement of Potential Participant Groups in Decision Making

Because APHIS did not see a need for conducting an environmental assessment before approving this introduction, there was no opportunity for public or external scientific involvement in this permit decision.

PETITIONS FOR DEREGULATED STATUS: FOUR CASE STUDIES INVOLVING SIX PETITIONS

Two Virus-Resistant Squash Petitions

Background

Virus-based diseases can sometimes pose important problems for crop production (Hadidi et al. 1998). Virus resistance may be transferred into a crop via conventional breeding methods but only if that resistance already exists in the crop or in a sexually compatible relative. Transgenic virus resistance provides an opportunity for disease resistance in crops whose close relatives are not resistant to the virus in question. Transgenic virus resistance can be obtained by introduction of part of the disease viral genome into the susceptible plant genome; in particular, expression of the viral coat protein (CP) often confers resistance (Powell-Abel et al. 1986, Grumet 1995).

Field trials of dozens of crop species with transgenic-based virus resistance have been conducted (see "Field Test Releases in the U.S.," Information Systems for Biotechnology online database: *www.nbiap.vt.edu*). As of April 2001, APHIS had approved six petitions for the deregulation of transgenic crops with virus resistance (see "Current Status of Petitions," APHIS website: *www.aphis.usda.gov/biotech/petday.html*). The deregulated crops transformed are papaya, potato, and squash. The case of deregulation of Upjohn/Asgrow's virus-resistant crookneck squash varieties (application numbers 92-204-01p and 95-352-01p) exemplifies how APHIS evaluated a number of different issues associated with the biosafety of a transgenic product.

Viral diseases are periodically an important problem for growers of squashes and other cucurbit crops (Desbiez and Lecoq 1997). These diseases include those caused by zucchini yellow mosaic virus (ZYMV), watermelon mosaic virus 2 (WMV2), and cucumber mosaic virus (CMV). Aphids act as the vectors of all three viruses.

Interestingly, squash varieties with genetically based virus resistance to WMV2 and ZYMV were developed almost simultaneously by both transgenic and conventional methods. In the same year, 1994, Harris Moran released a conventionally bred virus-resistant zucchini (Tigress) and Asgrow's transgenic virus-resistant yellow crookneck squash (ZW-20)—whose resistance was based on expression of viral CP genes—was deregulated (USDA 1994b, Schultheis and Walters 1998). Subsequently, Asgrow's second transgenic yellow crookneck squash (CZW-3)—containing viral CP genes to confer resistance to WMV2, ZYMV, and CMV—was deregulated in 1996 (USDA 1996). Conventionally created CMV-resistant marrow squash is commercially available from Thompson and Morgan (USDA 1996). As of January 2001, APHIS had received 66 notifications and permit applications for squash varieties with transgenic resistance for as many as five viruses (see "Field Test Releases in the U.S.," Information Systems for Biotechnology online database: *www.nbiap.vt.edu*).

APHIS's action on the Upjohn/Asgrow petition for ZW-20 squash received considerable comment from the public, both pro and con. By the time APHIS made its final decision, the agency had published three *Federal Register* announcements and conducted a number of public meetings. Feedback on the petition is detailed below under "Involvement of Potential Participant Groups." APHIS provided a detailed response to commenters who disagreed with its ruling in *Response to the Upjohn Company/ Asgrow Seed Company Petititon 92-204-01 for Determination of Nonregulated Status for ZW-20 Squash* (USDA 1994b).

The petition for CZW-3 did not generate controversy. APHIS received only a few comments, all favorable to the petition, in response to a single *Federal Register* announcement. Nonetheless, APHIS detailed its findings by covering much of the same ground as that for ZW-20 in *Response to the Asgrow Seed Company Petition 95-352-01 for Determination of Nonregulated Status for CZW-3 Squash* (USDA 1996).

Both APHIS response documents on the transgenic virus-resistant squashes considered a number of potential risks in some detail. Below is a highly abstracted overview of APHIS's arguments for finding "no significant impact."

Environmental Risks Considered by APHIS

Disease in the Transgenic Crop and its Progeny Directly Resulting from the Transgenes, Their Products, or Added Regulatory Sequences. Some of the DNA sequences used in transforming these squashes were derived from *Agrobacterium tumefaciens* (the agent of crown gall disease), but the disease-causing genes were removed. Likewise, the viral CP transgenes and additional viral regulatory sequences to control their expression were all derived from disease-causing organisms, but they and their products do not cause disease. CZW-3 also has a selectable marker for kanamycin resistance; APHIS did not consider whether that transgene and its products will cause disease CZW-3, presumably because the source of this gene is not a pathogenic organism. Thus, APHIS concluded that the transgenes, their products, or added regulatory sequences do not result in plant pathogenic properties (USDA 1994b, 1996).

Evolution of New Plant Viruses. For both ZW-20 and CZW-3, APHIS addressed the risks that other viruses would appear with altered host specificities (via transcapsidation, when one virus bears the coat protein of another; also known as "heteroencapsidation," "genomic masking," or "masked viruses") or evolve increased virulence (from recombination with virus-derived transgenes; Matthews 1991).

With regard to the first issue, APHIS pointed out that mixed infections by plant viruses are not uncommon, that these viruses are common viruses of squash, and that transcapsidation is already occurring in infected plants. Given that the amount of coat protein in the transgenic squashes is considerably less than that in naturally infected plants, the chances of transcapsidation are lower in transgenic plants than infected ones. Furthermore, APHIS pointed out that even if masked viruses (i.e., viral nucleic acids enrobed with a coat protein of a different virus produced by the transgenes of the plant) were produced, they would have biological properties identical to those produced in naturally infected plants (USDA 1994b, 1996).

With regard to evolution of virulence via recombination, APHIS concluded that:

> because the viral transgene is derived from virus that naturally infects the squash host, is synthesized in the same tissues as in the naturally-infected plants, is produced in less concentration than during natural infections, and if a recombinant was formed would have to be competitive with other squash-infecting viruses. APHIS believes that even if a recombinant virus did occur that [*sic*] this virus could be managed just like the numerous new viruses that are detected every year in the United States. (USDA 1996; cf. 1994b)

APHIS addressed two additional issues for its assessment of CZW-3: the release of subliminally infecting viruses (those unable to move from the initial site of infection) and synergy (the increased severity of symptoms from multiple infections; Matthews 1991). The agency addressed the concern that infection from a different virus may release subliminally infecting viruses. In the case of plants expressing CP genes, the worry is that those genes might facilitate movement out of the transgenic plants. But "since the CP transgenes in CZW-3 are all from viral strains that routinely infect the curcubit family, it is not expected [that] subliminally infecting viruses will present a problem any more serious than can occur in naturally infected squash plants" (USDA 1996). Synergy was not considered an environmental risk but rather an agronomic problem. "Asgrow inoculated CZW-3 with several common squash-infecting viruses. No synergistic symptoms were seen in infected plants" (USDA 1996).

In both response documents, APHIS concluded that the transgenic squashes should pose no greater risk of evolution of new viruses than naturally infected plants. In the second document, that conclusion was extended to cover the case of wild relatives that pick up the transgenic traits through introgression.

Increased Weediness in the Transgenic Squash Relative to Conventionally Bred Squash. For both ZW-20 and CZW-3, APHIS addressed the risk that the virus resistance genes would increase the weediness of yellow crookneck squash. Yellow crookneck squash is not listed as a common or troublesome weed anywhere in the United States; for example, it is not on the Weed Science Society of America's "Composite List of Weeds" (available online at *http://ext.agn.uiuc.edu/wssa/*). Squash volunteers occur adjacent to squash production fields and, if necessary, are controlled mechanically or with herbicides. They do not readily establish as feral or free-living populations (USDA 1994b, 1996).

For ZW-20, Upjohn/Asgrow supplied APHIS with data comparing the transgenic squashes with their nontransgenic counterparts, showing "no major changes in seed germination, cucurbitin levels, seed set viability, susceptibility or resistance to pathogens or insects (except ZYMV and WMV2), and there are no differences in overwintering survivability" (USDA 1994b). For CZW-3 the APHIS response document stated that "Asgrow has reported that there are no major changes in CZW-3 performance characteristics (except for resistance to CMV, ZYMV, and WMV2)" (USDA 1996). Given that the transgenic squashes would be expected to be grown in the same regions as squash is typically grown, APHIS concluded that "there is no evidence to support the conclusion that introduction of virus resistance genes into squash could increase its weediness potential. Many pathogen and insect resistance genes have been introduced into commercial varieties of squash by conventional means in the past without any reports of increased weediness" and noted that conventionally improved cultivars having resistance genes to viruses had already been developed. In both response documents, APHIS concluded that the virus resistance transgenes are unlikely to increase the weediness of yellow crookneck squash (USDA 1994b, 1996).

Impact on Non-target Organisms Other Than Wild Relatives. APHIS pointed out that both ZW-20 and CZW-3 transgenic squash plants have no direct pathogenic properties. The protein products of the transgenes are already present in high concentration in naturally infected plants, and the levels of cucurbitin—a naturally occurring plant defensive compound—are likely to be unchanged. Therefore, APHIS concluded that "there is no reason to believe deleterious effects on beneficial organisms could result specifically from the cultivation of" the new transgenic squashes. APHIS noted Upjohn/Asgrow taste tests for cucurbitin levels, but otherwise its conclusions were based on the fact that the coat proteins present in the transgenic squashes are already present in the environment in virus-infected plants. In the second determination, APHIS briefly examined the issue of whether insecticide usage might be reduced by the introduction of the CZW-3 but did not reach a conclusion.

Impacts on Free-Living Relatives of Squash Arising from Interbreeding. Most of the APHIS discussion in the response documents, particularly the 1994 one focuses on whether wild relatives could benefit from virus-resistance alleles, leading to the evolution of increased weediness. The effort was, in part, in response to several negative comments received after the three APHIS *Federal Register* announcements associated with ZW-20. Many of the comments questioning the decision did so because it marked an important APHIS precedent. As noted in Chapter 2, the sexual transfer of beneficial alleles from a transgenic crop to a wild relative might result in the evolution of a more difficult weed. This issue is perhaps the most widely discussed risk associated with transgenic crops (e.g., Colwell et al. 1985, Goodman and Newell 1985, Snow and Moran-Palma 1997, Hails 2000).

This case study has all three elements that could create such a risk—transgenes of a type that could confer a fitness boost in the wild, a sexually compatible wild relative, and the fact that the wild relative has been classified as a weed. If the crop mates with the wild relatives introducing virus resistance into wild populations and if the primary factor limiting the aggressiveness of wild populations is disease caused by the same viruses, introgression of the transgenes could result in increased weediness of the wild relatives. To obtain more information on the relevant biology of the wild relative, APHIS commissioned a report on the risks that might be posed by crop to wild gene flow by Hugh Wilson, an expert on cucurbit taxonomy and ecology. Wilson (1993) concluded that free-living *Cucurbita pepo* (FLCP) is a significant weed that might benefit from protection from ZYMV and WMV2. Key information on squash and its weedy North American relatives is summarized below with APHIS's conclusions and the committee's evaluation of those conclusions.

Yellow crookneck squash belongs to the species *Cucurbita pepo*. As discussed above, the squash itself is not a weed. However, squash freely crosses with wild weedy plants known as Texas gourd (originally classified as *C. texana* but now considered a subspecies of *C. pepo*) and as FLCP. An experiment by Kirkpatrick and Wilson (1988) demonstrated that squash and FLCP naturally hybridize freely; the crop sired 5% of the seed set by FLCP, growing 1,300m from cultivated squash. Hybrids between the crop and FLCP are fully fertile (Whitaker and Bemis 1964). Clearly, if the crop and FLCP grow in the same region, natural hybridization will occur, and crop alleles will readily enter the natural populations. Indeed, cultivated squash and FLCP co-occur in many regions of Texas, Louisiana, Alabama, Mississippi, Missouri, and Arkansas (Wilson 1993). APHIS concludes that natural hybridization will move the virus resistance genes from the transgenic crop to the wild populations (USDA 1994b, 1996).

FLCP is an agricultural weed in cotton and soybean fields. At one time it was one of the 10 most important weeds in Arkansas (McCormick 1977). APHIS contacted three weed experts for their opinions on the current status of FLCP as a weed. The three experts noted that FLCP plants appeared to be "less a problem" in 1994 than during the 1980s because of new herbicides not available in the 1980s and suggested that new herbicide-tolerant crops would "further expand the tools for effective control of FLCP plants" (USDA 1994b). APHIS concluded that "FLCP plants are not a serious weed in unmanaged or agricultural ecosystems [because] "the registration of new herbicides now allows effective management of these plants" (USDA 1994b). However, two of the three weed experts reported that FLCP is less of a problem; it is not clear how serious a weed they still consider it to be.

The key issue raised by the foregoing data is whether virus resistance genes will provide enough of a benefit that FLCP becomes a more difficult weed. Wild and cultivated *C. pepo* are susceptible to the same viruses (e.g., Provvidenti et al. 1978). To determine whether viruses limit the population size and number of FLCP, Asgrow conducted a survey in 1993. Fourteen FLCP populations (two in Arkansas, four in Louisiana, and eight in Mississippi—a severe drought precluded sampling in Texas) in nine locations were visited once (when plants were at maturity); no visual symptoms of viral infection were noted. Some of these sites were within a mile of cultivated squash. But it was not reported whether the nearby cultivated plants were infected with virus; that information would have shown whether viruses were present that year. A single plant was sampled from each population. Each plant was subjected to multiple analyses to check for asymptomatic viral infection, and all were found to be virus free. On the basis of these data and qualitative anecdotal reports from weed experts, APHIS concluded that "there is no scientific or anecdotal evidence to suggest that these viruses routinely infect FLCP plants" (USDA 1996) and that because "FLCP plants are **not** under significant environmental stress from viral infection, the selective pressure to maintain the virus resistance genes in natural populations of FLCP plants should be minimal" (USDA 1994b).

These conclusions warrant some discussion. One issue is the adequacy of the survey. The committee questions whether a single visit and the small sample of plants are adequate to determine whether a disease is among the important factors that limit population size, number, and niche. The APHIS documents reveal that these viral diseases vary tremendously in the crop from place to place and year to year. Should they be expected to do any differently in closely related plants? Is a single-year survey sufficient? Likewise, is a single visit per site sufficient? APHIS defends the survey as "an appropriate, adequate, and **proven** means for determining whether a plant is a significant natural host for a particular virus." The committee questions the phrase "appropriate, adequate, and **proven.**" Standard sampling procedures of plant epidemiology include sampling several plants per site, repeated visits over time, and a statistically derived basis for determining sample size (e.g., Campbell and Madden 1990). The committee does not see those standards in the survey defended by APHIS. In fact, the timing of the survey might have been important. The committee notes that Provvidenti et al. (1978) found that virus-infected wild plants die prematurely. If wild plants are especially vulnerable to the disease at an early age, by the time the survey was conducted it is possible that the viruses could have wiped out whole populations or that the only plants remaining in populations were those that, for whatever reason, escaped viral infection. The committee also notes that the drop in FLCP frequency reported by the weed experts interviewed by APHIS (see above) is coincident with the arrival and spread of ZYMV in the United States reported by APHIS (USDA 1994b).

Another issue is that of alternative methods for identifying whether virus resistance might provide a fitness boost for FLCP. Some comments suggested experimental approaches for measuring whether virus resistance would confer a fitness benefit to the wild plants. Two types of experiments were proposed:

1. Inoculating greenhouse-grown plants with ZYMV and comparing the response to infection of the following plants: FLCP, ZW-20, nontransgenic yellow crookneck squash, and F_1 and F_2 hybrids of FLCP with ZW-20 and nontransgenic yellow crookneck squash. The rationale is that a

change in the weediness potential should be apparent by comparison of plant fitness.

2. Experiments to exclude the virus from an array of genotypes to determine whether the virus impacts the fitness of FLCP. The rationale is that if the virus is a significant natural enemy, excluding it should result in a fitness boost in FLCP. The comments did not give details of the experiment in the reports to APHIS.

APHIS rejected the "experimental approaches suggested by commenters to determine the impact of movement of virus resistance gene to FLCP plants" as "flawed" (USDA 1994b). APHIS described the limitations of the proposed experiments at length. The first approach was criticized because the greenhouse is an artificial environment. "It is not uncommon for crop plants to be susceptible under controlled conditions to a widely prevalent virus yet [be] rarely infected by that virus under natural conditions." APHIS concluded that the survey "will provide more reliable information than a greenhouse experimentation based one." APHIS interpreted the second proposed experiment as one that would involve placing insect-proof cages over plants growing in natural conditions. APHIS criticized that presumed approach as not feasible as well as artificial because cages would exclude all insect pests of FLCP and all insect-borne viruses, not just the two in question.

It is notable that Asgrow itself conducted experiments similar to the type APHIS rejected. Regrettably, the sample sizes in those experiments were too small to draw any conclusions. APHIS also referred to another Asgrow experiment which showed that "FLCP × ZW20 hybrids do not appear to be strong competitors when growing in fields that have not been tilled to remove competing wild plants based on survival and seed set (data report 93-041-01)" (USDA 1994b). APHIS did not state to what the hybrids were compared. The committee does not necessarily support the proposed experimental methods but does question why APHIS is so critical of those methods to measure potential weediness while accepting, apparently without criticism, experimental data of the same general type from Asgrow.

The third issue is whether transgenes will be maintained in natural populations. The APHIS statement above that "the selective pressure to maintain the virus resistance genes" and others in the documents belies an assumption that crop alleles are maintained in natural populations only when they are beneficial. Population genetics theory has demonstrated that even very low levels of gene flow (two successful pollinations per generation) are sufficient to maintain a *neutral* crop allele at substantial frequencies in a natural population (Wright 1969). Given the hybridization rate of 5% that Kirkpatrick and Wilson (1988) observed, the committee predicts that neutral and beneficial crop alleles would persist in the FLCP populations. Indeed, at this rate of hybridization even somewhat *deleterious* alleles are expected to persist in wild populations (Slatkin 1987); that is, a 5% migration rate would maintain alleles that cause a 5% fitness drop. The maintenance of crop alleles in natural populations does not necessarily pose a risk; indeed, the flow of neutral, beneficial, and deleterious squash alleles into natural FLCP populations must have occurred in the past and may still be occurring. However, the APHIS expectation that only beneficial alleles will be maintained in natural populations under gene flow from crops reflects inadequate expertise in population genetics at the time the squash documents were prepared.

APHIS may well be correct that virus resistance would not result in increased weediness of FLCP. But its defense of the Asgrow survey, the "double-standard" for experimental data, and the flawed population genetics arguments are scientifically inadequate to support that conclusion.

Finally, APHIS makes a "substantial equivalence" argument in comparing the product produced by a transgenic method to one produced by conventional, sexual crosses. APHIS notes that because some nontransgenic virus-resistant squashes (see "Background" above) were available at about the same time, it argues that the risks of gene flow of virus resistance alleles from those conventional varieties should be equivalent to the transgenic products in their potential impact on FLCP (USDA 1994b, 1996).

APHIS does not have a mandate to consider the environmental impacts of transgenic plants outside the United States. Because of a negative comment suggesting that ZW-20 would find its way into Mexico, the APHIS response document associated with that regulated article states that "many crucial scientific facts explained in this determination for the United States also apply in Mexico" (USDA 1994b).

Environmental Risks Not Considered by APHIS

As noted in Chapter 2, crop to wild gene flow, transgenic or otherwise, may present an extinction risk to wild crop relatives. Preexisting nontransgenic squashes may already pose that risk to their wild relatives in the United States and Latin America. APHIS did not directly consider whether transgenic squash would have a negative impact on its wild relatives in the United States or Latin America, either in terms of posing an impact that might lead to their extinction (for example, *Curcubita okeechobeensis* is a federally listed endangered species) or in terms of changes in their genetic diversity. Relative to nontransgenic crops, a transgenic crop could pose an greater extinction risk to a wild relative or, more likely, alter its genetic diversity if it permits the crop to be grown in closer proximity to the wild plants, thereby increasing interpopulation hybridization rates and detrimental gene flow.

Involvement of Potential Participant Groups

On September 4, 1992, APHIS announced its intent to issue an interpretive ruling on the Upjohn/Asgrow petition that ZW-20 squash did not present a plant pest risk and would no longer be considered a regulated article in the *Federal Register* (57:40632). During the 45-day comment period, APHIS received 17 comments regarding its proposed ruling; seven were generally supportive of APHIS's proposed action, and 10 were not.

On March 22, 1993, APHIS published a second *Federal Register* notice (58:15323) requesting additional information on eight issues raised by commenters to the first notice. At the same time, APHIS commissioned Hugh Wilson of Texas A&M University, a cucurbit taxonomist and ecologist, to prepare a report (see above) related to issues raised in comments to the first *Federal Register* notice. There were 12 comments to the second notice; 10 were generally supportive of APHIS's action and two expressed serious reservations. After the close of the official comment period, APHIS received two additional letters urging the agency not to approve the petition.

On May 23, 1994, APHIS published a third notice in the *Federal Register* (59:26619–26620) announcing an environmental assessment (EA) and preliminary finding of no significant impact (FONSI) for comment at a public meeting and for written comment during a 45-day comment period. Two individuals, one in favor of the EA and FONSI and one against, spoke and provided written comments at the meeting. During the rest of the comment period, APHIS received 52 additional written comments. Twenty-nine comment letters disagreed with APHIS's proposal to approve the subject petition; 23 comments supported APHIS's findings in the EA and FONSI. The affiliations of the final set of commenters were as follows: private individuals (18), universities (12), agricultural experiment stations (11), public policy and public-interest groups (6), industry (2), associations (1), cooperative extension service (1), and federal research laboratory (1). About a third of the final set of comments were detailed and substantial. The affiliations of prior commenters were not published by APHIS.

The second petition for CZW-3 was not nearly as controversial. On February 2, 1996, APHIS published a *Federal Register* notice (61:3899–3900) announcing that the Asgrow petition was available for public review. APHIS received a total of four comments, all of which were favorable to the petition.

Soybean with Altered Oil Profile

Background

Introduced to North America from China, soybean (*Glycine max*) produces almost half the world's vegetable oil. The United States grows almost half the world's soybean crop (Singh and Hymowitz 1999). Soy is a highly versatile crop, grown for human food, animal feed, and industrial components. It is considered the world's foremost provider of vegetable oil and protein. Its major product, soybean oil, is used in the production of a wide range of food products, from margarine to salad oil and as an additive in hundreds of processed foods. The seed meal remaining after oil extraction is a nutritious (but incomplete) protein source. In addition, soy is made into tofu, soy sauce, meat substitutes (e.g., "veggie burgers") and industrial products such as soaps, diesel fuel, and oil-based carriers for other chemicals (Smith and Huyser 1987, Fehr 1989).

Like all vegetable oils, soybean oil is composed of a mixture of several fatty acids. The composition of fatty acids gives the various oils their properties. In developing seeds, fatty acids are synthesized in a known biochemical cascade. Starting with common nutrient building blocks, specific enzymes act to create simple fatty acids, such as the saturated lauric (C12:0) and palmitic (C16:0) acids. Other enzymes can then act on these fatty acids to result in more complex fatty acids, such as stearic acid (C18:0) and the monounsaturated oleic (C18:1) acid, the major constituent of olive oil. Each plant species produces enzymes to act on the fatty acids to synthesize a mixture of fatty acids characteristic of the seed oil for that species. Over half the soybean oil (~54%) consists of the polyunsaturated fats linoleic acid (C18:2) and alpha-linolenic acid (C18:3; ~8%). About a quarter (~23%) of the oil is the monounsaturated fatty acid oleic acid (C18:1). The remainder, about 15%, is saturated fatty acid.

The fatty acid composition of various oil crops has been subject to conventional breeding to create desirable or enhanced properties. Because the incidence of coronary heart disease correlates more strongly with the amount of saturated fat in the diet than with total fat intake (Zyriax and Windler 2000), oils that are low in saturated fatty acids are considered healthier. One example is the oil from linseed flax. Linseed oil is used primarily as an industrial drying oil in paints and varnishes but is also considered a "health food" because of the high proportion (over 60%) of omega-3 polyunsaturated fatty acid (linolenic acid, C18:3). Linolenic acid oxidizes rapidly, which is a good characteristic for paint but not for food because oxidation makes it rancid and unpalatable. Thus, too much linolenic acid causes oil to spoil easily. Using ethylmethane sulfonate-induced mutagenesis, plant breeders in Australia produced linseed flax strains with substantially reduced linolenic acid (Green 1986). Oil from this mutant is virtually identical to high-quality oils from safflower or sunflower. The mutant linseed varieties are now in commercial production in several countries.

Plant breeders were not as successful developing altered oil-quality soybean genotypes using mutagenesis, although mutant soybean oil varieties do exist (Brossman and Wilcox 1984, Kinney 1994). The exact mechanisms or genotypes of most mutations are unknown. In mutation breeding, plant material is exposed to a mutagenic agent, such as ionizing radiation, in the hope of obtaining a mutation with a desirable phenotype. Mutagenic agents alter the DNA of the subject organism, causing destruction or duplication of a gene or parts of genes. Plant breeders select genetically based phenotypes and rarely characterize mutants at a molecular genetic level. It is almost certain that any mutation resulting in a desired phenotype will also contain mutations at other less obvious genes. These conceivably could affect ecological fitness, nutritional or antinutritional composition, or other important characteristics. In practice, however, such accessory or pleiotropic effects are identified during the breeding evaluation process prior to commercial release.

To avoid some of the problems and uncertainties with mutation breeding, DuPont used transgenic technology to improve the soy oil quality by increasing the proportion of oleic acid from approximately one-quarter to over one-half of the fatty acids. This change was accomplished by inserting another copy of a soybean gene to interfere with the enzyme responsible for converting oleic acid to polyunsaturated fatty acids. Sometimes, introducing an extra copy of a gene results in a phenomenon called cosuppression, the simultaneous silencing of the activity of both the endogenous and the inserted copies of the gene. This enzyme is encoded by a gene called *Gm fad 2-1*, which catalyses the biochemical reaction converting oleic acid to linoleic acid. DuPont's strategy was to develop transgenic soybean lines in which the inserted soybean gene interferes with the normal activity of that enzyme through cosuppression, resulting in a buildup of oleic acid and a reduction in linoleic and other polyunsaturated fatty acids. DuPont inserted DNA, including the *Gm fad 2-1* gene, into soybean (cultivar Asgrow A2396) using the particle gun delivery method, regenerated transgenic soybean plants, and analyzed them for activity of the *Gm fad 2-1* gene.

In the DNA plasmid containing *Gm fad 2-1*, the gene was linked to the soybean seed-specific promoter of the beta-conglycinin gene and the transcription terminator region from the phaseolin gene of *Phaseolus vulgaris* (common bean). This plasmid included two additional genes from *E. coli*: the ampicillin resistance beta-lactamase gene (*bla*) with a bacterial promoter, which was used in selecting the initial DNA constructs in bacteria and the beta-glucuronidase gene (*uid-A*) with the CaMV 35S promoter, which served as a reporter gene in the initial selection of transformed soybean cells. Neither of these *E. coli* marker genes is active (i.e., they do not generate functional proteins) in the transgenic soybean plants under review. However, the nonfunctional DNA is present in every soybean cell.

A second piece of DNA, a plasmid carrying the dihydrodipicolinic acid synthase (*dapA*) gene from *Cornybacterium glutanicum*, also was included in the initial soybean transformation. When active, the *dapA* gene can increase free lysine content, another potential quality trait in soybean. However, in the soybean transformation event under review, two insertions occurred: one, locus A, consists of the *Gm fad 2-1* plasmid and the second, locus B, on a different chromosome, carries both the *Gm fad 2-1* plasmid and the *dapA* plasmid. Subsequent analysis showed the activity of the *Gm fad 2-1* transgene at locus A suppressed the endogenous *Gm fad 2-1* gene, resulting in accumulation of oleic acid to over 80% of the seed oil. But the activity of the *Gm fad 2-1* at locus B caused accumulation of the active enzyme, resulting in the reduction of oleic acid to less than 4% of seed oil. Because reduced oleic acid content was undesirable, conventional plant-breeding methods were used to eliminate the chromosome-carrying locus B (consisting of both the active *Gm fad 2-1* transgene and the *dapA*-containing plasmid). Then, three soybean lines carrying only locus A were selected and designated G94-1, G94-19, and G168. The soybean lines under review therefore lack the *dapA* gene (as well as the second *Gm fad 2-1*-containing plasmid). They accumulate oleic acid and do not accumulate additional lysine.

Besides the effects on oleic acid accumulation, DuPont noted changes in the phenotypes of these soybean lines. In addition to a substantial elevation of oleic acid (C18:1) and a reduction of linoleic (C18:2) acid, linolenic acid (C18:3) and palmitic (C16:0) acid, a saturated fatty acid, are significantly reduced. Trace amounts of a linoleic acid isomer also were detected in the oil. In the meal, beta-conglycinin a and a′ subunits of the storage protein were replaced by glycinin subunits. DuPont suggested that this change occurred because the recombinant gene in locus A, consisting of the beta-conglycinin promoter fused to the *Gm fad 2-1* gene, interfered with both the *Gm fad 2-1* gene and the endogenous beta-conglycinin genes. In its petition, DuPont argued, citing Kitamura (1995), that this protein subunit substitution is nutritionally advantageous.

DuPont conducted about 25 confined field trials with the transgenic soybean lines under APHIS notifications in the mid-1990s and an additional trial in Chile over the winter of 1995–1996. DuPont submitted a petition for an APHIS determination of nonregulated status for the soybean lines G94-1, G94-19, and G168, and APHIS approved it. In May 1997, APHIS issued an EA and FONSI on the environment from agricultural cultivation and use of the DuPont soybean lines (USDA 1997a). Much of the salient discussion and justification for this determination appear in APHIS's appendix to the USDA (1997a) document.

Environmental Risks Considered by APHIS

APHIS considered the genotypic and phenotypic characteristics of the transgenic soybean lines and how they might pose an environmental hazard in unconfined release. The primary issues include the following:

Potential for the Transgenes, Their Products, or the Added Regulatory Sequences to Present a Plant Pest Risk. APHIS concluded that "neither the introduced genes, their products, nor [the] regulatory sequences controlling their expression presents a plant pest risk in the soybean sublines" (USDA 1997a). Subsequent discussion outlines the basis for this assertion, detailing the composition of the plasmids, the origin and function of each gene segment, the method of insertion (particle bombardment), and the mechanism leading to observed elevated oleic acid content in the seeds. This part also describes the genetic status of the initial transformant, carrying two inserts (loci A and B) as determined by Southern blot analysis, and how DuPont used ordinary plant-breeding methods to select progeny homozygous for locus A but lacking locus B.

DuPont's analysis of Southern blots over four generations (R1 to R4) led APHIS to conclude there are two copies of the *Gm fad 2-1* gene (one endogenous, one inserted) and that they are transmitted to progeny in a normal Mendelian manner. Unspecified evidence was provided that the *dapA* gene (from the second plasmid and inserted into locus B) is absent.

APHIS concluded that introduction of the vector DNA does not present a plant pest risk in the subject soybean lines because there are no pathogenic DNA sequences present, plus the likelihood of nonsexual transmission of DNA to other organisms is exceedingly small and, even in that eventuality, would not constitute a plant pest hazard. Finally, even though the DNA constructs did use portions derived from known plant pathogens (*Agrobacterium tumefaciens* and CaMV), those segments were not derived from pathogenic sequences in the respective pathogens. DuPont provided unspecified evidence that no symptoms of either *Agrobacterium* or CaMV infections were noted during greenhouse and field trial monitoring of the soybeans. Furthermore, DuPont reported no difference in disease and insect susceptibility between the transgenic lines and the parental soybean.

Potential for Increased Weediness of the Subject Soybeans over Conventional Soybeans. APHIS concluded that the high oleic soybean lines had no significant potential to become weeds. APHIS compared weed characteristics (Baker 1965) to the crop itself, noting that "soybean does not possess characteristics of plants that are notably successful weeds" (USDA 1997a). (See discussion of Baker's list of weed characters in the next case study.) Citing evidence provided by DuPont, APHIS said the applicant had not observed any significant changes in such characteristics as seed production, germination, standability, overwintering capacity, or pathogen susceptibility that might affect their potential to become successful weeds.

Potential for Outcrossing to Wild Relatives. APHIS noted that there are no free-living close relatives of cultivated soybean in the continental United States but that there are some wild perennial *Glycine* species in the Pacific territories. However, soybean cannot naturally hybridize with those species. APHIS acknowledged that soybean naturally hybridizes with *G. soja*, a weed of northeast Asia (including Japan) but not found in the United States (Kwon et al. 1972, Holm et al. 1979). APHIS also noted that soybean is almost exclusively self-pollinating, so hybrids due to natural outcrossing between the subject lines and wild plants is very rare. Even if outcrossing were to occur, they argue, there would be no significant impact because the high oleic trait confers no selective advantage as it does not appear to contribute to enhanced ecological fitness or weediness characteristic to wild populations.

Potential Impact on Non-target Organisms. APHIS considered the possibility that the modified soybeans might have some deleterious effect on other organisms, including beneficial insects and rare or endangered species. APHIS determined that there would be no significant deleterious impact, citing the nontoxic nature of oleic acid and also that the genes and enzymes responsible for the modified phenotype are naturally present in soybeans. The higher concentration of oleic acid is not known to have toxic properties. Oleic acid is a very common food component (and the primary ingredient of olive oil), so there is substantial history of consumption and literature on nutritional and toxicological aspects.

Observations from field trials revealed no negative effects on non-target organisms, although specific parameters measured were not provided. APHIS also stated that, since there are no novel proteins in the transgenic soybeans, there is no potential for exposing other organisms to new, potentially harmful proteins. The expressed enzyme is common and well characterized in nature, suggesting no potential for harm to beneficial organisms, and the product, oleic acid, should not result in any harm beyond that caused by a high monounsaturated fat diet. No other potential mechanisms for harm to beneficial organisms were identified.

APHIS discussed the indirect metabolic alterations in the transgenic soybeans, including the presence of a linolenic acid isomer, reduction in beta-conglycinin subunits, and corresponding increase in glycinin subunits. The agency determined that the linoleic acid isomer, present in trace quantities (<1%) in the oil, and the alteration in the seed storage protein subunits should have no adverse effects. The glycinin subunits are common components of soybean and so are unlikely to cause harm. The linoleic acid isomer is also found in other common foods, including partially hydrogenated vegetable oils and human breast milk, so it is unlikely to be harmful in the transgenic soybeans.

Soybean naturally produces antinutritional components, notably trypsin inhibitors, phytic acid, and oligosaccharides. APHIS noted that the presence of antinutritional factors in the transgenic lines were similar to those in conventional soybeans. They also said the presence of DNA from the *uidA* and *bla* genes should not have any adverse effects, as they are not expressed in the plant and do not encode infectious agents. APHIS concluded that none of these indirect effects should have any negative impact on non-target organisms.

Potential Impact on Agricultural Commodities. Citing its authority to investigate the potential for plant pest effects, APHIS determined that the subject soybean lines, their components, and their processing characteristics have no indirect plant pest effect on any processed plant commodity.

Based on unspecified evidence provided by DuPont, APHIS concluded that the high oleic soy lines present no threat to raw or processed agricultural commodities. APHIS noted that DuPont consulted with the Food and Drug Administration concerning the safety of consumption of both the oil and meal of these lines by humans and other animals.

Potential Environmental Impacts from Growing the Modified Soybeans outside the United States. Executive Order 12114 (January 4, 1979) allows APHIS to consider the possible environmental impacts of cultivation of regulated articles in other countries. APHIS concluded that the subject soybeans will not have an adverse environmental impact when grown anywhere in the world because there are no significant differences from the parent line for any investigated parameter, except for the higher levels of oleic acid, the glycinin protein substitution, and the linoleic acid isomer. APHIS also noted that "all national and international regulatory authorities and phytosanitary regimes that apply to introductions of new soybean varieties internationally apply equally to those covered by this determination" (USDA 1997a).

Environmental Risks Not Considered by APHIS

APHIS considered the information provided by DuPont and others in reaching its determination, but there is little experimental data in the decision document or even the appendix to allow an independent critique. One recommendation of another National Research Council study was that "the quality, quantity, and public accessibility of information on the regulation of transgenic pest-protected plant products should be expanded" (NRC 2000c). This recommendation should be expanded to include all new plant products and should be implemented as soon as possible.

Some environmental questions remain. For example, temperate plant species (like soybean) tend to have a higher proportion of polyunsaturated fatty acids in the seed oil than do more tropical or subtropical species (like olive, where monounsaturated fatty acids predominate). Recent studies investigated the effects of fatty acid profile changes on the degree of cold tolerance in plants. Kodama et al. (1994, 1995) transformed tobacco to produce higher amounts of polyunsaturated fatty acids and noted an alleviation of cold-induced growth suppression. If polyunsaturated fatty acids help the seed survive cold winters, perhaps the transgenic soybeans, with such a deficit of polyunsaturates, would have reduced winter survival capacity. The data required to address this question might have been considered and investigated but were not included in the report. APHIS reported that DuPont provided evidence on overwintering but did not elaborate. Possibly, the transgenic soybeans suffered more winterkill than the nontransformed soybeans, but since this segment of the report dealt with the increased weediness potential, that fact might not have been considered important. Also, this question might have been considered irrelevant; that is, reduced overwintering potential of the transgenic soybeans would represent a reduced ecological fitness and hence no increased threat. It is unlikely that the modified-oil soybean changed overwintering capacity because the seed oil was modified, not the plasma membrane lipids, which are more relevant for cold tolerance. APHIS might have been aware of this fact and therefore did not think it necessary to mention it. But by not mentioning the possibility, APHIS leaves itself open to the charge that the agency overlooked it.

An interesting aspect of this case study is the prior presence in the market of nontransgenic soybean varieties with similar (high oleic acid content) attributes. The plant pest risk potential and differences between conventional soybeans and the transgenic varieties are uncertain, not because of incomplete information and regulatory scrutiny of the transgenic lines but because of almost complete lack of information on the conventional ones. The functional assumption is that conventional breeding methods, including irradiation, chemical mutagenesis, somaclonal variation, and other mechanisms of dramatic genetic disruption are safe and environmentally benign.

DuPont provided information on the molecular structure, though not the nucleotide sequence of the genetic modifications, of the transgenic soybeans. The difference between these soybean lines and regular soybeans is clear; the transgenic soybeans carry a tiny additional piece of DNA, an addition of less than one one-hundredth of 1% of the total DNA in the soybean.

Conventional mutant soybeans (Kinney 1994) manifest an unstable increase in oleic acid in the seed oil. The genetic mechanism by which this trait is achieved is unknown, but it must be the result of genetic modifications in the soy genome. Mutagenesis may have altered the amount of DNA, either by destroying portions of the genome or by causing a duplication (perfect or imperfect) of portions of the genome. In any case, it is likely that several genes were altered, not just those regulating oleic acid content. Any number of genes relating to environmental fitness, production of antinutritional compounds, or other undesirable consequences also may have been altered. But APHIS does not assess these new crop cultivars because, as noted above, the current trigger for regulatory review limits oversight to plants altered using rDNA breeding methods. This is not to suggest that new crop varieties developed solely by conventional breeding should be regulated as stringently as transgenic varieties currently are. Despite the genetic uncertainties and unknown consequences of conventional breeding, as noted in Chapter 1, real damage to the environment from new crop varieties is rare. Nevertheless, the additional knowledge of the genetic changes in transgenic varieties allows, in this particular case study, more confident and reliable predictions of environmental (as well as health and nutritional) effects than conventional varieties. There is no indication that the risks associated with these transgenic soybean lines differ in any material way from those of the same species with similar but non-transgenic based attributes. High oleic acid oil soybean varieties, regardless of how the varieties were derived, appear to present similar degrees of hazard.

Also, APHIS did not consider the effect of large-scale commingling of the high-oleic soybeans with regular soybeans, as that might adversely affect the quality of the processed commodity (soybean oil) but not the plant pest characteristics. The transgenic lines will be grown in a segregated identity-preserved manner to avoid commingling with regular soybeans. Any significant inadvertent mixing of the transgenic soybeans and conventional ones could be problematic because the resulting oil would be a blend of high and low oleic acids, the exact proportions depending on the ratio of the blend. The uncertainty of the final oil composition could adversely affect the value of the commodity oil product. While not an environmental risk, it is a potential impact on other agricultural commodities that are part of APHIS's charge to review. This is true of all identity-preserved crops being grown.

Involvement of Potential Participant Groups

The designated public comment period on the petition ended, with no submissions, on April 29, 1997. There was no indication from the documents of any other public involvement.

Two *Bt* Corn Petitions

Background

Insect pests of corn (*Zea mays* L.) comprise members of many different orders, including seed maggots, rootworms, wireworms, grasshoppers, flea beetles, aphids, and leafhoppers, which feed on kernels, roots, leaves, silks and vascular tissue (Hill 1983, 1987; Davidson and Lyon 1987; Straub and Emmett 1992; Steffey et al. 1999). Some leafhoppers and aphids are vectors for plant pathogens. The main Lepidopteran pests of field corn in the United States are borers (European corn borer, ECB, *Ostrinia nubilalis* (Hübner); southwestern corn borer, SWCB, *Diatraea grandiosella*, and corn earworm, CEW, *Helicoverpa zea* (Boddie) and defoliators (e.g., armyworms). Some of these pests are a chronic problem; others periodically account for severe yield losses (e.g., Ostlie and Hutchison 1997, Mississippi State University Extension Service 1999).

According to various extension service fact sheets for corn growers (Mississippi State University Extension Service 1999, Swanson 2000, University of Illinois 2001), management options for pest control include (1) early planting; (2) crop rotation; (3) resistant varieties; (4) varieties that attract natural enemies of target pests; (5) soil quality management; (6) noncrop management to reduce alternate hosts for pests, e.g., barnyard grass for armyworms, and to enhance naturally occurring predators and parasitoids; (7) conservation of natural enemies through intercropping or insectary plantings of noncrop vegetation; (8) augmentation (field releases) of natural enemies (e.g., *Trichogramma* spp. egg parasitoids); (9) destruction of corn stalks after harvest; and (10) insecticide treatments. Organically managed corn in the Midwest relies primarily on crop rotation for pest control (Swanson 2000). Price premiums compensate for lower yields under organic management. A relatively small proportion of the total U.S. field corn acreage is treated with insecticides, in part because sprays are often not a reliable or cost-effective means of controlling some of the major pests. For example, less than 5% of U.S. field corn acreage was sprayed for ECB before adoption of *Bt* corn (Gianessi and Carpenter 1999).

Genetically based pest resistance in corn was developed through the use of conventional breeding methods utilizing mechanical, chemical-repellent, or antibiotic properties of the corn plant. Commercial varieties with Lepidopteran resistance traits have been available since at least 1902 when Burpee Seeds released "Golden Bantam." Varieties with antibiotic properties have been developed by selecting for high levels of the toxin 2,4-dihydroxy-7-methoxy-1,4-benzoxazin-3-one (DIMBOA) and related hydroxamic acids, which vary widely in concentration in different corn plant tissues (Houseman et al. 1992). Structural traits such as highly lignified stalks have also been selected to serve for increased pest resistance. However, many of these varieties are variably or moderately resistant and may have relatively lower yields than susceptible corn lines. For example, DIMBOA levels in initially resistant varieties tend to decrease with plant age (Barry et al. 1994, Frey et al. 1997). Thus, some conventionally bred varieties resistant to ECB have been replaced by higher-yielding varieties with increased total yields despite heavier losses from ECB (Gianessi and Carpenter 1999).

Transgenic insect pest resistance can be obtained by transferring part of a bacterial genome (from various strains of *Bacillus thuringiensis*) into a plant genome. The genetic sequence encodes production of a toxic protein so that plant tissues become lethal to target caterpillar pests feeding on those tissues. At least three major opportunities are afforded by transgenic insect resistance traits in crops: (1) pest resistance in crops whose close relatives are not resistant to the target pests, (2) more effective pest control than pesticides and/or conventionally available resistance, and (3) the trait can be transferred into varieties showing excellent agronomic performance (e.g., Wiebold et al. 2000), thus avoiding the yield drag that sometimes accompanies conventional breeding efforts. For caterpillar pests of corn, all of these factors can be important. Therefore, starting in the 1980s, transgenic *Bt* corn lines were developed and field-tested for efficacy against target Lepidopteran pests, including ECB, SWCB, and, when toxins are expressed sufficiently in the corn ears, CEW.

Other crops have shown promise for *Bt*-based insect resistance. By October 2000, 25 petitions for nonregulated status (from Monsanto, Northrup-King, Ciba-Geigy, DeKalb, AgroEvo, and Calgene) had been submitted to APHIS for *Bt* crops (corn, potato, tomato, and cotton). Sixteen were approved for deregulation of transgenic crops with *Bt*-based insect pest resistance, seven petitions were withdrawn, and two were pending a decision. Field trials of dozens of other transgenic *Bt*-based insect-resistant plant varieties (including cotton, potato, rapeseed, poplar, broccoli) have been conducted (see "Field Test Releases in the U.S.," Information Systems for Biotechnology online database: *www.nbiap.vt.edu*).

For comparison, one of the earliest *Bt* corn petitions for deregulation by APHIS and one of the most recent are included in this case study (see "Current Status of Petitions," *www.aphis.usda.gov/biotech/petday.html*). Transformation Event 176 corn (Maximizer), with marker genes for resistance to the antibiotic kanamycin and tolerance to glufosinate, was the first *Bt* corn to be deregulated (USDA 1999). The petition for *Bt* corn Transformation Event CBH-351 (StarLink) was approved in 1998 (USDA 1998). These two events expressed different kinds of plant-produced toxic proteins. The kind of toxin expressed by Event 176 has been commercially available in formulations of bacterial insecticides, but Event CBH-351 expresses a Cry toxin that has not been available commercially in insecticides. Below are highly abstracted overviews and comparisons of the assessments by USDA-APHIS for its findings of "no significant impact." To the extent possible without access to the original research designs and test data submitted by the petitioning companies, this summary illustrates how APHIS evaluated a number of different issues associated with the biosafety of these transgenic products.

Environmental Risks Considered by APHIS

Disease Resulting Directly from the Transgenes, Their Products or Added Regulatory Sequences. For Event 176, two promoter sequences were used to allow high levels of Cry1Ab protein expression in both green tissue and pollen, which in combination were expected to be most effective in controlling ECB. Some of the DNA sequences used in transforming these plants were derived from cauliflower mosaic virus (CaMV), but the disease-causing genes of this virus were not involved. The marker gene *bar* from the soil bacterium *Streptomyces hygroscopicus*, which encodes for an enzyme resulting in glufosinate tolerance, was used as a selective marker. Plasmid pUC19 also harbors ampicillin resistance but is expressed only in a bacterium, not in plant cells.

Event CBH-351 was developed with two pUC19-based plasmids, which contained the modified *cry* gene and the *bar* gene, respectively. As in Event 176, the *bar* gene was cloned from *S. hygroscopicus*, and its expression results in whole-plant glufosinate tolerance. Event CBH-351 expresses modified Cry9c proteins, using sequences from *Bacillus thuringiensis* subsp. *tolworthi* (isolated from grain dust in the Philippines). Event CBH-351 uses regulatory regions from petunia and two plant pathogens, CaMV and *Agrobacterium tumefaciens*. In both cases, APHIS concluded that although pathogenic organisms were used in their development, the transgenes, their products, or added regulatory sequences do not result in plant pathogenic properties (USDA 1995, 1998). In the CBH-351 determination document, a discussion follows this conclusion, providing some information on the genetic constructs, including the number of insertion sites, the copy number and expression level of the transgenes, and evidence of stable inheritance of the transgenes in crosses of CBH-351 into several different genetic backgrounds. Unpredicted, unexplained effects are also mentioned, such as leaf striping on transgenic plants as compared to nontransgenic plants in Puerto Rican field trials and more predictable but secondary effects, such as reduction of stalk rot in transgenic plants.

In contrast to the document for CBH-351, the APHIS determination for Event 176 did not mention or evaluate descriptions of the event regarding pleiotropic effects or other unpredictable consequences of the random insertion of these genetic sequences, the number of copies, and stability of the transgenes. The determination for CBH-351 was more detailed than was the earlier determination, suggesting either that APHIS received more information, adjusted its evaluation procedures, reported more details in its summary, or some combination of the three.

Weediness of the Crop Plant Resulting from the Transgene and Associated Gene Sequences. For both Event 176 and Event CBH-351, APHIS addressed the risk that expression of the insect control protein might provide a selective advantage to the plant, sufficient to make it a plant pest. For Event 176, APHIS first compared the characteristics of nontransgenic cultivated corn with a list of ideal characteristics of weeds published by Baker (1965) and found little overlap. Though noting that some ecologists have criticized this list, APHIS relied on it anyway because no more broadly accepted suite of characteristics was available. Second, APHIS assessed the weed status of corn, consulting several weed compendia and found that corn was not listed in most of them. Third, APHIS considered the trait itself—insect resistance. APHIS concluded that since insect resistance was not among the weedy characteristics listed by Baker, that trait would not likely contribute to weediness in transgenic *Bt* corn. The herbicide tolerance trait of Event 176 was not perceived as one that could reverse the plant's nonweed status, primarily because glufosinate had not yet been registered for use in corn. Finally, "APHIS evaluated field data submitted by Ciba Seeds which specifically demonstrates that Event 176 corn is no more weedy than the non-modified recipient." These field data were not available to this committee, so it cannot comment on the adequacy of the study in supporting that conclusion.

APHIS used the same arguments for the later determination (USDA 1998) for Event CBH-351, citing Baker (1965), and noting that insect resistance was not listed as a characteristic of weeds. APHIS also found no differences between CBH-351 and its nontransgenic counterpart or a nontransgenic standard line in field tests conducted from 1995 to 1997 in 17 states and territories of the United States in *Bt* corn's ability to compete or persist as a weed. The CBH-351 determination pointed out that other corn genotypes with resistance to certain Lepidopteran pests, including ECB, have been in use for decades and have not been reported to cause increased weediness. Given these data and observations, APHIS concluded that the introduced genetic constructs and new traits, Lepidopteran insect resistance and tolerance to glufosinate, are not expected to release *Bt* corn volunteers from any constraint that would result in increased weediness. APHIS added that CBH-351 corn is still susceptible to other non-Lepidopteran pests and diseases of corn.

Although the committee believes that APHIS has come to a reasonable conclusion for this case, APHIS's analysis lacked scientific rigor, balance, and transparency. First, APHIS's statement that most weed lists did not include corn surprised the committee because corn was on the first list consulted. Corn is listed as a common or troublesome weed in the United States according to the Weed Science Society of America's "Composite List of Weeds" (available online at *http://ext.agn.uiuc.edu/wssa/*). Although the accompanying APHIS environmental assessment listed some of the authorities used to determine that corn is not a weed, the committee suggests that APHIS cite all sources it consults. The environmental assessment should then explain the reasoning for adhering to some but not other authorities.

Second, the herbicide tolerance trait in Event 176 corn was considered not to pose a hazard in terms of "weedier" free-living relatives because glufosinate had not yet been registered for use in corn and exerts no selection pressure for this trait in nature (USDA 1995). There was no discussion about herbicide-tolerant Event 176 corn volunteers persisting where glufosinate is applied, such as adjacent habitats or rotational crops. By the time CBH-351 was evaluated, however, glufosinate was registered for use on corn. The subsequent determination provided a more thorough discussion of the movement of this trait via gene flow and listed possible changes in agronomic practices needed to control resulting corn volunteers, including alternative herbicides or mechanical control measures (USDA 1998). Such herbicide-tolerant volunteers, however, were not considered weedier, even though they would require additional control measures. Later in the determination, however, APHIS implies that if a herbicide tolerance trait were to introgress into teosinte, the descendents would become weedier in the presence of glufosinate herbicide selection. Different conclusions for corn and teosinte descendants could be considered contradictory.

Third, APHIS relies extensively on Baker's list of weed characteristics in both determinations. Although an excellent heuristic tool for understanding weed ecology, Baker's list, in itself, is not an effective predictive tool. APHIS mentioned that some ecologists question the effectiveness of using the list for that purpose. If APHIS uses that list, the committee suggests that the agency include more discussion of such studies as Williamson (1994) and Perrins et al. (1992), which demonstrate that known weeds cannot be separated from nonweeds based on that list.

Another argument (used by APHIS in these determinations) is that the "lack of reported incidences" is evidence that something has not happened. For example, the determination for CBH-351 mentioned that increased weediness of corn due to an insect resistance trait has not been reported, as far as APHIS is aware. The committee agrees that if a phenomenon is dramatic and occurs over a short time period, it is likely to be noticed and even reported. That is, a lack of reporting of obvious phenomena may be a strong indicator that such phenomena have not occurred. However, a slight increase in fitness due to insect resistance is an invisible phenomenon, and any increase in weediness of corn due to that trait would likely be slow and subtle, especially since the conventionally bred traits were only moderately effective in conferring resistance.

To demonstrate that corn volunteers are more common as a result of conventionally bred ECB resistance, one would need a detailed comparison between, for example, the incidence of volunteer corn in the 1940s and the 1990s, including a way to isolate the resistance mechanism from other changes in corn production and management. Such an approach could be used to test for environmental effects of pest resistance factors. However, because a "lack of evidence" is used to conclude that there are no effects, there is no encouragement for such testing in the APHIS determinations. In general, use of the term "lack of evidence" can mean anything from "detailed, replicated, long-term experimental studies found no evidence" to "there are reasons to expect a problem but no one has tested this so there is a lack of evidence."

Finally, the committee found the argument that insect resistance cannot release a plant from ecological constraints to be both weak and inconsistent. Indeed, APHIS is the federal agency that evaluates and approves biological control applications for the importation of insects (including Lepidoptera) specifically to increase herbivore pressure and thereby reduce the incidence of certain weeds. The embedded assumption in the biological control of weeds is that the presence or absence of herbivory, by even one species, can determine whether a plant species behaves as a weed. Biological control was not mentioned in either environmental assessment. A stronger argument for why Lepidopteran resistance would not release corn volunteers would have included an explanation of why corn and its Lepidopteran herbivores are different from those plant-herbivore pairs that result in biological control of weeds. (APHIS cited Gould [1968] as saying that corn is incapable of sustained reproduction in feral populations.)

Impacts on Free-Living Relatives of Corn Arising from Interbreeding. APHIS considered the potential for gene flow from Event 176 transgenic corn and, should it occur, two of its possible consequences. Those consequences were increased weediness in wild relatives and population changes that would lead to reduced genetic diversity. Glufosinate resistance was not considered to pose a hazard in either regard, first because selective pressure for this trait would not occur either in natural habitats or cornfields, where glufosinate was not registered for use (USDA 1994a), and later because alternative management techniques for glufosinate-tolerant corn exist (USDA 1998).

First, however, APHIS discussed the possibility for introgression from transgenic corn into other corn cultivars via wind pollination. Gene flow would be reduced, according to USDA (1995), by low survival rates of corn pollen after 30 minutes. Some controversy exists now, however, about corn pollen lifetimes; the EPA (2000b) claims corn pollen is viable for several hours in good conditions. APHIS pointed out that maintaining an isolation distance between cultivars also reduces gene flow, but no particular separation distance is suggested. Additionally, APHIS noted that the primary use of corn kernels is food or feed rather than seed; thus, any hybrid seed resulting from cross-pollination with *Bt* corn would more likely be eaten than sown. APHIS did not discuss any indirect effects of cross-pollination on the food quality of recipient non-*Bt* corn. This eventually became a concern for the CBH-351 (StarLink) Event. As discussed in Chapter 2, there is disagreement about when gene flow to other non-propagating plants should be considered an environmental risk.

The analysis of Event 176 for impacts on corn relatives was not conceptually different from the analysis of corn's weediness. No other wild relatives, in the United States or abroad, were mentioned in this section of the determination despite the fact that sexual transfer of beneficial alleles from a transgenic crop to a wild relative that might result in a more difficult weed had been a widely discussed risk associated with transgenic crops prior to that time (e.g., Colwell et al. 1985, Goodman and Newell 1985, Darmency 1994, Kareiva et al. 1994). The environmental assessment for deregulation of Event 176 noted that hybrids between corn and *Tripsacum* weeds are sterile. For wild *Zea* (teosintes), which are native to Mesoamerica, the environmental assessment states that "in the wild, introgressive hybridization does not occur because of differences in flowering time, geographic separation, block inheritance, developmental morphology and timing of the reproductive structures, dissemination, and dormancy (Galinat 1988)."

The environmental assessment for CBH-351 discussed the possible avenues for transgene introgression into populations of corn's wild relatives to a greater extent. It mentioned that corn's wild relative, Eastern gamagrass, *Tripsacum dactyloides*, which is native to the United States, is grown as a new crop in the Midwest. The strong statement against the occurrence of introgressive intrageneric hybridization for Event 176 was altered for CBH-351 with less absolute language (e.g., "does not occur" was replaced with "is limited, in part, by several factors" citing Doebley [1990] as well as Galinat [1988]). The statement in the environmental assessment for Event 176 regarding the sterility of *Zea* × *Tripsacum* hybrids was revised for Event CBH-351 to the effect that resulting hybrids are often sterile or have greatly reduced fecundity. The latter information suggests that hybrids with some fertility might occur at a very low frequency if Tripsacum and field corn are grown in close proximity, such as in the Midwest. For both Event 176 and Event CBH-351, APHIS concluded that environmental impacts anywhere in the world should not be significantly different from those arising from the cultivation of any other variety of insect- and herbicide-tolerant corn.

The key questions posed by APHIS in the foregoing two analyses are: (1) Is it likely that the *Bt* endotoxin trait would introgress into at least one population of a wild relative of corn? (2) Can pest resistance genes against certain Lepidopterans provide sufficient benefit to result in making corn or its hybrids a more difficult weed? (3) Are *Bt* varieties similar enough in conventionally bred varieties to use the latter as a model for the former? Whereas the absence of selection pressure by herbicides was mentioned for the herbicide tolerance trait and weediness, no clear analysis of release mechanisms for wild relatives was provided by APHIS with respect to herbivore selection pressure.

APHIS's view of whether introgression of the transgene into wild relatives could occur seems to have changed from impossible for the first *Bt* corn determination to being unlikely in the United States and likely in Mesoamerica for the most recent determination. This shift in interpretation shows that APHIS adapts its arguments to take into account additional or overlooked scientific information as it becomes available. Certainly, compared to a crop like *Sorghum bicolor*, which has an abundant cross-compatible wild relative in the United States, corn has fewer avenues for gene exchange with wild relatives in this country. But it apparently has a slight chance given that populations of *Zea* and *Tripsacum* species are reported by APHIS (USDA 1998) to occur in the United States.

Morphologically intermediate plants between teosinte and corn appear to be spontaneous hybrids in Mesoamerica (Wilkes 1997), but they could also be crop mimics; actual hybridization and introgression rates are unknown (Ellstrand et al. 1999). Should there be crop-to-wild relative gene flow, in either the United States or Mesoamerica, introgression and maintenance of the transgene can occur even if hybridization rates are low. Despite APHIS arguments that a positive selection pressure must exist to maintain a transgene in a wild population, under modest gene flow pressure, *Bt*-based resistance could, in fact, be maintained at reasonably high frequencies in populations of wild corn relatives (see arguments above for virus-resistant squash) even if it were a neutral or deleterious allele. That is, relatively rare gene flow events provide an opportunity for persistence, perhaps leading to changes in the genetic diversity of natural populations (Ellstrand et al. 1999). Thus, the consequences of persistence of the transgene in natural populations are worthy of a serious assessment.

When considering whether transfer of a Lepidopteran resistance trait could allow an ecological release (i.e., an increase in population size, density, or range) of any wild relatives of corn, two further questions arise. Which Lepidopteran species frequently use wild and cultivated *Zea* and *Tripsacum* as host plants? Which of those species are susceptible to the *Bt* toxins? APHIS does not provide specific information on these questions. With respect to the range of Lepidopterans that feed on corn and/or its wild relatives, two contradictory statements and an extrapolation from a conference discussion comment are provided. In the determination document for CBH-351, APHIS stated that Cry9C protein has insecticidal activity against ECB and members of the families Pyralidae, Plutellidae, Sphingidae, and Noctuidae. Another statement in the same document states: "CBH-351 corn plants were generally indistinguishable from control corn plants for disease susceptibility and insect susceptibility except for tolerance to European corn borer, where a clear advantage was noted for CBH-351 corn" (USDA 1998). Perhaps these statements seem contradictory because in one case the plants were challenged with various Lepidopteran pests and in the other case plants were exposed to natural levels of pests; however, no such explanation is given in the APHIS documents.

APHIS cited Sánchez González and Ruiz Corral (1997) to suggest that teosinte is susceptible to most of the same pests and diseases (including Lepidopterans) that feed on *Zea mays*. The statement is actually less clear; in the conference discussion following the paper by Ruiz and Sánchez, Bruce Benz is quoted as saying, "In the case of insects, certain Coleoptera (*Macrodactylus murinus*) infest both *Zea mays* ssp. *diploperennis* and maize, although the impact seems to be different, while in the case of Homoptera and Lepidoptera, even though both species of plants may be infested, it seems that the damage tolerance of teosinte is lower than that of maize." Lepidopterans feeding on corn's wild relatives may indeed be primarily those that feed on corn; if so, they may be the very species that limit the ranges and aggressiveness of such wild grasses. However, few records are available in the most comprehensive Lepidopteran host plant database (Robinson 1999, Robinson et al. 2000). Thus, it is premature to make a conclusion or to suggest which species feed on which wild relatives.

Letourneau et al. (2001) list 376 Lepidopteran species recorded as feeding on *Zea mays*, two feeding on *Z. mexicana* (Schrad.;=*Z. mays* ssp.

mexicana), and no records for Lepidopterans feeding on other teosintes (e.g., *Z. diploperennis*). Of the 376 Lepidopteran species feeding on *Z. mays*, fewer than 15 are listed as having been tested for commercial *Bt* toxin susceptibility in the most comprehensive database (van Frankenhuyzen and Nystrom 1999). Most testing has been carried out on target pests; thus, little is known about which nonpest Lepidopterans are susceptible to any or all *Bt*-based proteins. Even less is known about whether these herbivores limit the population size of corn's wild relatives. Nevertheless, APHIS found it unlikely that potential introgression of Lepidopteran resistance will cause teosinte to become weedier (USDA 1998).

Whether or not conventionally bred insect resistance traits in corn and the *Bt*-based resistance trait inserted in CBH-351 corn are similar enough to make conclusions about the selective pressure of the *Bt* traits on feral progeny or hybrids with wild relatives is unknown. The genetic novelty of *Bt* toxins and herbicide resistance is likely to be outside the range of natural variation found in populations of wild relatives, whereas some of the traits responsible for causing Lepidopteran resistance in conventional corn hybrids may correspond to traits already present in populations of corn's wild relatives (e.g., tough stalks, low protein content, high DIMBOA content). Appropriate measures of environmental effects of conventionally bred resistant varieties are not provided by APHIS, though some may be available; at least some discussion of the kinds of resistance mechanisms present in conventional varieties and their relationship to the range of naturally occurring levels of resistance in wild relatives should be included in this discussion. This would aid in determining if the *Bt* proteins constitute fundamentally different mechanisms of resistance.

While APHIS might be correct that introgression of the *Bt*-resistance and glufosinate tolerance traits into feral corn and wild corn relatives is unlikely and that Lepidopteran resistance and herbicide tolerance would not result in increased weediness of these plants, APHIS's data on Lepidopteran pressure in natural populations, their population genetic arguments, definitions of weediness, and model of conventionally bred resistant corn varieties are inadequate to support such a strong conclusion.

Impacts on the Components, Quality, and Processing Characteristics of Corn Raising from the Event of Inserting Btk Insecticidal Protein Gene. APHIS concluded from extensive field tests that the Event 176 and Event CBH-351 corn plants had agronomic characteristics typical of nontransgenic corn such that, except for the desired effect of insect resistance, the introduced sequences did not confer any disease-specific property of the donor organism or any other plant pest characteristics.

Potential to Harm Organisms Beneficial to Agroecosystems. The *Btk* protein (Cry1Ab) expressed in Event 176 was tested by Ciba Seeds for toxicity to non-target organisms, as either an extracted leaf protein powder, intact pollen grains, or a bacterial cell paste, as well as in a comparative field trial. No information is given on the experimental design of any of these tests, except that the appropriate form of *Btk* protein described above was used for different non-target organisms, including earthworms, aquatic invertebrate *Daphnia magna*, and bobwhite quail. The field test focused on predators and parasites in three insect orders and included at least ladybird beetle larvae, a fly, and honeybee larvae. In the section on threatened or endangered species, APHIS summarized results of tests and observations on quail, the aquatic and soil-inhabiting invertebrates, a moth, and a number of butterflies. Bobwhite quail showed no adverse effects on feed consumption, body weight, or mortality when fed corn protein powder from *Bt* or isogenic non-*Bt* hybrids. Indirect effects of *Bt* corn on quail through its changes in its food supply were considered by APHIS to be negligible if the effect of *Bt* corn is restricted to the European corn borer (however, it is known that *Btk* affects many Lepidoptera). No data were mentioned on the toxicity of *Btk* proteins on endangered Lepidopterans, presumably because exposure to the toxins in nature was deemed unlikely. From these data, and data in the literature APHIS concluded that only a specific group of Lepidopterans should be affected and that the majority of quail's prey species should remain unaffected by Event 176 corn.

APHIS concluded that, because the specific receptors for *Bt* toxin in the midgut of target insects were not expected to occur in other invertebrates or any vertebrates, Event CBH-351, although not derived from a type of *Bt* available as a commercial formulation, should not adversely affect non-target taxa, including humans. Tests for effects on non-target organisms for Event CBH-351 (Cry9c) by AgroEvo (formerly Plant Genetic Systems, now Aventis) differed somewhat from tests on non-target toxicity for Event 176. Specifically, test organisms included adult honeybees, ladybeetles, earthworms, *Folsomia candida* springtails, bobwhite quail, and mice. Details on experimental design are not given, so a scientifically based review of APHIS evaluation of the data is not possible.

The fact that the committee cannot review the actual methodology and data summaries of any of the *Bt* corn testing is unfortunate for two reasons: (1) the committee's evaluation will be superficial, as if it were asked to review a scientific report but were given only the introduction and discussion sections without the methods or results, and (2) the committee's questions may be misguided because of incomplete information provided in the APHIS descriptions of tests and results. Nevertheless, the committee has several questions about what seem to be the methods used by the petitioners to test for environmental effects of the proteins expressed in Event 176 and Event CBH-351. It is curious that adult honeybees were tested, since adults both ingest pollen and feed it to their brood. APHIS justified the use of adult honeybees by noting that corn is not a nectar-producing plant, and so honeybees would visit it infrequently compared to other plants; thus relatively low concentrations of *Bt* pollen would be expected in adult pollen loads; the committee remains unconvinced that larval tests are unnecessary. Specific criticisms about the adequacy of *Bt* tests for lethal and nonlethal effects on natural enemies were discussed by Hilbeck et al. (1998a, 1998b). They noted, for example, that aphids were not an appropriate test prey for determining indirect effects of *Bt* proteins on natural enemies since the aphids may not ingest *Bt* proteins. Indeed, thorough testing for non-target impacts of *Bt* pollen might also include larvae of Lepidopteran species that might encounter corn pollen and feed on it more frequently in nature (Losey et al. 1999, 2001, Wraight et al. 2000).

The committee also questions the ability to detect community-level differences in the comparisons described. Field tests seem to have been conducted using a factorial design, with three replicated split plots of *Bt* corn and non-*Bt* isolines with or without pyrethroid insecticide application. Arthropod predator comparisons were based on 5 five-minute visual observations (biweekly) and seven weekly sticky trap samples per split plot. On three of seven dates, trapped predators were significantly more abundant in unsprayed non-*Bt* corn than on unsprayed *Bt* corn; no effect of *Bt* corn was found on the other four dates. Nor were differences found between plants that were and were not sprayed with pesticide. Information to decipher whether the trap samples showing higher levels of predators in non-*Bt* corn plots reflected predator colonization rates or resident predators emerging as adults from the corn (thus reflecting performance). Visual observations showed no difference in predator abundance between plots with and without *Bt* corn. Similar to the findings for Ciba Seeds field comparisons, the diversity of predators in experimental plots did not differ between *Bt* and nontransgenic lines. It is not clear what measure of diversity was used or at which taxonomic level. APHIS mentioned possible problems with the reliability of the field test due to small plot sizes (10 rows by 20 feet). Another problem might be the statistical power of the test to detect differences in predator diversity, especially if many of the arthropods were identified taxonomically at levels higher than species (genus or family).

Hoy et al. (1998) have pointed out that there could be complex indirect effects in crop fields that could cause shifts in the relative abundance of natural enemies. Also, indirect impacts of *Bt* corn on pests might occur. For example, a reduction in pyrethroid use for corn borers on *Bt* corn might minimize outbreaks of spider mites by allowing their natural enemies to survive. In contrast, minor pests may become more predominant as ECB is controlled and if foliar insecticides are reduced (Ostlie and Hutchison 1997). Community-level tests were restricted to above-ground species; given the persistence of *Bt* proteins in the soil from root exudates or after incorporation of plant material (Stotzky 2000, 2001), any evidence for no significant effect on communities of soil organisms should also be discussed.

Potential for an Adverse Impact on Threatened or Endangered Species. APHIS consulted the list of threatened and endangered species (50 CFR 17.11). None of these species feed on corn, so APHIS concluded that *Bt* corn would not affect these species. No mention of toxic effects of corn pollen on any sensitive Lepidopteran species was made for either Event 176 or CBH-351. Nor was there mention of any possible effects to non-target susceptible Lepidopterans that may be dependent on *Zea* host plants. The latter, though beyond the scope required, may be a proactive consideration for assessing possible environmental consequences should transgenes escape and affect population levels of non-target moths and butterflies, which are also plant pollinators (Letourneau et al. 2001). The possibility that threatened or endangered plants could rely on *Bt*-susceptible Lepidopteran pollinators was not mentioned. While this possibility may not be high, it is best to be thorough in these assessments.

Potential for an Adverse Impact on the Ability to Control Non-target Insect Pests. In the CBH-351 determination, APHIS briefly examined the issue of whether insecticide usage might be reduced by the introduction of *Bt* corn Event CBH-351 but did not reach a conclusion (USDA 1995, 1998). Perhaps because the APHIS assessments expected no adverse effects on natural enemies and possibly positive effects due to the curtailed usage of broad-spectrum pesticides, the notion of secondary pests is not discussed directly. If, in response to low levels of target pests, non-target insect pests increased and became secondary pests in *Bt* corn, those species would need to be controlled by alternative measures.

Effects of the Cultivation of Bt Corn on the Ability to Control Insects and Weeds in Corn and Other Crops. APHIS considered evolution of pest resistance to Cry proteins for Event 176. These proteins are similar to those used for ECB control in commercially available crystalline powder formations. Based partly on experimental demonstrations of resistance evolution in Lepidoptera to Cry toxins, APHIS predicted that resistant insects would probably evolve in response to *Bt* corn. However, a resistance management strategy was outside the scope of the determination.

APHIS summarized statements from CIBA Seeds—that *Bt* corn be used within the scope of an integrated pest management (IPM) strategy, that populations of ECB be monitored for resistance, that high-dose expression coupled with non-*Bt* corn refugia and development and use of new insect control proteins would delay the evolution of pest resistance, and that farmers be educated about resistance management strategies. APHIS suggested that, should these measures fail, resistant ECB populations might be controlled by agronomic practices such as rotation and alternate insecticides. The determination document concluded that evolution of resistance to *Bt*-based insecticides is a potential risk associated with Event 176 but that this risk was no greater than that posed by applying insecticides themselves. For Event CBH-351, which expresses a protein different from those in commercial formulations of bacterial sprays, similar arguments were tendered. Cross-resistance was considered unlikely due to separate receptors in some species, so the Cry9C protein was suggested as a useful alternative when resistance evolves to other *Bt* toxins (such as that in Event 176). APHIS concluded that CBH-351 should pose no greater effects in resistance evolution than the use of ECB-tolerant corn cultivars, chemical insecticides, or biological insecticides (USDA 1998). The possible consequences of herbicide tolerance in affecting weed control, addressed only for Event CBH-351, were predicted to be positive, allowing more choice among postemergent herbicides and no-till options.

Although APHIS makes some comments about resistance, the agency has apparently relied on the Environmental Protection Agency (EPA) to formalize and/or enforce resistance management plans, encouraging its consideration. The committee suggests that APHIS either indicate that resistance management is beyond its scope and not discuss it or provide detailed analysis of the practical issues of resistance management efforts in field corn. Otherwise, the outcome can be perceived as different levels of scrutiny between the EPA and APHIS.

Environmental Risks Not Considered by APHIS

APHIS did not directly consider whether transgenic corn would have a negative impact on corn's wild relatives in the United States or elsewhere, either in terms of changes in their genetic diversity or in terms of posing an impact that might lead to their extinction. This is likely due to the conclusions that hybrids with *Zea* or *Tripsacum* would rarely occur, especially in the United States. Mortality of non-target Lepidoptera should susceptible species ingest toxin-containing pollen on their host plants is not discussed. Although threatened or endangered Lepidoptera were considered, the link between threatened or endangered plants and *Bt*-susceptible Lepidopteran pollinators was not explored.

Public Involvement

The determination documents describe comments received on each of the two petitions (Event 176 and Event CBH-351) during the designated 60-day period after posting in the *Federal Register*. Most (2,271 of 2,309) were form letters (source not specified), and 2,307 either favored deregulation or endorsed the concept of an ECB-resistant corn variety. One letter pointed out that CBH-351 controls third- and fourth-stage ECB larvae. The two commenters expressing reservations were concerned about resistance management and the establishment of refugia of non-transgenic corn where the 176 Event would be grown. There was no indication from the documents of any other public involvement in APHIS's decision-making process.

Herbicide-Tolerant and Insect-Resistant Cotton

Background

Cotton production has historically relied on heavy use of both insecticides and herbicides. On a per-acre basis during the 1990s, the number of pounds of insecticide used in cotton was three to eight times more than in corn and about 100 times higher than in soybean (NRC 2000b). A number of key insect pests of cotton such as the tobacco budworm have evolved resistance to many insecticides. In the mid-1990s insecticide resistance threatened the economic viability of cotton farming in a number of areas of the United States (e.g., Luttrell et al. 1994).

Herbicide use in conventional cotton has been high and on par with that for other row crops such as soybean and corn. However, the available herbicides have been difficult for farmers to work with because of the limited time period for high efficacy and the limited spectrum of weeds killed by each herbicide. As a result, cotton farmers have often had to use multiple herbicides to control weeds. Any technology that increases the efficiency of weed control is of interest to farmers.

As discussed in the prior case study, transgenic cultivars expressing insecticidal proteins derived from the soil bacterium *Bacillus thuringiensis* (*Bt*) have been successful in limiting damage by a number of Lepidopteran insect pests (e.g., European corn borer, pink bollworm, tobacco budworm, cotton bollworm; Gould 1998). It was therefore not surprising that many cotton farmers whose livelihood was threatened by insecticide resistance embraced transgenic *Bt*-cotton, which caused nearly 100% mortality of the tobacco budworm.

Transgenic cultivars with herbicide tolerance and/or insect resistance were planted on over 40 million hectares in 2000 (James 2000), making these two crop traits the most widely used products of agricultural biotechnology in the world. Commercial sale of insect-resistant transgenic cotton began in 1996, and the sale of herbicide-tolerant cotton began in 1997 (USDA 1999; see also Biotech Basics 2001). Increasingly, cotton cultivars are being produced that have both herbicide tolerance and insect resistance. In 2000, one-third of all transgenic cotton in the United States had both traits (USDA-NASS 2001).

There are two approaches for gaining regulatory approval of a cotton cultivar that is both herbicide tolerant and insecticidal. The most common approach since 1996 has been to obtain regulatory approval for each trait individually. For example, a herbicide-tolerant cotton genotype is developed and a petition is sent to APHIS asking for deregulated status. A *Bt*-producing cotton genotype is developed separately and goes through the EPA regulatory process as well as a petition for deregulated status with APHIS. Once the herbicide-tolerant cotton and *Bt*-producing cotton are granted nonregulated status, APHIS has no authority over those plants. Because the herbicide-tolerant cotton is not in itself a pesticide, EPA has no authority to govern its sale (EPA does regulate the sale of all herbicides). Therefore, anyone with legal access to the deregulated insecticidal cotton germplasm and to the deregulated herbicide-tolerant cotton germplasm can cross the two types of cotton and produce a new cultivar with both traits by this conventional breeding technique. This multitrait (generally referred to as "stacked trait") cotton can then be commercialized without further regulatory oversight.

In the case study examined here (petition 97-013-01p for determination of nonregulated status for Event 31807 and Event 31808), Calgene took a different approach to developing a cotton plant with herbicide tolerance and insect resistance. Although not clearly stated in the APHIS environmental assessment (USDA 1997b), it appears that Calgene developed a single construct for insertion into the cotton genome that contained both a gene for bromoxynil tolerance and the *Bt* gene, Cry1Ac, for insecticidal activity. The breeding advantage for using a single construct with both genes tightly linked is that the probability of segregation of the two genes during backcrossing to other cotton cultivars is extremely low. Because the two genes are essentially inherited as a unit, Calgene had both traits reviewed simultaneously by APHIS. The committee selected this environmental assessment as a case study of multiple genes because it is a case in which APHIS examined a petition for a plant with two genes, each governing a different agriculturally important trait. Many petitions for deregulation involve plants with multiple transgenes. In most cases one gene produces the phenotype of commercial interest and a second gene acts as a selectable marker. In the case of virus-resistant squash, multiple genes are present for resistance to a number of viruses. This cotton case stands out because both genes have distinct commercial uses.

The environmental assessment and determination documents for this petition were relatively short. The assessment formally considered only two alternative actions: "no action," which would mean refusal to grant nonregulated status, or a determination of a "finding of no significant impact," which would result in complete deregulation. These alternatives contrast with other recent environmental assessments. For example, in the environmental assessment of a *Bt* corn petition that was also reviewed in 1997 (96-317-01p), three alternative actions are stated. The additional action listed is to "approve the petition with geographical limitation." No explanation was given in this case study's assessment about why only two options were considered.

Environmental Risks Considered by APHIS in Its Environmental Assessments and Determination Documents

Disease in the Transgenic Crop and Its Progeny Resulting from the Transgenes. Because the herbicide tolerance and *Bt* genes were inserted using *Agrobacterium tumefaciens*, and because a cauliflower mosaic virus 35S promotor and a chimeric 35S promotor were part of the inserted DNA, APHIS examined the potential for risk from these sequences that came from plant pest species. The potential for these sequences to result in risks was dismissed because the disease-causing genes were not present.

Potential Environmental Impacts. APHIS recognized the potential for transgenic cotton to cross with wild cottons in some parts of the continental United States but concluded that "none of the relatives of cotton in the United States show any definite weedy tendencies" (USDA 1997b). (APHIS acknowledged that judgment of weediness based on the 12 traits listed by Baker (1965) or subsequent modifications are "imperfect guides to weediness." (The utility of Baker's list as a regulatory guide is discussed at length in the previous case study and is not repeated here.) Furthermore, APHIS stated that gene flow to wild relatives would not be a problem because (1) "any potential effects of the trait would not significantly alter the weediness of the wild cotton; and (2) wild cotton populations have not been actively protected, but have in fact been, in some locations such as Florida, subject in the past to Federal eradication campaigns because they serve as potential hosts for the boll weevil" (USDA 1997b). The EPA, which has also reviewed transgenic *Bt* cotton, came to a different conclusion. EPA allowed the planting of cotton in all areas of the continental United States except southern Florida because of the presence of feral populations of *Gossypium hirsutum* that can cross with commercial cultivars. However, the EPA stated that commercial cotton is not presently grown in that area of Florida. The agency also found a risk of transgenic cotton crossing with wild cotton in Hawaii. In this case the agency was specifically concerned with the risk that hybridization of the transgenic cotton with wild *G. tomentosum* could threaten that species' biodiversity and put restrictions on all but isolated breeding nurseries. The EPA noted that nontransgenic cotton would pose a similar threat but is not regulated (EPA 2000a).

Potential Impacts on Non-target Organisms. APHIS concluded there is no reason to believe that transgenic cotton lines would have deleterious effects on non-target species, based in part on EPA's finding that "foliar microbial pesticides indicated no unreasonable adverse effects on non-target insects, birds, and mammals" (EPA 1995). Also, APHIS argued that "invertebrates such as earthworms, and all vertebrate organisms, including non-target birds, mammals and humans, are not expected to be affected by the *Btk* insect control protein because they would not be expected to contain the receptor protein found in the midgut of target insects" (USDA 1997b; see also BOX 4.1).

The comparison of *Bt* cotton with *Btk*-sprayed pesticides is not appropriate in this case because the *Btk* pesticides degrade very rapidly in the field, due in large part to ultraviolet light exposure, while the *Bt* toxin in cotton is expressed constitutively and is tilled into the soil. Furthermore, no information is given to indicate whether the Cry1Ac toxin produced by the plant is a protoxin, as in the pesticide, or if it is an activated toxin that could have different ecological impacts (see discussion in Chapter 2).

Potential Impacts on the Development of Insect Resistance to the Btk Insect Control Protein. APHIS considered the issue of insect pests evolving resistance to the *Bt* toxin. The environmental assessment indicated that the EPA's active resistance management program should delay the onset of resistance. APHIS also concluded that if resistance to *Bt* does evolve in insect pests, the ability to control the insects will not be reduced because conventional insecticides will still be available.

At the time this assessment was written, the tobacco budworm had become highly resistant to pyrethroid insecticides in major cotton-growing areas, and the cost of chemically novel replacement pesticides was about triple the cost of pyrethroids. APHIS did not discuss the fact that one of the major pests affected by *Bt* cotton is the corn earworm (*H. zea*). That species feeds on many vegetable crops and is treated with *Bt* sprays by organic farmers. One of the factors that led to developing resistance management programs for *Bt* crops was concern that in the absence of *Bt*

BOX 4.1
The Mysterious Ecological Role of *Bt* Toxins

Over 500 scientific journal articles have been published on *Bacillus thuringiensis* since 1999 (*www.WebofScience.com*). Of these, only a handful discuss topics related to the natural ecology of this bacterium. An examination of the older literature reveals a similar trend. Most ecological studies tend to examine persistence of the bacterium in the soil (e.g., Addison 1993) or competition between *B. thuringiensis* and related species (e,g,, Yara et al. 1997). Most of the toxicological testing of *B. thuringiensis* isolates have pragmatically focused on pest insects (see Schnepf et al. 1998). Some papers examined toxicity to non-target organisms, including collembola, honeybees, and daphnia, as part of the process for regulatory approval (e.g., Sims). Most of these tests indicate that the common *B. thuringiensis* toxins are specific to small taxonomic groups of insects, and there is therefore a tendency to conclude that insects and *B. thuringiensis* have coevolved with each other (see Yara et al. 1997).

Indeed, there is no basis for such a conclusion. As a case in point, many of the commercialized *B. thuringiensis* toxins are considered specific to Lepidopteran larvae (Schnepf et al. 1998). However, general knowledge of the ecology of these larvae and this bacterium indicates that they rarely come in contact. *B. thuringiensis* is considered a soil bacterium and is rapidly killed when exposed to ultraviolet from direct sunlight. In order to be toxic, *B. thuringiensis* must be ingested. Most Lepidopteran larvae feed on leaves, fruits, flowers and plant stems. The few that feed on plant roots only ingest soil, and the bacteria in it, as a contaminant of their diet. Many Lepidopteran larvae pupate in the soil, but the prepupal stages in the soil do not feed. Furthermore, epizootic of *B. thuringiensis* in Lepidopteran populations are rare. The only habitat where these larvae and bacteria could commonly come in contact is in grain bins where a small number of Lepidopteran species live.

If Lepidopterans are an unlikely natural host for *B. thuringiensis*, is there some other more likely host? It certainly seems unlikely that this bacterium would use over 10% of its protein to make a toxic crystal of protein unless it had some function. One candidate for a host that has received minimal attention is the bacteriophagous nematode. As the name implies, these nematodes eat bacteria, including *B. thuringiensis*. It has been known for over 10 years that some *B. thuringiensis* isolates are toxic to *C. elegans* (Feitelson et al. 1992) but only recently have studies begun to look at other nematodes (Marroquin et al. 2000, Griffitts et al. 2001). Given that bacteriophagous nematodes are one of the most diverse groups of soil invertebrates, there is at least a reasonable expectation that *B. thuringiensis* has evolved in interaction with these organisms. Of course, until more studies are done on the ecological interactions of *B. thuringiensis* and soil-dwelling organisms, it will not be known what is the most common host or food of *B. thuringiensis*.

Without this knowledge, our ability to develop tests to examine non-target effects of *B. thuringiensis* toxins will at least be inefficient and at most totally misguided. In order for APHIS to develop more rigorous environmental assessments, it would be helpful to accumulate knowledge about the natural ecology of *B. thuringiensis*.

organic farmers would have no means to control this insect. The environmental assessment does not comment on whether the return to the use of conventional insecticides would cause environmental problems. As indicated in the previous case study on *Bt* corn, it would seem best for APHIS to consider this issue in more depth or to completely defer to EPA authority.

Environmental Risks Not Considered by APHIS in its Environmental Assessments and Determination Documents

Potential Impacts on Non-target Organisms. *Bacillus thuringiensis* is a soil-dwelling organism that would rarely seem to come in contact with foliage- and fruit-feeding insects. *Bt* protoxin created by this bacteria must be ingested before its insecticidal properties can be activated. Many Lepidoptera pupate in the soil, but Lepidoptera with soil-dwelling feeding phases are very rare. Based on the lack of interaction between the bacteria and Lepidopteran-feeding stages, there is no obvious ecological or evolutionary explanation for *B. thuringiensis* producing a Lepidopteran-specific toxin. Presumably, the bacteria produce endotoxins for another purpose, but this purpose has not been determined (see BOX 4.1). APHIS presented no data on tests that the applicants might have conducted on impacts of the truncated *Bt* toxin on organisms in the soil, including microbes and nematodes that could interact ecologically with the *Bt* bacterium and its toxin.

Potential to Cross with Wild Species in Some Geographic Areas in the United States. As stated in the "Background" to this case study section, the environmental assessment presents only two alternative responses to the petition for a finding of nonregulated status. APHIS did not mention the option of approving the petition with geographical limitations, even though this option was presented in other APHIS environmental assessments. In the current assessment, APHIS makes the decision to grant complete approval of the petition. The transgenic cotton lines under consideration were deregulated throughout the United States.

Impacts of Commercialization of Transgenic Cotton on Environments Outside the United States. Other APHIS environmental assessments discuss concerns about the impact of a transgenic crop approved for use in the United States being planted in other countries. One example is the squash case study. The environmental assessment of Calgene's transgenic cotton mentioned the existence of wild cotton in Mexico but does not really assess potential impacts on those species. Furthermore, this environmental assessment did not consider the fact that the specific *Bt* toxin gene under review might be useful for resistance management in the United States but might also facilitate resistance in pests that occur beyond U.S. borders. At least one cotton pest, *H. zea*, is known to move between Mexico and the United States each year (Raulston et al. 1986, Pair et al. 1987). Therefore, inappropriate planting of *Bt* cotton in Mexico could select for resistant pest individuals that would then migrate to the United States.

Interactions among Multiple Transgenic Traits. APHIS treated herbicide tolerance and production of the *Bt* insecticide as two separate traits. It did not consider that there might be interactions between the two traits that could have a detrimental effect.

Integrated pest management (IPM) emerged in the late 1950s as an effort to put pesticide use on a more ecological footing. One of the tenets of IPM is that natural processes can be manipulated to increase their effectiveness, and chemical controls should be used only when and where natural processes of control fail to keep pests below economic-injury levels (NRC 1996). Even with crops that have only *Bt* toxin genes, it is difficult to follow IPM guidelines because seed must be purchased in the spring before pest abundance can be predicted (Gould 1988). When two traits are combined in a single cultivar and it is impossible to purchase cultivars with only one of the two traits, farmers are forced to buy a cultivar with both traits even if they need only one for their farming operation. In the case of the stacking of herbicide tolerance and *Bt* toxin production, a farmer who needs herbicide tolerance may end up planting cotton with the *Bt* trait, even if the densities of the *Bt* target pests on the farm do not warrant control with the *Bt* trait. While it is difficult to determine how many farmers have specifically begun to use cotton with the *Bt* trait based on their desire to use herbicide-tolerant cotton, interviews with North Carolina farmers indicate that it may be over 20% (Bacheler 2000, North Carolina State University Extension, personal communication). This approach to the use of a pest control tool is clearly not an appropriate way to achieve the goals of IPM.

In addition to negating progress in adopting IPM farming methods (NRC 2000b), overuse of pesticidal crops due to a lack of seed choice could lead to more intense selection for *Bt*-resistant pest strains. The EPA has developed regulations to delay the evolution of *Bt*-resistant pest populations. In 1998 the EPA's Scientific Advisory Panel recommended that resistance management for *Bt* crops must include the following two components: (1) the transgenic plants must produce a high enough dose of toxin to kill partially resistant individuals (this dose was set at 25 times the dose needed to kill susceptible individuals) and (2) enough nontransgenic hosts must be planted on each farm to produce 500 susceptible pest individuals for each resistant individual produced in the *Bt* crop.

In the case of cotton, EPA-registered transgenic plants do not produce a high dose of *Bt* toxin for the cotton bollworm, so a large proportion of partially resistant individuals could survive on the *Bt* cotton. This registration was not appropriate according to the EPA Scientific Advisory Panel, and current requirements for on-farm, non-*Bt* acreage are not expected to produce the desired 500:1 ratio. Only a substantial increase in the refuge acreage could ameliorate this problem. While the EPA may not increase the on-farm refuge requirements, concern over resistance evolution in *H. zea* due to an inadequate resistance management program was reemphasized by the most recent EPA Scientific Advisory Panel (EPA 2000b). If the stacked trait cottons such as the one approved by the environmental assessment discussed here are commercially successful, they could increase regionwide adoption of *Bt* cotton, further accentuating the risk of rapid evolution of *Bt* resistance in *H. zea*.

This case study identifies only two negative environmental effects that could be caused by the interaction of two transgenic traits. If outcrossing to weedy relatives was more of a problem with cotton, the interaction between herbicide tolerance genes and *Bt* genes could exacerbate an additional risk—the transfer of the *Bt* trait to noncrop plants. A *Bt* gene inserted into a crop along with a herbicide tolerance gene could be transferred to wild relative populations much faster than in cases where the *Bt* gene was inherited separately from herbicide tolerance. A potential scenario is as follows: Pollen for the stacked trait cultivar crosses with a weedy relative in a cotton field. The next year progeny from the cross as well as other individuals of the weed species germinate in the cotton field. The farmer sprays bromoxynil. This kills most of the weeds without the herbicide tolerance gene, but those with the gene increase in frequency. Although the *Bt* gene confers no direct advantage with regard to survival against herbicide spray, there is a major increase in the frequency of weeds with the *Bt* gene because it is linked to the herbicide tolerance gene. This results in a large fraction of weeds that are now protected from insect feeding. If APHIS reviews transgenic plants with weedy cross-compatible relatives in the United States, such as canola, with stacked herbicide tolerance and insecticidal genes, it would definitely need to consider this interaction.

Public Involvement

The APHIS environmental assessment indicated that the agency received no responses to its *Federal Register* announcement of this petition for deregulation. The committee is unaware of any other attempt to involve the public in this specific assessment.

CONCLUSION

This chapter has reviewed case studies of the three primary APHIS regulatory pathways for field release of transgenic organisms as well as a representative sampling of the vast array of transgenic species, phenotypes, and molecular mechanisms designed to obtain those phenotypes. In many cases the committee simply reports, without much comment, how and with what information APHIS made a specific decision. The committee has little to add in those cases. In certain cases, it has pointed out situations in which APHIS might have improved its assessments. The committee has supplied substantial supporting text to explain how those improvements might have been made. While it is recognized that a few of those suggestions benefit from hindsight, most of the suggestions are based on scientific information available, but not utilized, at the time of assessment. The opportunities for improvement of assessment provide a context for the committee's recommendations in the next chapter.

5

Analysis of APHIS Assessments

As indicated in Chapter 1, regulatory agencies charged with assessing the safety of transgenic plants face a daunting task. This is partly because environmental risk assessment for transgenic plants is new and partly because the social context in which regulatory decisions about transgenic organisms must now be made is dramatically different from the social context in which these agencies are accustomed to working.

The Animal and Plant Health Inspection Service (APHIS) began its involvement in the regulation of transgenic organisms in the mid-1980s. APHIS's regulatory system has improved substantially since that time. As pointed out in the case study of two *Bt* corn petitions for nonregulated status (see Chapter 4), the scope of environmental issues addressed and the degree of rigor with which they were addressed both increased between 1994 and 1997. Furthermore, development of a notification process that focuses on plant ecology was an important step in effectively streamlining the field-testing process. The learning process at APHIS has not come without missteps, but the agency seems to use those missteps as opportunities for further improvement. The present analysis is designed to facilitate additional improvements in the APHIS system, which will be necessary to meet future challenges of assessing potential environmental risks of a large number of diverse and novel transgenic products. In analyzing the APHIS regulatory process, the committee searched for problem areas. The committee's criticisms here are not meant to be an indictment of the system but rather a means to help improve an already functioning system. It is hoped that APHIS personnel will find this analysis useful in the spirit it is intended.

APHIS's regulatory process has never led to the release of a transgenic plant that clearly caused environmental damage. However, without systematic monitoring, the lack of evidence of damage is not necessarily lack of damage. Furthermore, based on the questions raised in Chapter 4 regarding the need for a stronger scientific basis of APHIS analysis, the committee recommends that the APHIS decision-making process could be made significantly more transparent, thorough, accurate, and scientifically robust by enhanced scientific peer review, active solicitation of public input, and development of determination documents with more explicit presentation of data, methods, analyses, and interpretations.

As discussed in Chapter 2, there are two roles for risk assessment—technical decision support and creation of legitimacy in the regulatory process. The analysis here is divided into two main sections that relate to these two roles. First, the committee analyzes how APHIS has involved the public in the development of its risk assessment process and its specific rulings. Then, the technical approaches taken by APHIS to support its decisions are analyzed.

ANALYSIS OF PUBLIC INVOLVEMENT

The issue of public involvement in APHIS's decision making is complex and must be accorded a somewhat summary, even ancillary, role in the present analysis. A thorough analysis would require a book of its own! This issue is an important one because public involvement in the regulatory decision-making process is desired on at least three counts (NRC 1996):

- Public involvement in government rule making is required by basic principles of democracy. Government authority ultimately rests on the consent of the governed, and it is desirable for public agencies to find appropriate ways to ensure that decisions are consistent with this principle.
- Opportunities for public involvement can broaden the basis of information on which regulatory decisions are made, improving the quality of decision making.
- Research on environmental risk indicates that public confidence in environmental policy making is particularly sensitive to the opportunity for concerned citizens to be involved in the decision-making process.

Currently, APHIS policies for public involvement conform to a fairly narrow interpretation of those required by the Federal Administrative Procedures Act. It is useful to summarize involvement at two distinct levels:

- First, who is involved or has input into decision making in the notification, permit, and petition processes?

- Second, who is involved or has input when policy for review of these decisions in being established?

With regard to both questions, APHIS may choose to solicit advice from sources deemed useful by agency staff (although the short turnaround required for notifications may limit the time to obtain that advice). Also, the widest level of public input presently solicited by APHIS is achieved through publication in the *Federal Register* of intent to deregulate, issue permits, or alter internal APHIS procedures. APHIS generally receives comments on these *Federal Register* notices for a limited period and is required to issue responses to all comments that express disagreement with intended action.

External Input into the Decision-Making Process

Table 5.1 summarizes the current range of external participation in APHIS's deliberative process for characterizing and evaluating risks asso-

TABLE 5.1 APHIS Involvement of Potential Participant Groups in the Process of Commercializing a Transgenic Plant or Its Products

	Applicants	External Experts	Critics	Consumers	Users
Petition for Non-regulated status					
Hazard identification	I	I	N	N	N
Risk measurement	I	I-N	N	N	N
Making a decision[a]	F	F	F	F	F
Risk management	I-N	N	N	N	N
Permit for release					
Hazard identification	I	N	N	N	N
Risk measurement	I	N	N	N	N
Making a decision[b]	I	I	I	I	I
Risk management	I	N	N	N	N
Notification					
Hazard identification	I	N	N	N	N
Risk measurement	I	N	N	N	N
Making a decision	I	N	N	N	N
Risk management	I	N	N	N	N

The process for involving interested or affected parties is designated as F, formal process; I, informal process; N, no process available.

[a]Notification of an impending decision must be published in the *Federal Register*, requesting comments from interested parties.

[b]Issuance of a permit is under the authority of the Secretary of Agriculture and does not require publication in the *Federal Register*. When permits are announced interested or affected parties can request the related decision documents.

ciated with regulated transgenic plants. There are three possible paths to commercialization in which external input could be considered:

- An applicant can petition for nonregulated status, which means the transgenic plant can be released without regulatory oversight. (See Chapter 3 for details of the petitioning process.) At present this avenue is the most common pathway to commercialization. As of the end of 2000, APHIS had received a total of 73 petitions for deregulation. Of these, 52 petitions had been approved, 19 had been withdrawn, two were incomplete, and none had been denied. APHIS works with applicants to clarify the standards for approval. As a consequence, petitions are improved, sometimes substantially, prior to an APHIS decision. The lack of denials could be related to this clarification process, which among other functions serves to inform applicants of the likelihood of approval of petitions.
- An applicant can commercialize the products of a regulated article (but not the live plant itself) under the permitting system. (See Chapter 3 for details of the permitting process.) The permitting system is also used to accumulate data needed for commercialization through the first path. However, based on the relatively small number of field tests conducted under permit, this approach is rarely used. In the year 2000, only 36 field tests were performed under permit.
- A final path for developing commercial products from a regulated article is through use of the notification system. One drawback to this approach is that APHIS may, at any time, require permits for applicants who are growing a regulated article under notification. (See Chapter 3 for details of the notification process.) Notification is also currently the primary route for field testing to gather information needed for commercialization through the first path. In the year 2000, almost 900 field tests were conducted under notification, and several products have been commercialized.

APHIS publishes its intent to deregulate or permit (if it conducts an environmental assessment for the latter) in the *Federal Register*. Comment periods are summarized in Chapter 3. In the case of notifications, the agency does not publish each notification separately or in a published list. Instead, it maintains an updated list of all submitted notifications (*Federal Register* 1992). APHIS periodically publishes a *Federal Register* notice announcing the list's availability (but apparently has not done so for several years). On request, the agency provides the list directly to interested parties. APHIS has made arrangements with the National Biological Impact Assessment Program to make the notification list available online (see "Field Test Releases in the U.S.," Information Systems for Biotechnology database: *www.nbiap.vt.edu*). Note that there is essentially no opportunity for either general public comment or external scientific review on notification decisions prior to a decision being made.

Finding 5.1: The notification process involves no public input.

Beyond publication in the *Federal Register*, the legislative authority under which APHIS regulates transgenic plants does not stipulate formal mechanisms for involving external participants. However, the agency has developed guidelines for petitioning for nonregulated status and for submitting notifications. These guidelines provide for deliberative interactions between APHIS and the parties seeking to commercialize the product (see Chapter 3 for details). APHIS sometimes uses informal processes to contact outside experts to solicit information (see Chapter 4, squash case study), but it also has the authority to convene such groups formally to provide external advice.

External Input into the Establishment of Policy

Table 5.2 summarizes previous involvement of potential participant groups in APHIS policy-making processes, including developing risk analysis procedures and procedures for commercializing a transgenic plant or its products. Three of the primary processes are listed.

As discussed in Chapter 3, APHIS once used a subcommittee of the U.S. Department of Agriculture's Agricultural Biotechnology Research Advisory Committee (USDA-ABRAC) to provide advice for initiating the notification process. The agency has also held a number of national and regional meetings to discuss its policies with stakeholders.

The rationale for APHIS risk management activities is determined largely by statutory authority. Because the U.S. Coordinated Framework for the Regulation of Biotechnology was developed under statutes put in place before the advent of transgenics, no public debate or congressional testimony specifically relating to the environmental risks or public confidence issues associated with transgenic plants occurred in connection with the creation of APHIS statutory authority.

Effectiveness of Efforts to Solicit External Input

The processes summarized in Tables 5.1 and 5.2 are somewhat effective in ensuring democratic decision making and improving the quality of APHIS's decisions, as noted earlier. Certainly, publication in the *Federal Register* in compliance with the Federal Administrative Procedures Act is intended to provide at least some degree of democratic legitimization of the general rule-making process. Although theorists of democracy and governmental procedure continue to debate the adequacy of this approach to public participation, debate transcends APHIS policies and the appropriate focus of this report. In addition, APHIS has been somewhat successful at using public input to improve the quality of its decisions, as

TABLE 5.2 Public Involvement in Policy Making for Risk Assessment and Management

	Applicants	External Experts	Critics	Consumers	Users
Permit for release (1987)					
Hazard identification	N	N	N	N	N
Risk measurement	N	N	N	N	N
Making the decision to establish system[a]	F	F	F	F	F
Notification system and petition for nonregulated status (1993)					
Hazard identification[b]	I	I	I	I	I
Risk measurement	I	I	I	I	I
Making the decision to establish system[c]	F	F	F	F	F
Expansion of notification system (1997)					
Hazard identification	N	N	N	N	N
Risk measurement	N	N	N	N	N
Making the decision to establish system[d]	F	F	F	F	F

The process for involving interested or affected parties is designated as F, formal process; I, informal process; N, no process available.
[a]Proposal was published in the *Federal Register*, where anyone could make comments.
[b]During the 1993 proposal, APHIS co-convened an expert panel to review the proposal. APHIS also presented the proposal at several public meetings during which interested or affected parties could comment. These procedures were not repeated during the 1997 proposal.
[c]Proposal was published in the *Federal Register* where anyone could make comments.
[d]Proposal was published in the *Federal Register* where anyone could make comments.

noted earlier in the virus-resistant squash case and the involvement of the ABRAC. However, the third desired outcome of public involvement—improving public confidence in the decision-making process—has not received sufficient attention.

Effectiveness of External Input on Specific Decisions

APHIS's informal procedures for consulting with outside scientific experts and the record of *Federal Register* notices, public comments, and APHIS responses to comments suggest that the agency's approach has been useful in some specific cases for assembling technical information needed to exercise its regulatory authority regarding permits and petitions. Comments on a number of permits have elicited detailed technical responses from APHIS staff and on several occasions have resulted in requests for additional data or additional studies on the likely risks posed by issuance of a permit (see deregulation case study of virus-resistant squash in Chapter 4). The effectiveness of the public comment period for this purpose depends on an informal network of people and organizations that monitor the *Federal Register* and generate or recruit comments from people having relevant expertise.

The committee notes that external input through the *Federal Register* publications, whether from scientists or the public at large, has dwindled over the years. In information provided by APHIS, the committee found that they received a total of 378 comments for the first 10 petitions they considered (from 92-196-01p to 94-319-01p); the greatest number was for the second petition involving virus-resistant squash (92-204-01p, see Chapter 4). In contrast, for the 10 most recently approved petitions (from 97-287-01p to 99-173-01p), APHIS received a total of 11 comments for 3 of them, and no comments for the others. It is possible that the decline in comments is due to improved APHIS decision making, or a decrease in interest on the part of external scientists and the public. In the case of decreasing numbers of negative comments, another possibility is that frustration with the process may have resulted in declining public involvement in this specific process. Indeed, information provided by members of the public interest community confirmed that their perceived lack of responsiveness of APHIS to the comments they provided during deliberations over the transgenic squash petition (92-204-01p) marked a "watershed" for them in which they felt their efforts proved a "waste of time" and that in the future their efforts were better spent on activities other than writing comments to APHIS. No matter what the actual cause, APHIS may be losing potentially valuable public input. The lack of input from outside scientists in precedent-setting decisions is especially problematic. In contrast, the Environmental Protection Agency (EPA) often convenes formal scientific advisory panels to provide information that will help improve the technical rigor of its decisions.

The *Federal Register* comment process has several weaknesses for the purpose of eliciting public involvement. First, *Federal Register* notices are often quite technical and not written in a manner that is accessible to the lay public. Second, when public comments have been issued regarding questions about confidence in the regulatory process for biotechnology, APHIS's responses have largely been perfunctory. For example, early in the history of permits a number of comments were submitted expressing the view that the U.S. regulatory approach was incomplete or had gaps. The APHIS response was either to simply note disagreement with these opinions or to reply that they were not relevant to the particular decision at hand. These responses are adequate within a narrow interpretation of administrative procedures, but APHIS could have sought alternative methods of involving the public. As noted above, the agency has held public meetings to inform the public about its policies, but such meetings reach a very small audience.

Recommendation 5.1: For precedent-setting decisions involving permits and petitions, APHIS should actively solicit external scientific review.

External Input into the Establishment of Policy

As reviewed in Chapter 3, the present regulatory system used by APHIS was established in a series of policy-setting decisions. Starting in 1987 with establishment of the permit system, with its risk-based process regulatory trigger, several subsequent decisions excluded certain transgenic organisms from regulation. Two significant policy-setting decisions were the 1993 decision to establish a notification system and a process to petition for nonregulated status (the deregulation decision) and the 1997 decision to expand the notification system. APHIS did not use any formal processes (i.e., processes that are part of APHIS written policy) beyond the use of the *Federal Register* to announce 60-day public comment periods for involving the public in these decisions. For the 1993 decision, the agency used informal mechanisms to solicit public input, which included several public workshops and a scientific review by a subcommittee of ABRAC to evaluate the scientific merits of the 1993 proposal for a notification system. Increased utilization of a system of external scientific review of important policy decisions could help in securing greater public confidence in the regulatory process.

Recommendation 5.2: For changes in regulatory policy, APHIS should convene a scientific advisory group to review proposed changes.

Input of the general public in the process of policy development has been limited to the same mechanisms used in soliciting input on specific regulatory decisions. Actively broadening the basis for public involvement in environmental decision making is a difficult and potentially expensive proposition. Among the strategies that have been used by other agencies are advisory committees with representation from a number of self-identified advocacy groups, and public hearings. Outside the United States other approaches for distributing information and eliciting responses from the interested public have included consensus councils, blue ribbon committees, standing committees, external contracts, and public jury approaches. Some European countries use standing committees of public-sector scientists or mixtures of public- and private-sector scientists to provide wider deliberations on critical issues. Others contract with public-sector scientists to provide detailed critiques of materials entering into the decision process. The 2001 revision of European Union's Directive 90/220 specifies that applicants perform the risk assessment (unlike in the United States, where a government agency performs the risk assessment). This allows the European regulatory community to solicit evaluations of these assessments using external scientists in ways that would be unwieldy for APHIS.

None of these efforts at involving the public can be undertaken without considerable cost in terms of both money to actually carry out the exercises and staff time and potentially needless delays in the regulatory process. Arguably, the effect of efforts undertaken outside the United States to involve greater segments of the public has been to considerably broaden the basis of decision making beyond that of the scientific disciplines deemed relevant to the interpretation and measurement of environmental risk by APHIS staff. Whether this other approach to decision making reflects greater sensitivity to alternative perspectives, including consumer and citizen desires, or an intrusion of inappropriate nonscientific viewpoints regarding what should be a science-based regulatory process is the subject of continuing debate.

Finding 5.2: There is a need to actively involve more groups of interested or affected parties in the risk analysis process while maintaining a scientific basis for decisions.

TECHNICAL ANALYSIS OF APHIS OVERSIGHT

There is no question that the APHIS regulatory processes that have evolved to handle transgenic plants have resulted in a system that allows for thousands of field tests and dozens of deregulations of transgenic plants while at the same time allowing for scientific scrutiny of the regulated articles. Nonetheless, there are opportunities for improvement of this system.

This section starts with general comments relevant to all APHIS decision-making processes—notification, permit, and petition. It then focuses on each process separately. The most important APHIS decisions regarding transgenic plants are those made during the notification and deregulation processes, because of the hundreds of field releases of transgenic plants under notification that occur every year and the fact that deregulation removes a transgenic plant from any further APHIS oversight. Because of the major differences between the notification and the petition for deregulation processes, each of these processes is considered individually after making a few general comments about the overall APHIS regulatory process. Comments also are provided on the third, now relatively rare, permitting process. That commentary is placed between notification and petition because the permitting process shares certain features with both. It is noted that the permitting process may become more common as it is the only route to commercialization of products from plants intentionally grown to produce pharmaceutical compounds.

General Comments and Concerns

Geopolitical Scale

The scope of APHIS oversight is limited to whether and how transgenic plants are moved and released in the United States. Although APHIS determinations sometimes include consideration of environmental impacts of deregulated genetically modified organisms outside the confines of the United States (see the case studies in Chapter 4), the agency is under no obligation to do so. Indeed, relevant scientific data may be unavailable for many countries that are centers of diversity for wild relatives of a transgenic crop. Yet these locations might be the very places where environmental impacts might occur due to transgene flow into those populations. Also, once a transgenic plant is deregulated, descendants of that plant may find themselves intentionally or unintentionally transported far beyond the borders of the United States, to radically different environments. A transgenic crop variety developed to fit into U.S. systems of agriculture may cause changes in the agricultural systems of other countries that would cause environmental degradation.

One can imagine an argument being made by certain stakeholders, that if the U.S. government found a plant to be safe, that judgment should be good enough for a country without the resources to conduct its own environmental analysis. That would be wrong. Just because APHIS finds a transgenic plant to have no significant impact in the United States is not a guarantee that it will not have an impact elsewhere. To its credit, APHIS has held biosafety meetings in a number of developing countries to inform policymakers of the potential environmental effects of transgenic plants.

Finding 5.3: APHIS's environmental assessments should be interpreted as being confined to the evaluation of effects occurring in the United States.

CBI in APHIS Decision Documents

Many people, including those generally supportive of biotechnology, decry the apparently large amount of data and information in submissions marked "CBI" (confidential business information). Under this CBI stamp, all manner of data are hidden from public view and even from independent scientific scrutiny. This form of business information protection is not unique to the biotechnology industry and those who monitor the applications for approval of new pesticides by the agrochemical industry are faced with similar challenges. This committee sometimes found that it could not provide an independent scientific assessment of APHIS rulings because of the broad use of CBI. Clearly, businesses have the right and the need to protect sensitive information from their competitors. But APHIS does not have a mechanism to distinguish what is truly competitive information (and worthy of confidence) from less sensitive information that might be of value in the public discourse and open to scientific analysis. The lack of such a mechanism may be related to issues much wider than the APHIS regulatory system. However, its absence creates several contradictions. Without a transparent mechanism to evaluate and judge confidential data from applicants, APHIS appears to accept without question an applicant's assertion of what is CBI. Public credibility is eroded when the same information marked CBI in APHIS documents is not considered CBI and is open to public inspection in other jurisdictions, such as Canada or Europe. It must be understood that U.S. standards for what constitutes CBI are very broad. A company can claim that certain information is CBI as long as it can show there is any possibility that disclosure of the information could directly or indirectly harm its business. In this legal sense APHIS may have little power to limit CBI. Another interpretation of the degree of CBI in documents sent to APHIS by applicants is that the agency is not working to provide as much information as possible to the public. Regardless of the accuracy of either of these interpretations, the extent of CBI in these documents makes public assessment of APHIS decisions extremely difficult.

Finding 5.4: The extent of CBI in registrant documents sent to APHIS hampers external review and transparency of the decision-making process.

Resistance Risk and Non-target Effects

For pesticidal plants, both APHIS and EPA examine risks of non-target effects and the potential for pests to evolve resistance to the pesticidal component of the plant. The committee examined APHIS's approach to assessing these risks in case studies described in Chapter 4 and by examination of the general guidelines to applicants (Chapter 3). In these chapters the committee notes that there is sometimes a lack of rigor in these guidelines and assessments. Often APHIS documentation refers to documents written by the EPA on these risks. If both EPA and APHIS conducted detailed, independent analyses of these risks, this could serve as a system of checks and balances. However, this does not seem to be the case. In some assessments it is clear that the EPA examinations are more comprehensive than those of APHIS. Therefore, it is not clear that anything is gained by having applicants submit information on non-target effects and resistance risk analysis to both agencies.

> **Recommendation 5.3: For pesticidal plants APHIS should either increase the rigor of assessments of resistance risk and non-target impacts, or it should completely defer to the EPA, which also assesses these risks.**

Technical Analysis of the Notification Process

The total number of notification decisions is now several thousand, and as detailed in other parts of this report and in the Office of Science and Technology Policy (OSTP) case studies (OSTP/CEQ 2001) it is a highly streamlined process that occurs without external input from either scientists or stakeholders. Furthermore, information available to the public from acknowledged notifications is limited because of CBI.

The information provided by the notification applicant generally appears to be sufficient for APHIS to make sure that the regulated article does not activate any of the triggers that would require the applicant to apply for a permit. (However, an example given below indicates this may not always be the case.) If a transgenic plant is eligible for oversight through the notification process and the applicant complies with the performance standards, environmental impacts should be minimal (especially if the notification is for a small-scale planting; see below). Of course, this implies that the applicant understands the performance standards and that APHIS personnel have sufficient time to carefully examine the notification application.

A plant under notification must not be a noxious weed, according to a federal weed list, but plants that are on the weed lists of states, regions of the United States, or other countries are allowed to be grown under notification, provided they are not recognized as a weed in the location of the field site. Relying on only a single weed list to determine whether a plant is a weed or invasive might not capture all species that are considered weeds or invasives, especially if they are important problems in a restricted locale or ecosystem. For example, in the case study of *Bt* corn (see Chapter 4) APHIS indicated that it did not find corn listed as a weed, while the first weed list the committee used indicated that corn is a weed. Reliance of APHIS on either its weed list or the committee's could result in an inappropriate decision. There is excellent scientific evidence showing that the Baker criteria are inadequate ex ante predictors of weediness. Yet APHIS continues to use these criteria in decision making.

There are no acreage limits for a crop grown under notification or permitting. While the typical acreage is small and the largest plantings are far less than 1,000 acres, the fact that plants under notification and permitting can be grown for commercial purposes may, in the future, motivate considerably large-scale plantings. There is nothing to prohibit a single planting of 1000 acres or more under notification. Alternately, one could imagine an applicant filing 20 different notifications in a single year for separate plantings of the same transgenic genotype in different or contiguous fields of 50 acres each, resulting in a total of 1,000 acres. Because many ecological effects are scale dependent and more likely to occur at larger spatial scales, these considerations raise concerns about acreage limitations.

Finding 5.5: There is no acreage limit for transgenic plants grown under notification or permit.

As the scale of release increases, so does the challenge for confinement of the transgene. For example, larger plantings of a species broadcast pollen farther and at higher rates than smaller stands (Handel 1983 and references therein). Also, as the scale of release increases, it is more difficult to check for compliance to the performance standards mandated for plants grown under notification. Unlike plantings under permit, which are visited annually, only a subset of field sites under notification are inspected annually.

Finding 5.6: The combination of no limit to scale of planting under notification and permit and the opportunity for commercialization of products under notification and permit may motivate large-scale plantings that may erode compliance to performance standards.

Transgenic plants that are toxic to non-target species cannot be tested under notification. Such plants are considered toxic to non-target species if the plant or its seeds are toxic to non-target species in the field and must be field tested using the more stringent oversight of the permitting process. If the toxic compound were intended for use as a pesticidal substance, it would be regulated by EPA no matter how small the acreage planted. If the toxic compound were intended for use as a pharmaceutical, it could not be regulated under notification, and may require oversight by the Food and Drug Administration. Intent is determined by the applicant with a declaration of intent or no such declaration. Although an applicant does not now intend either pesticidal or pharmaceutical use, in the future an applicant may declare such intent. Criteria based on intent have little effective regulatory scope and have the additional disadvantage of possibly misleading the public into believing that some types of transgenic plants are being closely regulated when in fact they may not be. Thus, a transgenic plant producing a toxic compound with pesticidal or pharmaceutical properties may be regulated under APHIS notification if the plant or its seeds do not harm non-target species in the field. Toxins that harm non-targets in storage or toxins in plant parts that harm non-

BOX 5.1
Avidin

Transgenic maize that produces the glycoprotein, avidin, has been grown commercially under the APHIS notification procedure since 1997 (Hood et al. 1997). The avidin is extracted from the corn and is used commercially for a number of purposes, including medical diagnostic procedures. Concern has been raised about the appropriateness of having avidin-producing corn included among transgenic crops that merit a low level of oversight through the APHIS notification system.

The avidin molecule binds to the coenzyme biotin with high affinity and results in biotin inactivation. Because biotin is involved in the basic metabolism of all organisms, avidin can act as a general toxicant (Hood et al. 1997, Kramer et al. 2000). However, if excess biotin is provided to an organism, the effects of avidin can be ameliorated in at least some cases (Kramer et al. 2000). Avidin is produced in chicken eggs (Livnah et al. 1993) and has never been implicated as a cause of any human illnesses (Kramer et al. 2000) or allergies (Langeland 1983). Short-term toxicity tests with mice were negative (Kramer et al. 2000). However, long-term biotin deficiency can be harmful to hamsters (Watanabe 1993) and mice (Baez-Saldana et al. 1998). Furthermore, while not shown to cause human allergies, avidin and a related protein, streptavidin, do elicit immune response in humans (Subramanian and Adiga 1997, Meyer et al. 2001).

Avidin has been known to have "antivitamin" effects on insects since Levinson and Bergmann (1959) did their early research for the Israeli Army Medical Corps. The more recent literature indicates that at least 26 insect species have been shown to be killed or chronically impaired by low doses of avidin (Levinson et al. 1967; Bruins et al. 1991, 1992; Morgan et al. 1993; Kramer et al. 2000; Markwick et al. 2000; Sedlacek et al. 2001).

Because ProdiGene, the company involved in production of corn-based avidin, is not growing avidin-producing corn for the pesticidal effects of the molecule, the EPA does not regulate the product or any field testing. APHIS regulations indicate that a transgenic plant is not appropriate for testing or commercialization under the notification system if the transgenes "encode substances that are known or likely to be toxic to non-target species known or likely to feed on the plant." However, according to the user's guide (APHIS 1997b), the definition of "toxicity to non-target species" is restricted to only those non-target species can be regulated by notification. For example, toxins in pollen that harm non-target species could be regulated by notification.

As indicated above, it is often difficult to ascertain many details about a tested plant from the published notifications. The committee did, however, find one case where enough information was available to result in questioning the rigor with which the restrictions on the notification process are enforced. This case is presented in BOX 5.1 as an example, with the caveat that the committee did not have access to all of the information available to APHIS. The committee concluded that the plant tested had the potential for toxicity to a broad array of organisms both in the field and after the plant was harvested. Based on this information, the commit-

target species that feed on the plant, not dispersed plant parts, such as seeds, pollen, or plant residue. Even using this restricted definition, it is difficult to understand how the avidin-producing corn could qualify for commercialization under notification because the corn kernels are toxic to many diverse insect species (Kramer et al. 2000). Allergenicity is not one of the characteristics considered in accepting a notification package. While avidin has no history of causing allergies (Langeland 1983), the notification procedure used by APHIS does not require any reporting about potential allergenicity, and any such risks would not be considered as a part of a notification package. In other words, even if avidin were allergenic, the applicant could grow and commercialize the product under the APHIS notification system without any additional oversight.

Based on conversations with John Howard of ProdiGene and APHIS personnel, avidin corn appears to be well contained in current commercial plantings. All of the avidin corn is planted by a single contractor on less than 5 acres of land. Furthermore, biological containment of the avidin transgene is reinforced by its toxicity to specific corn tissues. Avidin-producing plants that express high levels of avidin are male sterile, presumably due to the toxicity of the avidin to pollen-producing tissues of the corn plant.

These containment methods are more stringent than required under the APHIS notification procedure. Under the notification system, an applicant only has to show that progeny from any gene flow to other crop plants will not survive in the environment. Processing corn for animal feed or human food is one way suggested by APHIS to ensure that these crop plants will not survive in the environment. Such procedures clearly will not ensure that risky transgene products will not enter human foods if the transgenic crop is commercialized under the APHIS notification system. It is worth noting that the first field tests of avidin-producing corn were carried out in 1993 based on an APHIS permit that had been applied for in 1992, before the notification system was established.

While there do not seem to be any unacceptable environmental risks in the current production of avidin corn, this case points to a lack of rigor in the notification process that could lead to problems with future transgenic plants. For example, it would be possible, under notification, to grow thousands of acres of a transgenic crop that produced a substance that was allergenic or toxic to livestock or humans after seed was harvested. There is a clear need for APHIS to reassess its notification process given the novel products likely to be produced by transgenic crops in the future.

tee does not understand why APHIS has repeatedly allowed field testing of this transgenic plant under the notification system. The fact that the substance coded for by the transgene has the potential to be an allergen was of special interest. Examination of APHIS guidelines revealed that plants with allergenic properties can be grown under notification. The committee questions the wisdom of allowing such plants to be grown under the streamlined notification system. Regulation of novel compounds under notification raises the more general question of how APHIS determines that a compound is not toxic to non-target species in the field, given the limited information that must be provided by applicants. The committee's comments on the APHIS procedures are not meant to imply that there is any risk in this specific case where the applicant seems to have taken extra steps to ensure human and environmental safety.

Finding 5.7: It appears that a transgenic plant with toxic properties to non-targets was grown to create a commercial product under the notification process.

APHIS personnel must process the notification applications very rapidly (see Chapter 3). The total number of APHIS personnel available to process notifications, permits, and petitions appears to be insufficient. Currently there are about 10 permanent APHIS biotechnology evaluations staff. From the time the committee began its study in July 2000 until July 2001, APHIS always had staff vacancies, and staff turnover rates appear to be high. This situation results in a heavy per-person workload. The heavy workload coupled with no public feedback may detrimentally affect the rigor of the determinations.

The number of APHIS personnel who conduct field visits to sites under notification and permit is also small. This includes the permanent APHIS Biotechnology, Biologics, and Environmental Protection (BBEP) staff plus APHIS field personnel who are not all trained to understand the implications of the evaluations they are making. (To the credit of APHIS, it is clear that these field personnel are, on average, better trained today than they were five years ago. Turnover in field personnel makes maintaining adequate training levels difficult.) To maintain compliance with performance standards, sufficient numbers of appropriately trained personnel must be allocated to visit field sites. Only a subset of notification sites receives an inspection visit. In field seasons such as 2001, when APHIS field inspectors have an emergency priority (of inspecting for foot and mouth disease), few may be available to visit notification sites.

Recommendation 5.4: Because of the large number of field tests conducted, resources for compliance monitoring are necessary to maintain a suitable number of well-trained APHIS field inspectors.

Technical Analysis of the Permitting Process

As noted earlier, the permitting process is currently a relatively rare APHIS decision-making process for transgenic plants, with only a few dozen such decisions per year. Prior to introduction of the notification process, the permitting process was the only regulatory mechanism for conducting field trials, so the total number of permit decisions associated with field tests numbers over a thousand.

Similar to the petition process, permit applications contain more than the simple list required for notification—that is, they provide some scientific information and analysis. The information provided by the applicant is primarily for the purpose of assuring appropriate confinement and disposal of the transgenic plants.

Most permit applications, like all notification applications, are not listed in the *Federal Register* during the process of APHIS assessment unless they include an Environmental Assessment as required under the National Environmental Protection Act. From 1996 to the present, APHIS appears to have conducted environmental assessments for about 6% of the permit applications processed. If APHIS does not issue an environmental assessment reporting on its environmental risk assessment, there is no potential for feedback from stakeholders except the applicant. And even when an assessment is issued, feedback from interested and affected parties could be frustrated by information that is limited because of CBI.

Plants grown under permit share certain characteristics with plants grown under notification. There are no restrictions to the acreage under which they may be grown, and they may be grown to produce nonliving commercial products. As noted elsewhere in this report, it is anticipated that transgenic plants intentionally grown to produce commercial pharmaceutical products will be grown under permit. Thus, problems associated with the scale of the plantation discussed for notification may also apply to plants grown under permit. However, while APHIS personnel visit only a subset of notification field sites, all permit sites are visited.

The committee's findings and recommendations under the section on notifications generally apply to the permit process as well.

Technical Analysis of the Petition Process

If APHIS approves a petition for deregulated status, deregulation is absolute. APHIS generally indicates that it cannot "conditionally," "temporarily," or "partially" deregulate. For example, APHIS cannot deregulate an article but require monitoring of it. Similarly, APHIS deregulates a regulated article and all of its descendants. As noted in Chapter 3, transgenes can be intentionally moved by crossbreeding into other varieties of the same species or even other species. Because the plants are fully deregulated, one can imagine that the transgenes could be used for purposes other than originally intended. For example, without any further APHIS consideration, a selectable marker transgene that confers herbicide tolerance in a deregulated article could be transferred to another variety or another species to create a new variety to be sold as a herbicide-tolerant product. As a hypothetical example, if a male-sterile poplar tree with a *Bt* gene were deregulated, it would be possible to cross the *Bt* gene into a male-fertile poplar without further regulation.

Although APHIS generally indicates that it cannot "partially" deregulate an article, there is some inconsistency in the agency's documents. For example, in the case of the 1997 *Bt* herbicide-tolerant cotton petition discussed in Chapter 4, the APHIS assessment formally considered only two alternative actions: "no action," which would mean refusal to grant nonregulated status or a determination of a "finding of no significant impact" which would result in complete deregulation. APHIS found no significant impact in this case and deregulated this article in the entire United States. This resulted in a ruling that differed with one by the EPA that restricted the growing of cotton in some areas of Florida and in Hawaii. This is a problematic outcome because the U.S. Coordinated Framework for the Regulation of Biotechnology indicates that all regulatory agencies should regulate transgenic plants with similar scrutiny.

Finding 5.8: APHIS and EPA made different decisions on the planting range of *Bt* cotton in the United States.

The use of two alternatives in the APHIS assessment of this cotton petition contrasts with an APHIS environmental assessment of a *Bt* corn petition that was also reviewed in 1997 (96-317-01p). In the corn assessment, APHIS presents three alternative actions. In addition to the two actions listed in the cotton case, the additional action listed is to "approve the petition with geographical limitation." If APHIS had used this approach with cotton, it could have provided a ruling more similar to that of EPA.

Finding 5.9: APHIS appears to be inconsistent in reporting its authority to deregulate on a geographic basis in the United States.

The need for a case-by-case assessment of the environmental effects of transgenic plants has been emphasized repeatedly (NRC 1987, 1989, 2000c) because even the same transgenic trait in two different plants may pose different concerns. Because each plant/trait/environment combination is different (see Chapter 2) and because our understanding of genetics and ecology is still developing, it is important for APHIS to resist a rigid "checklist" approach suggesting specific data needed for petitions. While such data requirements might be convenient from the regulator's viewpoint, this approach could be perceived as onerous busywork from the applicant's perspective, and as an indication of bureaucratic rubber stamping from the activist communities' perspective.

Beyond simple identification information, the only information provided to APHIS should be useful in identifying and determining risk. The data required of APHIS should be associated with the product (phenotype), not the process. Details of the molecular biology of the transgenic organism are helpful only when they define the phenotype of that organism in the context of exploring a potential risk. For example, it is important that APHIS be provided with the complete sequence of expressed transgenes in order to provide assurance that only the coding sequences of interest will be expressed in the transgenic plant. Evidence to date indicates that molecular gene transfer techniques often give complicated insertions of the transgene into the genome. Transgenics can be selected, however, that give Mendelian ratios and behave genetically in a normal single-gene fashion. Nevertheless, instead of a straightforward insertion of an unmodified DNA sequence, the transgene may be inserted in a rearranged manner, as multiple copies, or both. Recent studies have shown that host DNA of unknown origin may separate transgenes or parts thereof (Takano et al. 1997; Kohli et al. 1998, 1999; Pawlowski and Somers 1998; Jackson et al. 2001). An open reading frame (ORF) can be created (Somers, 2001, University of Minnesota, personal communication) at some point in the process of transfer or insertion.

The possibility exists that an ORF could lead to expression, although the likelihood is remote that such an RNA or subsequent protein product, if produced, could have any negative consequences. As cloning technologies improve, however, the ideal would be to clone all transgene components and be certain that no unexpected gene products would be produced. At present, cloning the complex insertion(s) is not a trivial matter. However, it would seem prudent to encourage the sequencing of inserts whenever possible and to include such information for review by the appropriate regulatory agency. Striving for a gene transfer technology that provides simpler inserts is important for the future. Agrobacterium-mediated gene transfer may provide simpler insertions than particle bombardment, and new ideas on how to further reduce disruptive events in the transformation process are emerging (Koprek et al. 2001). Gene replacement technologies may become common in the future and perhaps alleviate the concern of producing new ORFs. Of course, spontaneous gene mutation happens at a low frequency, so any sequence can occasionally be modified to produce a new ORF. How often this happens in nontransgenics is not known.

This committee recommends that as much transgene sequence information as is reasonably feasible be reported as part of an application. Because the transgene region might be quite complicated, a protein product may or may not be produced, and the product may be inconsequential, any recommendation to require sequencing must remain flexible until much easier DNA sequencing and other technologies are available. For these reasons, the committee recommends that APHIS should require reporting of full DNA sequences of transgenes as they are integrated in the plant genome unless the applicant can provide scientific or technological justification not to do so.

Recommendation 5.5: APHIS should require reporting of full DNA sequences of transgenes as they are integrated in the plant genome unless the applicant can provide scientific or technological justification not to do so.

APHIS's guidelines for applications for determination of nonregulatory status provide a suggested format for applicants to follow. Some of the features of this format are described in Chapter 3. While APHIS does not exactly give the applicant a checklist, it does come close to that in some places. In reading the guidelines this committee was often unsure of why some kinds of information were requested at all and was also unsure about what kind of evidence would be needed to respond to some of the questions asked. For example, the document states that:

> Applicants must report any differences noted between transgenic and nontransgenic plants that are not directly attributed to the expected phenotype. Differences observed could include changes in leaf morphology, pollen viability, seed germination rates, changes in overwintering capabilities, insect susceptibilities, diseases resistance, yield, agronomic performance, etc. Applicants must also note the types of characteristics that were compared between transgenic and nontransgenic plants and found to be unchanged.

While this paragraph offers some general guidance, what is missing is information on the standards of evidence. Does an assessment of overwintering capability mean performing quantitative tests in a dozen distinct habitats for three years, or does it simply mean not noticing any major change in overwintering while conducting yield trials? While leaf morphology could be measured in the process of typical agronomic trials, measuring disease resistance could require special testing with a set of disease organisms under the environmental conditions best suited for a disease outbreak of each organism. The applicant could use common sense in determining what to do or could consult with APHIS personnel. The problem is that the answer to the question may depend on the APHIS scientists' own value system and on their knowledge of the phenotype in question. Of course, the importance and likelihood of unexpected changes in disease resistance will depend on the plant being examined.

A similar concern about evidential standards is raised in the case study of virus-resistant squash presented in Chapter 4. Here the question was how to determine if gene flow of the virus resistance-conferring gene from the transgenic variety to weedy relatives of squash could change their level of weediness. In this case it was clear that regulatory personnel and external experts differed in what they considered sufficient evidence.

The four case studies present in Chapter 4 also indicate that the evidence considered to be sufficient differs among petitions. As indicated above, some variation is explained by the case-specific nature of the risk assessments and the APHIS learning process from 1994 through 1997. However, some of the variation in rigor is hard to explain scientifically.

Finding 5.10: From the committee's assessment of APHIS guidelines and case studies, it appears that the agency does not provide a sufficient guide for evidential standards to applicants preparing petitions for deregulation.

It also is not clear from the committee's analysis that APHIS personnel have a system for matching evidential standards to the potential level of hazard and risk. Development of such evidential standards is not an easy task and may be best accomplished by multiple external experts in specific areas. As indicated in Chapter 2, a consensus of multiple external experts is likely to be more rigorous than the expert judgment of regulatory personnel because disagreements among external experts typically lead to more robust risk assessments. Furthermore, no single person or small groups of people is likely to have expertise in all of the areas needed to assess the risks of transgenic plants.

The heavy workload of APHIS personnel was discussed earlier. Regarding expertise among APHIS-BBEP personnel, the committee's assessment of their professional backgrounds indicates an appropriate number of the personnel with training in molecular biology and related fields, but too few with any formal training in ecology or population genetics. Greater diversity in areas of expertise could be helpful to this group.

Finding 5.11: APHIS is understaffed, and the committee questions the match between the scientific areas of staff training and staff members' responsibilities.

Recommendation 5.6: APHIS needs to improve the balance between the scientific areas of staff training and the job responsibilities of the BBEP unit by increasing staff and making appropriate hires.

Even if the APHIS staff were larger and more balanced in expertise, an external scientific peer review process would still likely raise and address issues that would be missed by the staff. Peer review could help assess whether the data presented by an applicant were adequate to address specific risks or whether more data were necessary. This would be especially important when novel traits were being assessed.

A bona fide analytical-deliberative process should involve risk identification and data evaluation by the applicant, regulators, and interested and affected parties. In many ways the squash case study exemplifies the best analytical-deliberative approach of the case studies presented. The risks are identified by the applicant and addressed. APHIS's multiple *Federal Register* announcements and requests for external expert advice created an iterative process of analysis and deliberation. The fact that the process was controversial is, in part, a product of public involvement. The committee has identified more unanswered questions for the corn and cotton case studies discussed earlier, even though they are more recent than the squash case study because, in part, they were not as controversial and were processed without public or external scientist involvement. Apparently, with more stakeholder involvement and external expert information, fewer questions are left unanswered.

The information necessary to determine whether a transgenic plant poses an environmental risk is like a moving target. Experience and information accumulated with time have changed our collective judgments of environmental impacts. For example, seven years ago when the first *Bt* corn petition was reviewed, APHIS did not consider adequately the potential effects of pollen from *Bt* corn to harm non-target Lepidoptera (see Chapter 4). Publication of a paper showing that *Bt* pollen was a hazard to monarch butterfly larvae aroused great public concern (Losey et al. 1999). Recent studies now show that, although *Bt* pollen is hazardous to monarch butterfly larvae, the risk is likely to be low but more research is still needed (BOX 2.1).

Changes in both scientific information and social values affect the amount of attention given to specific risks. As discussed in Chapter 1, society's view of the interaction between agricultural and nonagricultural systems has been changing, and more attention is now focused on these interactions. One issue that has recently been getting more attention is the effects of large-scale planting of transgenic crops. There is concern that some environmental effects would not necessarily be detected in small-scale field tests. Indeed, there is a general problem of implicitly extrapolating from the results of small-scale field trials on the order of tens of acres to potential effects of large-scale commercial "grow outs" on the order of millions of acres. The committee did not encounter questions about such changes in scale in APHIS assessments, but large-scale changes are the very kind of environmental changes that make up the primary impacts of modern agriculture (see Chapter 1).

Finding 5.12: APHIS assessments of petitions for deregulation are largely based on environmental effects considered at small spatial scales. Potential effects from scale-up associated with commercialization are rarely considered.

The focus on small-scale field and laboratory tests is evident in the types of tests typically conducted to assess impacts on non-target organisms. For crops that produce *Bt* toxins, the general types of tests used to assess non-target effects are laboratory toxicology tests similar to those used to assess conventional insecticides. The organism being tested is exposed to the toxin itself (or a closely related toxin) at concentrations 10 to 10,000 times higher than the organism will experience in the field. If the right organism is tested using an appropriate method of exposure (but see Hilbeck et al. 2000, NRC 2000c, Marvier 2001), such testing can be valuable. However, these elevated-dose, acute toxicity tests can miss the effects of a chemical on biological processes other than the one that causes the acute toxicity. For example, testing of many conventional pesticides at maximum tolerated doses did not reveal their estrogenic activity, which can be seen at much lower doses.

Finding 5.13: The "toxicology-type" risk assessments used by petitioners are useful but not sufficient to assess the non-target risks of pesticidal crops.

In addition to laboratory testing, there have been some short-term, small-scale field tests that have examined the effects of transgenic crops on invertebrate biodiversity. It would be possible for applicants to conduct more comprehensive field evaluations of transgenic plants for environmental effects prior to petitioning for nonregulated status than is currently done. A number of the case studies in Chapter 4 point out specific ways that these studies could be improved. However, even more comprehensive and relevant precommercialization field testing would still be limited to detecting effects on organism abundances and field characteristics that occur on small-time and spatial scales.

Therefore, there is a need for a system of testing and monitoring after commercialization, when large-scale plantings begin. Development of such postcommercialization assessments is important, complex, and likely to be quite expensive. The next chapter is devoted to examining the conceptual and practical aspects of developing such postcommercialization assessments.

Changes in APHIS Oversight over Time

As emphasized repeatedly throughout this report, the regulation of transgenic organisms is relatively new, and APHIS should be commended for being the first regulatory agency in the world to develop a regulatory framework for the oversight of transgenic organisms. Being first, however, also has its disadvantages because there were no prior regulatory models specific to transgenic crops from which to adapt the system of oversight. Consequently, APHIS-BBEP has had to adapt its procedures at the same time it was creating them—a challenging and sometimes unsettling process.

For example, during 1987, the first year of operation under the newly approved regulatory system, only a handful of permits were received, and each could receive considerable attention from the staff. By 1991, APHIS-BBEP was receiving literally hundreds of permit applications, which greatly stressed the agency's technical capacity, and the number of petitions was anticipated to increase in the future. Moreover, APHIS-BBEP found that the majority of the applications concentrated on corn, soybean, cotton, potato, tomato, and tobacco. In addition, it was found that confinement procedures required for each of these crops had many similarities. Consequently, in 1992 APHIS-BBEP proposed the development of a notification system so that these common applications, which had already undergone an environmental review under permitting, could receive less oversight. This system was implemented in 1993 and is the cornerstone of the APHIS regulatory process. In 1997 they expanded this notification system, which allows many transgenic plants to be planted without any environmental review. The committee discussed some of the scientific concerns and inconsistencies associated with this expanded process. This notification system provides great regulatory flexibility for APHIS and in principle has considerable scientific validity. The 1993 system, in particular, is an outstanding example of how a regulatory agency can learn from experience and adapt its regulatory procedures.

As discussed in Chapter 2, transgenic crops have been assessed in the United States by APHIS assuming a static non-adaptive risk analysis framework, without systematic regard to the potential for mistakes, in either management or risk assessment. Certainly the system is designed to limit mistakes, but they do occur, and their significance is unknown outside the specific context in which they occurred.

The committee discussed later in Chapter 2 several models that could be used to formally supplement the management framework in APHIS's risk assessment practices. One such model—the fault tree-analysis—could be used to assess risks that could occur through failure of the current oversight system. Conducting such an analysis would provide APHIS with considerable information to improve its regulatory procedures and could contribute to a more formal methodology for a learning system. One caveat regarding these models is that the cost of such an analysis is unknown, but might be high.

Recommendation 5.7: APHIS needs to formalize its learning process.

CONCLUSION

The best regulatory decision making depends on using the best information available. The committee has identified two interweaving factors that could increase the amount of information available for the regulation of transgenic plants. First, it has identified that the flow of information to and from external scientists during the APHIS decision-making process could be improved if external input were more actively sought by the agency and if the impediments to flow of data by "confidential business information" were reduced. Second, the committee perceives that the small APHIS staff and their high workload precludes the opportunity to develop the flexibility and breadth necessary to deal with the complex, diverse, and evolving challenges of a growing workload, new products, new environmental questions, and increase in the spatial scale of commercial production.

6

Postcommercialization Testing and Monitoring for Environmental Effects of Transgenic Plants

INTRODUCTION

As indicated in Chapters 2 and 5, short-term experiments and general characterization of plant traits may not pick up all environmental effects of transgenic crop plants. It is therefore important to conduct postcommercialization testing to determine if the precommercialization testing protocols adequately assessed risks (i.e., validation of precommercialization decisions). It also is important to set up long-term, postcommercialization monitoring programs to record trends in predicted effects, and to detect effects that were not predicted by precommercialization testing. Taken to an extreme, postcommercialization testing and monitoring could be prohibitively expensive, and if not carefully conceived, could lead to collection of uninterpretable data. This chapter explains the logic behind validation and monitoring programs, describes the status of current environmental monitoring programs, and makes suggestions for the general structuring of validation and monitoring programs. It is beyond the scope of the committee to offer suggestions for development of specific programs for postcommercialization testing and monitoring.

Postcommercialization testing or validation programs are an essential part of any quality control program. Whether it be an automobile part or a new hotel service, the producer does testing after commercialization to make sure that the precommercialization quality control program was effective. Precommercialization testing for environmental effects of new crop varieties is a new endeavor, so follow-up validation testing seems essential.

Monitoring of past and present status or trends in quantity and quality of a resource has often proven essential in making decisions regarding the management of that resource. As a very familiar example, a storeowner makes ordering decisions based on monitoring of inventory, and on factors such as past experience with consumer behavior, upcoming holidays, and time for the delivery of goods. Management abilities and predictive power for anticipating future needs improve with experience (or as data accumulate). Over time, changes in consumer preferences, and changes associated with technological advances in the delivery systems affect how storeowners manage their inventories. Models used to forecast climate, weather, or performance of stocks, are based on the sophisticated use of data on key parameters accumulated over time. Models improve as data series become longer, however the intrinsic degrees of uncertainty in these systems will never allow perfect predictions so monitoring will always be needed. In the economic world, productivity of workers, unemployment rates, stock indices, mortgage rates, and other variables are monitored and this information is used to predict and manage the local, regional, national and international economy. In contrast, we do not have reliable statistics or long-term monitoring indicators for most of the nations' biological resources (NRC 2000a).

The charge to this committee as it pertains to environmental monitoring is to (1) evaluate the need for and approaches to environmental monitoring and validation processes and, if deemed necessary, to include recommendations for postcommercialization monitoring of transgenic plants and (2) provide guidance on the assessment of non-target effects, appropriate tests for environmental evaluation, and assessments of cumulative effects on agricultural and nonagricultural environments for transgenic plants.

THEORETICAL JUSTIFICATION FOR MONITORING AND VALIDATION AFTER COMMERCIALIZATION OF TRANSGENIC CROPS

Precommercial risk analysis has several inherent weaknesses. In general, small-scale precommercialization field experiments are not sensitive enough to detect anything but large effects. For any such experiment there will be some limit to what can be detected, and this limit will be rather high because the natural variability from one experimental replication to another is large. For example, in estimating the yield of a corn variety—a commercially important agronomic trait—it is necessary to run several hundred-yield trials to detect significant increases in yield. In the U.S. Corn Belt, corn yields may be greater than 150 bushels per acre. A variety that yields five additional bushels per acre is a significantly more productive one, which at today's depressed corn price of about $2 per bushel on a 500-acre corn farm would provide an extra $5,000 in gross income. This is only a 3% increase over previous varieties and yet requires hundreds of trials to demonstrate conclusively.

There is no environmental risk of any agricultural technology that will be tested in hundreds of experimental field trials prior to commercialization. Small-scale field trials will readily detect order-of-magnitude differences in an ecological effect, but smaller differences will be difficult to document. To illustrate this problem, a prospective power analysis (Oehlert 2000) can be conducted on published insect density data to model detection of potential non-target effects of transgenic crops. The population density of a pest insect—the European corn borer—was measured in three environments replicated four times each (Andow and Ostlie 1990). To detect statistically a 10% difference in population density among the experimental environments given the observed variation in density among replicates, it would be necessary to examine 134 replicates of the experiment. Thus, unless the effect of the transgenic crop were considerably larger (a twofold difference in population density could be detected with 10 replications), a small-scale field experiment is unlikely to detect 10% population reductions in non-target species. Yet ecological effects of 10% can be significant. For an endangered species, a 10% difference in survival (note that the committee is speaking of "survival"—not "density") could mean the difference between recovery and extinction. Thus, a small experimental effect does not imply a small ecological effect, and the magnitude of a significant ecological effect may be no different from that of a significant commercial effect.

Postcommercialization testing is a typical component of most quality control programs and is appropriate for determining whether the precommercialization tests of transgenic plants were effective. Precommercialization testing of transgenic plants has involved the adoption of testing protocols used for other types of products such as synthetic pesticides. Because it is not clear that these protocols are completely appropriate for transgenic plants, postcommercialization validation testing is especially important. If evidence is collected over time that confirms the accuracy of precommercialization testing protocols, the need for validation testing would decrease, but due to the uniqueness of new products its utility is unlikely to disappear.

Even if the precommercialization risk analysis process is validated so it effectively identified and mitigated all order-of-magnitude ecological effects, low-probability events and low-magnitude effects would likely escape detection and management. As the transgenic crop is commercialized to larger spatial and temporal scales, it may become possible to observe smaller and less frequent ecological risks as long as there is a system of monitoring to look for them.

One type of potential adverse ecological effects of transgenic crops that may be difficult to detect is their potential for invasion of neighboring ecosystems. Short-term, spatially limited field trials are poor predictors of environmental impacts of invasions (Kareiva et al. 1996). Small-scale field trials conducted as part of a permit or petition for deregulation of a transgenic crop will ultimately have little predictive power regarding potential ecosystem effects or potential for invasiveness. In an exercise to explore the predictive power of a large dataset comparing potential invasiveness of transgenic and nontransgenic canola over three years in the United Kingdom (Crawley et al. 1993), Kareiva et al. (1996) found that results varied depending on how many sites and years were incorporated into the analyses. Years were more important than sites, but the magnitude of errors often exceeded 100%, making use of long-term data imperative to increase predictive power. Kareiva et al. (1996) conclude, "we have so little faith in models and short-term experiments regarding predictions about invasions, that we advocate extensive monitoring of any introduced [transgenic crop] with any ecologically relevant traits (such as disease resistance, herbivore tolerance, and so forth)."

A second reason that ecological monitoring is needed after commercialization is that ecosystems are complex. As noted above, this complexity stems from interannual variations and indirect effects. Because laboratory and small-scale field experiments do not adequately replicate all the interactions that occur in an ecosystem, the only way to observe the full range of ecological effects of a transgenic crop is to observe it in actual ecosystems. Some of these effects cannot be predicted beforehand, so ecological monitoring will be necessary to detect any adverse ecological effects.

Social science provides an additional rationale for postcommercialization monitoring. From a social-psychological perspective, a compelling reason to monitor is that the public wants it. Rigorous monitoring assures the public that their concerns are being addressed seriously and reassures the public that scientists are being careful to keep the risks low. While it would be irresponsible to develop a large monitoring program for the sole purpose of reassuring the public, ignoring public concerns is also irresponsible.

Finding 6.1: There are several compelling theoretical arguments for the need for ecological monitoring and validation after commercialization of genetically modified organisms. These needs range from ecological to social and from the specific to the general. It is likely

that different kinds of monitoring will be needed to meet these various needs.

POSTCOMMERCIALIZATION VALIDATION TESTING

As indicated above, postcommercialization testing should be used to determine the effectiveness of the precommercialization risk assessments. Each of the four major categories of risk associated with transgenic plants that are examined before commercialization can be reexamined after commercialization (these are risks associated with movement of transgenes, impacts of the whole plant through escape and through impacts on agricultural practices, non-target effects, and resistance evolution). The specific approaches for validation will depend on the risk examined, and the intensity of testing may depend on the extent of the potential risk. Validation testing must always be based on testing specific hypotheses related to the accuracy and adequacy of precommercialization testing.

In some past cases, precommercialization tests have identified potential risks and restrictions that have been imposed on commercialization to limit these risks. One example is the risk of insect evolving resistance to *Bt*-producing transgenic crops. In this case precommercialization testing demonstrated that pest species had the genetic capacity to evolve resistance to *Bt* toxins (Tabashnik 1994), and in some cases the frequency of specific resistance genes was estimated (for example, *Bt*; Bentur et al. 2000). During the precommercialization period a number of biological attributes of target pests were estimated, and these estimates were used to develop simulation models that projected the dynamics of resistance evolution in the pest (Alstad and Andow 1995). Results from the simulation models *and* other sources were then used to determine conditions for commercialization that would limit the risks of resistance evolution (EPA 1995). At the time of commercialization there were still uncertainties regarding some of the key parameters used in the simulation models, and incorrect parameter estimates could result in inappropriately strict or lax restrictions on commercialization. Two types of postcommercialization testing could improve the match between risks and restrictions. First, further research to better estimate biological parameters of the system could increase the rigor of model predictions, and second, monitoring of changes in the frequency of resistance genes in the pest could be used to determine if evolution was proceeding at an acceptably slow rate. If information from parameter estimates or monitoring indicated that the precommercialization decisions about restrictions were inappropriate, adjustments to the conditions could be made.

In the case of risks due to non-target effects, there are limits to precommercialization tests because they are restricted to testing a small set of organisms over a short time period, on a small spatial scale (Chapter 5). Once the transgenic crop is marketed there is an opportunity for large-scale, multi-year testing to determine the adequacy of the precommercialization testing. Comparisons of the population dynamics of non-target organisms in large (about 100 acre) paired fields (transgenic versus near isolines) and surrounding natural habitats in a number of locations would be one approach to such validation testing. If such paired fields were repeatedly planted to the same crop variety for a number of years it would be possible to examine population trends in non-target organisms that would be impossible to examine at a smaller spatial and temporal scale. The committee recognizes that even the 100-acre scale would be insufficient for examining organisms that move long distances. For such organisms it would be necessary to utilize epidemiological approaches that examined relationships between the intensity of planting of a specific crop variety in a locality, and the population dynamics of these non-target organisms. At this level, a gray area develops between what might be considered a postcommercialization validation test and a monitoring program. While this committee treats validation testing and monitoring separately in this report because of the distinct and complex type of infrastructure needed for monitoring, it is important to recognize that these two approaches are used to accomplish broadly similar goals.

Recommendation 6.1: Postcommercialization validation testing should be used to assess the adequacy of precommercialization environmental testing. This validation testing must always involve testing specific hypotheses related to the accuracy and adequacy of precommercialization testing.

Recommendation 6.2: Postcommercialization validation testing should be conducted at spatial scales appropriate to evaluate environmental changes in both agricultural and more natural ecosystems.

A funding mechanism for postcommercialization testing will need to be established. It would be preferable to have such testing involve public sector scientists to help alleviate real or perceived conflicts of interest. Although validation testing is costly, the recent postcommercialization testing for impacts of *Bt* corn on monarch butterflies (Chapter 2) demonstrates that such testing is feasible.

The USDA Biotechnology Risk Assessment Research Grants Program is small ($1.9 million for fiscal year 2000) (USDA 2001a) and is directed primarily toward small-scale hazard identification. The objectives of this program expanded recently to include risk issues associated with commercialization. The USDA's Initiative for Future Agriculture and Food

Systems (IFAFS) program has established a section concentrating on Biotechnology Risk Management ($3.4 million for fiscal year 2000, of which approximately $2.75 million is dedicated to Ecological Risk Management; USDA 2001b), but this may not be a recurring grant program within USDA. The budgets for these programs are small and the needs are so diverse, that many deserving issues are not funded. Validation testing grant proposals could fall between the main objectives of both of these programs.

Present public research programs, such as the Biotechnology Risk Assessment and Risk Management Programs will need to be expanded substantially to meet this need for postcommercialization testing. These grants may need to last longer than the average grant since many effects may not be seen until the crop has been grown for several years at the commercialized scale. In addition, these programs need to identify, clearly, program objectives that encourage research related to long-term monitoring. Such postcommercialization validation testing with well-crafted controls can provide solid experimental research data to form the scientific basis for decisions. The peer review process associated with such grants plays an important role in ensuring proper design of experiments. Some validation testing requires that the same location be examined for a long period of time and this may require specific targeting.

Recommendation 6.3: Present public research programs, such as in Biotechnology Risk Assessment and Risk Management, will need to be expanded substantially.

STATUS OF LONG-TERM ENVIRONMENTAL MONITORING IN THE UNITED STATES

A multitude of individuals and private, local, state, and federal agencies manage the country's abiotic (soil, water, air, nutrients, etc.) and biotic (organismal diversity) resources. Effective resource management relies on accurate and timely information on the extent, condition, and productivity of these resources and how they respond to management. Although historic calls for monitoring in natural areas such as in the national park system were made as early as the 1930s (Wright et al. 1933, Wright and Thompson 1935), it took until 1990 before the National Park Service initiated a natural resource inventory and monitoring program in select parks to aid in fulfilling its mission (Woodward et al. 1999). Many individuals and agencies are involved in long-term monitoring and research. Many formal censuses of animal species are regional in scale and short term in duration. Annual waterfowl surveys conducted to set hunting seasons and bag limits are a notable exception. Waterfowl populations were declining rapidly at the end of the nineteenth century and for their protection the Migratory Bird Treaty Act became law in 1918 (Anderson and Henny 1972). Over time the methods used to estimate annual production and harvest became more sophisticated and now include aerial surveys of breeding and wintering grounds, banding of individuals, reporting of harvests, and use of computer modeling (Shaeffer and Malecki 1996, Shaeffer 1998). Data collected in the summer are immediately used to develop harvest regulations for the fall, and this approach has proven extremely successful for managing waterfowl.

Most state natural resource, natural heritage, and fish and game departments conduct censuses of game species (fish, birds, and mammals). All officially endangered species are periodically censused. Other examples of long-term datasets include the Christmas Bird Count, the North American Breeding Bird Survey, and the Annual 4th of July Butterfly Count. However, a review of long-term programs (Woodward et al. 1999) found that vague objectives, a piecemeal selection of indicators, and poor linkages between monitoring projects and the decision-making process plague such programs. In addition, lack of consistent standards for data collection across agencies makes it difficult to increase statistical power by combining data.

Monitoring programs associated with agriculture have a more extensive history. One of the oldest programs is the National Agricultural Statistics Service (NASS), which has collected agricultural statistics for over 160 years. NASS's charge has evolved over time, and today it collects a substantial and diverse set of agricultural data including crop acreage, farm expenditures, harvest yields, livestock inventories, and environmental data such as land usage and chemical use on those lands.

The Dust Bowl of the 1930s and loss of valuable land to erosion triggered the beginning of coordinated data-gathering efforts on the status of agricultural lands in the United States by the U.S. Department of Agriculture's Soil Conservation Service (now the Natural Resources Conservation Service, NRCS). Over time the inventory has expanded from the original purpose of monitoring soil erosion in the 1930s to include monitoring of clean water, prime farmland, wetlands, wildlife habitat, and environmental effects of agriculture. At present the NRCS, (through the Rural Development Act of 1972, the Soil and Water Resources Conservation Act of 1977, and other supporting legislation) is charged with assessing the status, condition, and trends of soils, water, and related resources on nonfederal lands in intervals not to exceed five years (Nusser and Goebel 1997). These data now form part of the National Resources Inventory (NRI) program. A constant feature of the sample surveys is the use of numerous measures and the reliance on detection over time through repeated visits to the same points. The longitudinal survey encompasses a network of 800,000 sampling points in primary sampling units (PSUs) across the United States. Each PSU is a 64.75 ha (160 acres) in area, and 0.8 km (0.5 miles) at each side (some PSU sizes vary; see Nusser and Goebel 1997 for details). Within each PSU, three sampling points are selected according to a restricted randomization procedure. For the comprehensive five-year NRI survey, sampling rates across the country generally range from 2 to 6% of the land area, but sampling rates increase for special studies. Statistical techniques and data collection protocols have evolved as inventory goals have become broader and more sophisticated. In recent years the NRI has conducted special studies to investigate such topics as changes in wetlands, erosion rates, and changes in field practices. Multidisciplinary teams using remote sensing techniques, and geographic information systems are now used for data collection.

Long-term monitoring activities of the NRI, however, have not been able to provide sufficient or relevant data for effective natural resource and environmental assessments or management (Goebel 1998). NRI data do not provide information about the presence or abundance of any plant or animal species (other than crop species) and thus fall short of a comprehensive assessment of natural resources (see BOX 6.1).

Despite the information available from the NRI and NASS, there are substantial gaps in current monitoring data from agricultural systems in the United States (The Heinz Center 1999). These gaps are particularly evident with regard to many of the ecological interactions that are important for assessing the long-term productivity and sustainability of agricul-

BOX 6.1
Information Currently Contained in the NRI

- Soil characteristics and interpretations (such as slope, depth, land capability class/subclass, prime farmland, salinity or acidity, flooding frequency)
- Earth cover (such as trees, shrubs, grass)
- Land cover and use (such as crop type, grazing, recreation)
- Erosion (such as sheet and rill, wind)
- Vegetative conditions (such as range condition and species, wetlands)
- Conservation treatment needs (such as erosion control, irrigation management, forage)
- Potential for cropland conversion
- Extent of urban land
- Habitat diversity
- Cover maintained under the Conservation Reserve Program (where applicable)

tural practices. For example, there are no consistent or comprehensive data available on crop losses to pests and disease or on pesticide resistance. There are little if any consistent or comprehensive data on soil qualities such as salinity, organic matter, and compaction. The same can be said about the status and trends of unmanaged pollinator populations that are vital to many crops. These gaps and many others will need to be addressed if monitoring of new agricultural practices and their impacts, such as the introduction and use of genetically modified crop varieties, are to be seriously considered. To study associations between the intensity of use of a transgenic plant and changes in biological indicators, monitoring will need to be conducted at a county or township level in order to gather statistically useful data.

Finding 6.2: Our ability to assess the impacts of large-scale planting of transgenic crops is hampered in part by the lack of baseline or comparative data on the environmental impacts of agriculture.

Finding 6.3: The data provided through the NRI or the NASS are not sufficiently detailed to allow an independent assessment of the environmental effects of transgenic plants. Data need to be available on at least the county or township level to become meaningful. More detailed information is preferable to be able to make inferences about trends in areas with or without commercialization of transgenic plants.

Long-term monitoring programs of federal agencies were often established in response to legislative mandates (Endangered Species Act, Clean Water Act, Biomonitoring of Environmental Status and Trends Program, and the Environmental Monitoring and Assessment Program), yet information about biological diversity or individual species is patchy at best. For example, in its latest strategic plan, the U.S. Forest Service (USDA 2000) has identified species inventory and monitoring as key components to its decision-making processes for natural resource management plans. The improvement and integration of information systems, data structures, and information management as well as the development of indicators and monitoring protocols are expected to lead to improved stewardship. A milestone for fiscal year 2006 is the establishment of measurable objectives and monitoring programs for populations, habitats, and/or ecological conditions for threatened and endangered species, other species for which there are viability concerns, and other management indicator species/focal species. Of the 19 species explicitly mentioned to be included in monitoring programs are eight bird, five fish, three mammal, two plant, and one frog species or subspecies. The strategic plan recognizes that it may take many years before trends resulting from management strategies can be separated from population fluctuations caused by other influences.

Two informative examples of the power and usefulness of monitoring come from the need for sustainable management of fisheries and waterfowl species, which has resulted in the design of standardized long-term monitoring programs. The National Marine Fisheries Service spent some $28.8 million on ship time to collect fisheries data (excluding personnel and analyses costs), $3.9 million for recreational monitoring, $9.2 million on observer programs (with another $10 million subsidy from private industry), and $2.8 million on a vessel monitoring system program in 1999 alone. The overall data collection expenditures approximate $45 million annually, while the total (commercial and recreational) value of fisheries' harvests, including economic effects, are estimated at $45.7 billion annually (NRC 2000a).

The other example is that of the U.S. Fish and Wildlife Service's Office of Migratory Bird Management, which has developed a continental monitoring program to provide information for sustainable harvest of waterfowl. As indicated earlier, each year management decisions such as season dates and harvest limits are set based on models incorporating information on breeding pairs, weather, and the age structure of populations (Shaeffer and Malecki 1996, Shaeffer 1998). Hunters (via their purchase of duck stamps) pay for many of these monitoring programs, and spend about $1 billion for hunting and related activities. These two programs illustrate that monitoring with specific goals to (1) provide information to set harvest limits and (2) provide information about the response of the resource to management actions can be a powerful tool to manage resources. In particular, the waterfowl monitoring efforts illustrate the use of conceptual models for evaluating factors regulating waterfowl populations (such as snowfall for Canadian geese on breeding grounds or population structure and precipitation for mallards; Shaeffer and Malecki 1996, Shaeffer 1998).

Long-term monitoring provides the necessary data to develop indicators and to model annual recruitment that are then easily measured and used in management decisions. Long-term monitoring data also provide baseline information necessary to evaluate whether changes in management or other environmental variables result in population changes. However, for most biological resources such long-term data do not exist.

Finding 6.4: The United States does not have in place an adequate environmental monitoring program for agricultural and natural ecosystems that would permit assessment of the status and trends of the nation's biological resources.

Selection of Appropriate Variables to Monitor

The real challenge in developing a long-term monitoring program is in determining what are the most important variables to monitor and how to monitor them in a cost-efficient manner. The natural world is itself variable and populations of organisms fluctuate widely in abundance and distribution even in relatively short time spans. Examples of such fluctuations are the cycles of many forest pests such as tent caterpillars or the lemming cycle. The ability to identify and separate ecological effects attributable to chemical, physical, or biological perturbations from inherent variability is central to characterizing effects and estimating risks in ecological systems. The frequency and amplitude of such oscillations make the development of sampling designs to detect population trends due to other factors challenging. "Choosing from the myriad of potential parameters to monitor is truly a daunting task for managers and biologists. Literally, any biotic or abiotic feature of an ecosystem can be monitored. Moreover, changes in natural resources can manifest at spatial scales from the individual to the landscape, biologic scales from genetic to community, or temporal scales from a few milliseconds to millennia" (Woodward et al. 1999).

Arguing with confidence that environmental conditions are better, worse, or just different owing to changes in crop varieties or cultivation practices is impossible unless natural oscillations (due, for example, to fluctuations in climate) can be separated from changes in the abundance of species and ecosystem function caused by chemical, physical, or biological stressors. The identification of natural fluctuations is complicated because information is lacking about causes and consequences of population oscillations for most organisms. Recent evidence on large population fluctuations caused by the North Atlantic Oscillation and the El Niño Southern Oscillation in birds, insects, and mammals (Sillett et al. 2000, Mysterud et al. 2001) demonstrate the importance of climate variability and long-term data collection. Historical information about diversity or disturbance can also be valuable in explaining the response of different communities to the same perturbation—that is, the invasion of a species, a drought or flood, and so forth. The sequence in which disturbances occur may lead systems to very different succession trajectories (Fukami 2001), again highlighting the need for long-term data collections.

Another challenge in determining what to monitor comes from the potential for strong indirect effects in ecosystems. Ecosystems can be viewed as aggregates of populations functioning together—however, ecosystems also have unique features (spatial structure, diversity, etc.) and function (nutrient cycling, food web relationships, etc.) that must be considered when assessing changes associated with the commercialization of transgenic plants. Outcomes of perturbations are difficult to predict due to the high degree of indeterminacy in the strength and direction of ecological interactions (Attayde and Hansson 2001). A broad message from community ecology in recent decades is that indirect interactions among species are pervasive in natural communities (Holt and Hochberg 2001). Such interactions are often quantitatively as important as direct trophic or interference interactions. It is therefore possible that a transgenic plant that is toxic to a single soil insect species could affect the diversity of soil microbe species. Addition or removal of individual species have both been found to cause dramatic changes in communities, and recent evidence suggests that even species with "weak" interactions (as opposed to keystone species) may play a significant role in stabilizing communities (Berlow 1999). Indirect interactions arise because most species live in a complex web of interactions, and in principle this makes it difficult to predict the response of even well-understood systems to environmental change (Yodzis 1988, Polis and Strong 1996). The prevalence of indirect effects in natural communities limits the accuracy of even the most basic predictions (e.g., the addition of a predator will reduce the population of prey; Attayde and Hansson 2001).

A central goal for postcommercialization monitoring is to understand (and potentially predict) the outcome of species interactions in agricultural and natural communities. However, understanding and predicting the outcomes of species interactions require knowledge about which species are dynamically coupled, either indirectly or through intermediate species (Attayde and Hansson 2001).

DEVELOPMENT OF MONITORING PROGRAMS FOR TRANSGENIC CROPS

The effects of commercialization of transgenic crops can range from a change in the state or dynamics of an organism, a population, or an ecological system. These effects can be characterized as either direct or indirect. Direct effects involve interactions of the transgenic plant with another organism (e.g., direct killing of an insect pest feeding on a *Bt* crop). Indirect effects are interactions that result from direct effects. For example, herbicide treatment of resistant crop varieties is intended to reduce weed infestation (a direct effect), but reductions in weeds may reduce populations of species feeding on the weeds (an indirect effect).

Many different types of metrics can be used to assess the impacts of transgenic plants or invasive species on populations and ecosystems (Parker et al. 1999). This committee recommends that a two-part approach be used to assess potential environmental impacts of transgenic crops.

The first element of this approach involves *trained-observer monitoring* involving technical personnel in agricultural and natural areas management. Trained observers are frequently in the field and could be important in detecting environmental effects. Their observations would enter into a nonmonitoring process of validation. Laboratory and field testing would be conducted to validate scientifically the observations made by the observers. If the effect is validated and shown to be possible to occur in the field, it will be possible to generate scientific hypotheses for how the effect would be observed in the field. The second leg of the monitoring system would be *long-term monitoring* programs using bioindicators.

Recommendation 6.4: The committee recommends a two-part approach to postcommercialization monitoring of transgenic crops: (1) *trained-observer monitoring* involving technical personnel in agricultural and natural areas management, and (2) *long-term monitoring* programs using appropriate indicators (see below).

TRAINED-OBSERVER MONITORING

Need

Although some environmental effects of specific transgenic crops might be predicted, many effects (intermittent, low-magnitude, or cumulative effects) may remain undetected during precommercial field trials, and it might be possible to detect them only after scale-up to commercial plantings. A system of monitoring based on trained observers can be used to help detect such effects. Monitoring for novel or acute ecological effects, such as detection of a new nonindigenous invasive species, has relied on such a system for several decades (Kim 1983). Detecting new effects of transgenic plants, such as a pathogen of herbicide-tolerant soybean or a decline in bird populations in areas planted with certain transgenic crops, would also require a network of trained observers who have an incentive for detecting such unexpected effects.

One concern that emerges from lessons learned from introduction of the ornamental plant known as kudzu (*Pueraria montana*) into the United States is that widespread planting may increase the likelihood of invasiveness. Ironically the initial spread of this species was linked to the commercial seed trade—plants were grown as ornamentals as well as forage crop (Winberry and Jones 1974, Mack 1996) in addition to its use for erosion control. Good surveillance and monitoring are essential for detecting early invasive tendencies and for allowing early eradication or control measures should they become necessary.

Logistics

Such a network of trained people would be prohibitively expensive to maintain for the sole purpose of detecting the effects of commercialization of transgenic plants. However, trained professional and volunteer networks already exist in agricultural and natural areas, and it would be useful to determine how to integrate monitoring with preexisting networks.

In the United States, the Agricultural Extension Service and the Animal and Plant Health Inspection Service (APHIS) have provided a detection network for invasive species. In addition, crop consultants and farmers themselves may notice environmental changes associated with transgenic crops. Building and maintaining the capacity to detect new acute ecological effects is paramount to an effective monitoring system. In natural areas, resource professionals in federal and state agencies (e.g., Fish and Wildlife Service, state departments of natural resources, natural heritage programs, National Park Service, Bureau of Land Management, Department of Defense) already assess and inventory or observe native species and ecosystems. The same is done in civil society organizations (such as the Nature Conservancy stewards, Ducks Unlimited, Trout Unlimited, land trusts) and by professional and volunteer naturalists (hikers, birders, entomologists, botanists organized to form exotic pest plant councils, for example, native plant societies, etc.). Any deviations or unusual occurrences can be reported and then verified.

A critical need for assessing whether such reports are associated with the commercialization of transgenic crops will be access to detailed information (at the township level at least) on the spatial and temporal patterns of planting of specific transgenic varieties (see below). Systematic monitoring in areas planted with the same variety is needed to gather data used to build an understanding of processes and changes in the field and surrounding natural areas. Ecological effects with a low frequency of occurrence will probably occur in a spatially heterogeneous pattern. Monitoring for these effects can be improved by taking this spatial heterogeneity into account. For example, because the probability of a change being detected is likely to correlate spatially with the amount of transgenic crop planted in an area, spatial maps of planting density can lead monitoring personnel to the optimal monitoring locations.

Needed Training

Many of these networks of people are already stretched thin by their present responsibilities, so it is important that any additional activities associated with detecting potential environmental effects of transgenic crops be integrated into their ongoing activities. For example, the Agricultural Extension Service maintains offices in many counties. These extension service personnel have many responsibilities and spend considerable time observing agriculture. They have little spare time, and it would be ineffective to mandate that they spend additional time specifically looking for environmental effects of transgenic crops. But because they are trained observers of agriculture, it is possible that while they are conducting their normal activities they could be on the watch for potential environmental effects of transgenic crops. This may be facilitated by a short half-day or one-day workshop on the potential environmental hazards of transgenic crops, so that these extension service personnel can understand better the types of effects that might occur. Crop consultants are another group of trained observers in agriculture. It may be possible to add a training module to the short courses they routinely take that would prepare them to observe environmental effects of transgenic crops.

LONG-TERM MONITORING AND THE USE OF BIOINDICATORS

Need

The release of transgenic plants is often compared to the introduction of nonindigenous species (Kareiva et al. 1996, Marvier and Kareiva 1999). Lessons from invasion biology and management of nonindigenous plants suggest that the reasons for differences in invasiveness among species are poorly understood (Williamson 1996), and that invasiveness may evolve or be delayed (Blossey and Nötzold 1995, Ellstrand and Schierenbeck 2000), often in the form of a "lag" phase between initial introduction and explosive spread (Williamson 1996). Ultimately, we have little power to predict which species will be a successful invader, which ecosystem may be particularly vulnerable to invasion (Williamson 1996, Williamson and Fitter 1996, Lonsdale 1999), or what the impact of invasions will be (Hengeveld 1999). In fact, for some of the most well-known invasive bird species, such as European starlings and the House sparrow, multiple introduction events were necessary before populations became established (Williamson 1996). Lag times in invasive species may be explained by (1) inherent factors associated with population growth and range expansion, (2) changes in environmental conditions favoring the introduced species, and (3) genetic factors (Crooks and Soulé 1999). Moreover, time-lagged effects are particularly evident after establishment of exotic perennial plants compared to annuals, and in association with range expansion of the invaders (Williamson 1996). Overall, short-term experiments (even if conducted over multiple years or a decade) cannot substitute for long-term time series observations because of the potential for lag times and the general unpredictable outcomes of altered ecosystem interactions. Long-term monitoring to assess potential changes associated with commercialization of transgenic crops is essential. The inherent difficulty in predicting indirect interactions and cumulative or synergistic effects makes the use of long-term monitoring essential in assessments of the environmental effects of new technologies.

Monitoring Transgenic Crops

Systematic monitoring of the spatial and temporal patterns of an area planted with different transgenic varieties will provide a basis for all other monitoring efforts. Ecological effects with a low frequency of occurrence will probably occur in a spatially heterogeneous pattern, and the probability of occurrence will be predicted to be proportional to the area of the transgenic crop planted. Information on the planting pattern will permit use of epidemiological methods (Waggoner and Aylor 2000) to evaluate reports received from the trained-observer monitoring system. Without this systematic monitoring data, it will not be possible to separate coincidental anecdotes from real ecological trends. The information should also be used to plan long-term monitoring sampling plans to optimize effort allocation. Moreover, in any analyses of long-term monitoring, indicator data need to be linked to the spatiotemporal planting pattern.

Annual reporting of this planting pattern is essential. The spatial pattern should be reported at as fine a spatial scale as possible because environmental effects are likely to be localized and aggregation of spatial data may eliminate the correlation between the occurrence of the crop variety and the environmental effect or may induce artificial correlations. The committee suggests that spatial resolution of planting patterns to a 6 × 6 mile grid may be sufficient, but spatial statisticians should develop the appropriate sampling scenario. Because different transformation events associated with the same trait can have different environmental effects (e.g., transgenic corn varieties Event 176 vs. *Bt*-11/Mon 810), planting patterns should be reported separately for the various transformation events, instead of aggregated across the type of trait within a crop.

Recommendation 6.5: The committee recommends that any post-commercialization monitoring program that is adopted should include monitoring of the spatial distribution of transgenic crops.

Monitoring Using Biological Indicators

The committee recognizes the difficulty in providing sufficient financial and taxonomic expertise to monitor more than a fraction of the biota occurring in agroecosystems and natural areas. Indeed, for transgenic crops, the role of monitoring in relation to the specific indicators and our capacity to monitor requires additional deliberation (BOX 6.3). The committee endorses the development of ecological indicators as proposed by the National Research Council (NRC 2000a) for both agricultural and nonagricultural environments. The development of indicators is based on the assumption that monitoring an indicator is more cost effective and accurate than monitoring individual processes or species. Using indicators will simplify what is communicated from such programs.

The NRC (2000a) has developed several criteria for evaluating ecological indicators. The criteria recognize that some ecological indicators have been less useful than anticipated because they have not been clearly linked to underlying ecological processes or because the data requirements are overly complex and extensive. The criteria provide a framework for evaluating indicators to assess the potential importance of a proposed indicator, its properties, its domain of applicability, and its limitations:

1. *General importance.* Does the indicator provide information about changes in important ecological processes?

2. *Conceptual basis.* Is the indicator based on a well-understood and generally accepted conceptual model of the ecosystem to which it is applied?

3. *Reliability.* What is the evidence that the indicator is reliable?

4. *Temporal and spatial scales.* Does the indicator inform us about national, regional, or local ecological changes? Are the changes measured likely to be short-term or long-term? Can the indicator detect changes at the appropriate spatial and temporal scales without being overwhelmed by variability?

5. *Statistical properties.* Are the statistical properties of the indicator (accuracy, sensitivity, precision, robustness) sufficiently understood that changes in its values will have clear and unambiguous meaning?

6. *Data requirements.* How many and what kinds of data are needed to estimate the indicator and to detect trends in the indicator?

7. *Skills required.* What technical and conceptual skills must the collectors of data for an indicator possess?

8. *Data quality.* Will sampling and analytical methods be documented sufficiently that future researchers will be able to understand how the indicator was estimated?

9. *Data archiving.* Data, physical samples necessary to recalibrate the entire dataset, and analytical models should be archived so that interested parties have access to the information.

10. *Robustness.* In an ecological sense, is the indicator relatively insen-

sitive to expected sources of interference, such as external ecological perturbations, and technological change in monitoring capacity?

11. *Internal compatibility.* Is the indicator compatible with any other indicator being developed or used by other nations or international groups?

12. *Costs, benefits, and cost effectiveness.* What is the cost to develop, implement, and refine the indicator? Benefits are more difficult to estimate, but what are the expected ones? Is there another method that can provide similar information at lower cost?

Details for the justification and rationale for each indicator are contained in the earlier NRC report (2000a) and the arguments will not be repeated here. For some of the recommended indicators, data are already being collected through the NRI; others, particularly the monitoring of biological diversity, will need further elaboration. This committee is not charged with and will not provide detailed guidance on organismal groups/taxa for monitoring. However, the committee encourages further research into the potential use of indicator or umbrella species to assess ecosystem health or changes associated with the commercialization of transgenic plants (see BOX 6.2). The USDA and many other federal and state agencies are committed to increasing the monitoring of biological and natural resources. A scientifically rigorous design modeled after the NRI and with cost sharing among agencies should reduce overall costs for such a program. More research is needed to identify organisms and bio-

BOX 6.2
Indicators of the Nation's Ecological Capital

The Committee to Evaluate Indicators for Monitoring Aquatic and Terrestrial Environments (NRC 2000a) chose the following as indicators of the nation's ecological capital:

- Total species diversity
- Native species diversity
- Nutrient runoff
- Soil organic matter

The committee chose as indicators of ecological functioning or performance:

- Carbon storage
- Production capacity
- Net primary productivity

The committee chose as indicators for agricultural ecosystems in particular:

- Nutrient use efficiency
- Nutrient balance

logical processes that are especially sensitive to stresses and perturbations. Future research may suggest better indicators, but their performance and reliability need to be carefully evaluated before implementing them nationally.

The committee suggests that the development of indicators used as "common currency" should be an open democratic process involving agencies, industry, and other stakeholders. Only if there is agreement of what to monitor will the results be accepted.

Recommendation 6.6: There should be an open and deliberative process involving stakeholders to establish criteria for environmental monitoring programs.

The committee follows the arguments for a common currency developed by the indicators committee (NRC 2000a):

> Indicators are more likely to be useful if they are understandable, quantifiable, and broadly applicable. They are likely to command attention if they capture changes of significance to many people in many places. Although indicators of local effects are not without value, they must be aggregated into some composite indicator if they are to serve broad policy purposes. Indicators are most policy-relevant if they are easily interpreted in terms of environmental trends or progress toward clearly articulated policy goals, and if their relevance is made clear (Landres 1992). In other words, indicators that convey information meaningful to decision makers and in a form these decision makers and the public can understand are more likely to be observed and acted on. Indicators are also more likely to be influential if they are few in numbers and capture key features of environmental systems in a highly condensed but understandable way. The manner in which data are aggregated to yield a small number of general indicators should be clear, especially to those who wish to understand how the indicators were developed. The reason for choosing indicators, and the selection criteria, should also be clear (Landres 1992).
>
> Any objective ecological indicators should be expressed numerically, so that results can be compared with those of indicators in other places and times. For the indicators to command attention and be used, the data and calculations they are based on must be credible. The choice of what indicators to use and how to define them is necessarily somewhat subjective, but the procedures for measurement and calculations associated with a particular indicator, once defined, must be clearly specified, repeatable, and as free of subjective judgments as possible. Where they are unavoidable, the sources of subjectivity should be defined and identified (Landres 1992, Susskind and Dunlap 1981). For example, the Consumer Price Index and the percent of people unemployed are calculated by well-defined rules that have been agreed on, regardless of a person's view about the value of full employment or low inflation or even the

validity of these indices. Debates about these numbers do not involve who calculated them. Similarly, ecological indicators need to be based on calculations that are well defined and agreed on.

In addition to being based on credible measurements and calculations, the choice, motivation, and interpretation of indicators should be publicly trusted for them to be of greatest use. That means that the people and organizations who produce the indicators should be generally trusted (Greenwalt 1992). The committee cannot specify the best methods for achieving this goal, but notes that in at least some cases separating the responsibility for preparing indicators from responsibility for carrying out policies based on them seems to enhance trust in the indicators. For example, the Bureau of Census has no policy-making responsibility; so, despite recent political arguments about the validity of sampling as opposed to counting everyone, the population estimates produced by the Bureau are usually trusted. Similarly, the National Weather Service has no responsibility for environmental policies, and so, beyond some scientific questions about the nature and placement of its instruments, its statistics are generally widely respected and trusted. The importance of public trust in the indicators is even more critical if ecological indicators are to be used as input for a national assessment of the state of the nation's ecosystems.

Crucially important in the data collection process is that the data satisfy standards of timeliness, level of detail, accuracy, accessibility to users, coverage or completeness, and credibility of the data collection and management processes using the data. Because data collections and use will be at various spatial scales, data system designers must allow for demands among users at these various scales. Any system must be able to deal with the needs of users to work at different spatial resolutions and different degrees of timeliness. For the National Marine Fisheries Service, credibility is a major concern. Many stakeholders mistrust data that are collected and analyzed by the same agency that makes policy recommendations, conducts stock assessments, and enforces fishery regulations. These multiple responsibilities create mistrust about the collection and use of fisheries data (NRC 2000d).

Recommendation 6.7: The committee recommends that an independent body separate from APHIS be charged with the development of a monitoring program. This monitoring program/database should allow participation by agencies, independent scientists, industry, and public-interest groups. The database depository should be available to researchers and the interested public.

Establishment of a database depository would build stronger confidence and allow access to more data if collection techniques were standardized. Major advances in monitoring could be achieved by establishing new computing and communications capabilities and by increasing integration and standardization of data management on local, regional and national bases.

Expertise for developing statistically reliable designs is available within the federal government or at universities. For example, the Resources Inventory Division of NRCS develops policy, procedures, and standards and provides guidance for natural resource data collection efforts by NRCS. The division also ensures that NRCS's data collection efforts are coordinated with other federal, state, and local, government agencies. The Federal Geographic Data Committee (FDGC) coordinates the development of the National Spatial Data Infrastructure (NSDI). The NSDI encompasses policies, standards, and procedures for organizations to cooperatively produce and share geographic data. The 17 federal agencies that make up the FGDC are developing the NSDI in cooperation with organizations from state, local, and tribal governments; the academic community; and the private sector in an effort to improve the quality of the collected information while minimizing the burden for individual agencies through the use of information technologies.

Recommendation 6.8: The establishment of long-term monitoring efforts is recommended to permit assessment of potential environmental changes associated with the commercialization of transgenic plants.

RESPONSES TO MONITORING

Need

The purpose of monitoring (using bioindicators, general surveys by trained personnel, and long-term assessments) is to allow timely and informed responses to changes in agricultural and natural ecosystems (positive or negative) associated with the commercialization of transgenic plants. Without long-term monitoring, informed management decisions are impossible to make because of uncertainties about the causes of shifts in abundance of individual species or of ecosystem function. The longer the data series from a monitoring program, the easier it is to distinguish true ecological relationships from fluctuations due to natural variability. A tradeoff between accuracy and utility develops because the longer we wait for accurate assessments, the less likely it is that responses can be developed to "restore" ecosystem function or rebuild species populations. When there is a considerable lag time before impacts become detectable, the problem of timeliness of response is exacerbated (Byers and Goldwasser 2001). Examples of such delayed impacts are extinction

BOX 6.3
Role of Monitoring

The question of what to monitor depends on the larger purposes that monitoring is intended to serve. With respect to the environmental effects of commercialization of transgenic crops, it is important to bear in mind certain points discussed in previous chapters. First, as discussed in Chapter 2, the focus should be on the environmental effects of transgenic crops rather than the transgenes themselves. These effects depend on the phenotypic characteristics of the crop rather than the fact that transgenes have been moved into the crop through rDNA techniques. Furthermore, even if transgenes move into other crops and wild relatives, such movement may or may not have further effects on such factors as crop performance, food quality, and non-target organisms. As such, the key questions that must be addressed in discussing the need and potential for monitoring have to do with our ability to detect such effects should they occur.

Social science research suggests that some members of the public may regard an unintended movement of the transgene *itself* as an environmental effect (Gaskell et al. in press). This is especially the case for focus groups conducted outside the United States, where participants are likely to interpret effects on farming practices in environmental terms. Thus, events that complicate production of crops that meet organic marketing standards are likely to be interpreted by some as environmental effects, even if they do not involve detectable effects on non-target organisms or ecosystem functioning. Others will regard the unintended presence of transgenes in a commodity as a direct economic or social effect, rather than an environmental effect, in that they affect the value of farmers' conventional and organic crops and possibly consumers' ability to find products that satisfy their preferences.

A system to test crops for the presence of transgenes, either in the field or at various points in shipment, is technically feasible. One could thus monitor nontransgenic crops for the movement of transgenes. The question is whether this is an objective that is consistent with the goals of environmental monitoring. There are both scientific and social value judgments that bear on this question. Given a clear set of criteria for the ecological health of conventional or organic agroecosystems, it might be possible to evaluate whether the presence of specific transgenes in particular farming systems is likely to have detrimental effects. There also may be specific cases, such as organic growers' use of *Bt* toxin, where monitoring for specific outcomes, such as increased resistance to *Bt* among crop pests, might prove beneficial to producers. In general, however, there are no scientifically agreed upon parameters for evaluating the ecological health of agroecosystems (Peck et al. 1998). Thus, the argument for undertaking such a system of monitoring involves an appeal to economic and social values or an interpretation of environmental effects that is broad enough to encompass many impacts that most Americans would characterize in social or economic terms.

While technically feasible, detecting low levels of gene flow into crops or noncrops may be challenging. Gene flow by pollen varies with a tremendous number of parameters, including pollen vector behavior, compatibility of the populations involved, and relative sizes of the source and sink populations (e.g., Levin 1981; Ellstrand et al. 1999). Likewise, seed dispersal can depend on many factors. Typically, the dispersal of pollen and seed follows a leptokurtic curve (Levin and Kerster 1974), such that the great majority of the propagules disperse a very short distance (on the order of meters), but the curve has a long tail (on the order of hundreds or thousands of meters). Of course, the sites to be sampled will be in that tail region of dispersal. The events in the tail are rare enough that they will not be uniformly distributed and are especially subject to idiosyncratic factors (e.g., wind direction) that will increase their patchiness in space. For monitoring to be effective, samples must be large enough to detect rare and patchy events (Marvier et al. 1999). Although flow into noncrop species may be reasonably interpreted as a necessary (but not sufficient) threshold for certain environmental effects to occur, monitoring gene flow may be of considerably less value for detecting environmental effects than direct monitoring of nontarget species or ecosystems themselves.

Scientific considerations alone do not provide a sufficient basis for general monitoring programs that would attempt to keep track of transgenes that may move either to conventional and organic crops or to noncrop plants. One should not rule out the possibility of future transgenic crops that would warrant such an approach on scientific grounds. Some of the scientific, social, and economic factors that could be considered in evaluating whether the movement of transgenes beyond the cultivars in which they are commercialized should be monitored are summarized below.

SHOULD FEDERAL AGENCIES MONITOR FOR THE MOVEMENT OF TRANSGENES?

Pro

- The identification of no transgenes in a sample might be economically beneficial to growers who seek to sell their "transgene-free" products to consumers who demand them.
- The spatial distribution of transgenes in nontransgenic crops would provide baseline information on the incidence of transgene movement in the environment.
- The incidence of transgenic plants "out of place" could test whether current cultural and genetic mechanisms for containing transgenes—motivated by concerns for protecting the environment and intellectual property—are effective.
- The monitoring activity would expand the universe of values reflected in monitoring activities and might be particularly satisfying to those who interpret movement of the transgene as itself an environmental effect.
- Monitoring could identify the loss of security of intellectual property.
- Monitoring for the unintended presence of transgenes or their products could be useful in avoiding or redressing any associated economic, legal, or public health effects.

Con

- Even an extensive monitoring system occasionally might miss transgenes and give false negatives.
- Unless bulk samples are processed or screening costs decrease, the costs of screening could be very high relative to the benefits of the monitoring program. Presently, that ratio is unknown and hard to estimate.
- Because the movement of transgenes, in itself, is not an environmental hazard, monitoring for the sake of measuring this parameter might create public confusion.
- Screening will likely depend on PCR technology. This technology is very sensitive to tiny amounts of DNA. Thus, if materials used for screening are inadequately cleaned or if the appropriate controls are not used, tiny amounts of contamination may yield false positives.
- Good intentions with screening can go wrong. A good example is that of efforts in the last century to screen African American schoolchildren for the hemoglobin S allele, which, in homozygotes, results in sickle cell anemia. This compulsory program, without appropriate public education, led to a number of ethical, social, and legal problems (NRC 1994). Therefore, screening for transgenes could stigmatize both transgenic and conventional crop production.
- Additionally, the costs of any required monitoring might prove so expensive for small growers that it might put them out of business.
- The identification of transgenes in fields where they were not intentionally grown would open up the possibility of lawsuits for protection of intellectual property.

"debts" due to habitat fragmentation or invasion of nonindigenous species, global climate change, and groundwater contamination with nitrate and pesticides (Holden et al. 1992, Phillips et al. 2000). Although monitoring may detect some unexpected effects so that action may be taken to prevent or ameliorate those effects, in other cases monitoring may detect those effects so late that environmental damage may be irreversible (e.g., extinction).

Finding 6.5: Monitoring cannot substitute for rigorous precommercialization regulatory risk assessments.

Long-term monitoring programs are expensive and labor intensive, and require standardization of monitoring units and verification of data collected. Society will benefit only if timely responses become part of the framework for evaluating the environmental impacts of transgenic plants. One category of responses to a finding of environmental hazard from a monitoring program would be the option to reassess the environmental risks associated with a specific transgenic plant implicated in the finding. Another category of responses could be the adaptive integration of knowledge gained from monitoring one transgenic cultivar in developing more appropriate evaluation protocols prior to commercialization and postcommercialization monitoring systems for future cultivars.

In general, risk analysis can improve safety using a combination of prerelease assessment and management and postrelease monitoring and mitigation. At present, APHIS has no formal process for responding to data gathered through environmental monitoring. However, some of the responses described below have already occurred to some extent (e.g., research to assess the potential risk of *Bt* corn pollen on monarch butterfly larvae) and have influenced the activities and regulatory processes of APHIS. A summary follows of the types of responses to monitoring that are essential to the objectives underlying the different types of monitoring approaches described and recommended above.

Currently there are no provisions for an APHIS regulatory response to the detection of environmental hazards after deregulation of a crop (but see Chapter 7). One of the underlying assumptions in recommending postcommercialization monitoring is that a feedback system in the regulatory process will be implemented, to respond to data accumulated during long-term monitoring. These responses may be informational or regulatory.

Examples of Responses

Validation testing could have a direct effect on regulation if, for example, testing revealed a substantial increase in the frequency of a *Bt* toxin resistance gene and a federal agency implemented a change in the recommended resistance management plan. Validation testing could be used more broadly to test for causal relationships between the distribution of transgenic crops and their hypothesized environmental effects after commercialization. For verification of precommercialization risk assessment testing, this type of testing would provide feedback on effects at larger spatial scales that would either confirm the validity of current risk assessment procedures or suggest that they should be modified. In the latter case, a formal response to the results of validation testing would be a change in the type or extent of precommercialization testing that an applicant would need to conduct before a permit would be issued or a petition approved.

Long-term monitoring of as many appropriate indicators as is reasonable (including the exact location and acreage of transgenic crops planted each year) might provide critical information on the spatial association between the intensity of use of a transgenic plant and specific environmental effects. Indicator monitoring, coupled with the potential detection of an unexpected effect through a network of observers, will allow the detection and interpretation of spatial relationships between the distribution of specific events and environmental change. Accounting for event location then allows "epidemiological"-style research methods to detect and associate specific environmental effects with specific transgenic crops. Another type of feedback or response to data resulting from indicator monitoring would simply be the provision of spatial distribution data to design additional, more focused monitoring efforts. By using spatial data we can detect quickly which changes are occurring near the use of transgenic crops and so may be resulting from these introductions. Finally, the historical placement of the various events will provide baseline data and trends against which change is assessed as part of the follow-up from potential detection of unexpected effects by observers. Of course, any regulatory response, such as disallowing the continued use of a specific transgenic crop associated with detrimental environmental consequences, would be aided by data about the status of bioindicators at the location of those events.

Any adverse effects detected through monitoring by trained observers would likely be correlational or anecdotal; it must be verified with experiments, tested in field environments to demonstrate that it could happen in such environments, and/or stimulate additional, more focused monitoring to establish the cause of these effects and their potential connection to planting of transgenic crops. Several examples are provided above in the discussion of the trained-observer monitoring system.

Thus, the typical appropriate response to effects reported by trained observers would be a formal verification process to ensure that the results of observer monitoring activities are acted upon. Of course, in some cases the observed effect may be so dramatic that it would justify immediate action, which would later be followed by a verification process. In addition to the various feedback functions described above, there should be serious consideration given for a process that allows clear regulatory responses to findings from environmental monitoring. For example, one of the possible responses should be to disallow continued planting of the transgenic crop. Such a response could consist of a two-step process. First, the problem should be identified, described, and a probationary warning issued that specifies an observation period during which, if the problem continues, planting will be disallowed. Second, during that observation period, mitigation efforts could be attempted and evaluated. At this time, the burden of proof should shift, so that one must prove that mitigation is successful or planting will be disallowed. Such a response to environmental disruption would require regulatory decision making to match different levels of risk or types of hazards to the timing of the response, mitigation and evaluation periods, the mitigation goal required, and the degree of evidence required. Under the present USDA regulatory system, such matching of a measured regulatory response to the degree of identified risk is not possible.

Recommendation 6.9: The committee recommends that a process be developed that allows clear regulatory responses to findings from environmental monitoring.

CONCLUSIONS

Sustaining the quality of the nation's ecological and natural resources requires effective management of those resources (Grumbine 1994, Olsen et al. 1999). Effective resource management relies on accurate, timely, and complete information on the extent, condition, and productivity of those resources. To obtain this information, federal, state, and local agencies have established ecological and natural resource monitoring programs (Olsen et al. 1999). Monitoring is also used in the identification and definition of environmental problems yet to be recognized or that may emerge in the future.

Major monitoring programs such as the NRI and the NASS provide valuable long-term datasets relevant to agriculture, but they are not sufficiently detailed or focused to allow an independent assessment of the environmental effects of transgenic plants. Currently the environmental monitoring of agricultural and natural ecosystems in place in the United States is inadequate for assessing the potential impacts of commercialized transgenic crops.

To be able to do so, the committee recommends that a two-part approach be used to assess the potential environmental impacts of transgenic crops after commercialization: (1) *Trained-observer monitoring* would involve technical personnel in agricultural and natural areas management making and recording observations; and (2) *long-term monitoring* would help identify and distinguish patterns of biotic and abiotic variation in natural and agricultural ecosystems from impact events due to transgenic crops. In order for the findings of such monitoring efforts to be useful, a clear and coordinated regulatory response must be in place.

7

The Future of Agricultural Biotechnology

The application of biological sciences in agriculture has become increasingly prominent in the past decade. Genes were first inserted into corn using molecular techniques in 1989, and by the late 1990s farmers were growing millions of acres of transgenic corn. Clearly, the science of biotechnology for agriculture is in its infancy, yet it shows an influence beyond its years.

The previous chapters review what is known about the environmental impact of commercialized transgenic crops and approaches for monitoring that might be adapted to screen for their unanticipated effects. One key finding is that particular phenotypic characteristics of a given transgenic plant determine its likely environmental interactions; the fact that recombinant DNA methods were used in its development only indirectly affects these interactions by influencing the phenotypic characteristics of the transgenic plant. Indeed, the significance of biotechnology for environmental risk resides primarily in the fact that a much broader array of phenotypic traits can be incorporated into crop plants than was possible about a decade ago. As such, our experience with the few herbicide-tolerant and insect- and disease-resistant varieties that have been commercialized to date provides a very limited basis for predicting questions needed to be asked when future plants with very different phenotypic traits are assessed for environmental risks.

This chapter is divided into three major sections. The first includes an overview discussion of some new kinds of transgenic crops and a selective discussion of some environmental risk issues that may be associated with the next generation of transgenic crops. The second section focuses on policy, beginning with a general discussion of the context in which environmental risk from transgenic crops should be framed and then moving on to specific topics that may arise as policies for the next generation of transgenic crops evolve. The final section is a discussion of some research needs to address future issues.

THE NEXT TRANSGENIC CROPS

This section describes anticipated future transgenic crops, including some expected to be commercialized in the next couple of years, others that may reach commercial status on a midterm horizon sometime during the next decade, and others that are mere twinkles of ideas for transgenic crops that will require research breakthroughs before they can reach fruition. As discussed in Chapter 2, it is not possible to characterize the environmental hazards that may be associated with all such crops in advance of knowledge about their phenotypic characteristics and the agricultural ecology of the settings in which they will be grown. However, the second part of this section offers a preliminary discussion of some representative environmental risk issues that may be associated with these new transgenic crops.

An Inventory of New Transgenic Crops

The first commercially produced transgenic crops were based on single-gene traits. Among these was the "Flavr-Savr" tomato, which used gene silencing to inhibit the expression of an enzyme involved in fruit ripening (Kramer and Redenbaugh 1994). The Flavr-Savr tomato was not a commercial success, but the technology was effective because the fruit not only had a slow rate of ripening but also was less susceptible to pathogen infection. Other early transgenic products were based on traits influencing agronomic performance (i.e., pathogen, insect, and herbicide resistance). The rapid and broad use by the American farmer of glyphosate-resistant soybeans and *Bt*-expressing cotton and corn attests to the commercial success of these transgenic crops (James 1998, USDA-NASS 2001).

Based on the successes of these initial transgenic crops, research laboratories throughout the world are now studying a wide variety of traits/genes that could greatly expand the spectrum of products from such plants. As was true of the first genetically engineered crops, the rate at which new transgenic traits can be expected to appear in the future depends largely on the number of genes encoding them. So traits controlled by single genes, or traits that can be reduced or eliminated by the loss of expression through gene silencing of a single gene or group of related genes, are likely to become the first available. The next logical step for expansion is the integration of single-gene traits. Crops with a set of single genes for a number of distinct traits already are in use. We can only speculate about the number of genes controlling the development and architecture of plants and the various physiological processes impacting yield. It seems safe to assume, though, that genetically complex traits will require additional years of research to understand, let alone express and regulate in a genetically engineered crop species. Nevertheless, complex traits, including those controlling adaptation to abiotic stresses, such as drought and salinity, flowering and reproduction, and hybrid vigor, are being actively investigated, and it would not be surprising if some of these could be regulated in crop plants by genetic engineering within the next 5 to 10 years.

These products have the potential to not only improve agronomic performance but also increase the nutritional value of grains consumed by humans and livestock, eliminate allergens and antinutritional factors, improve the shelf life of fruits and vegetables, and increase the concentration of vitamins and micronutrients found in seeds, creating healthier foods (Abelson and Hines 1999). While there are many different kinds of transgenic crops under development, most aim to address one of four broad social needs: improved agricultural characteristics, greater adaptation to postharvest processing practices, improved food quality and other uses of value for humans, and better mitigation of environmental pollution. Indeed, so many traits currently under investigation could become incorporated into transgenic plants that space limits consideration here to only a few.

Improved Agricultural Characteristics

Among the transgenic traits near commercial release are new *Bt* genes that provide protection against additional types of insect pests. One of these is a gene that protects corn against corn rootworm damage (Kishore and Shewmaker 1999). It is estimated that damage to corn roots by this pest result in losses approaching $1 billion annually. By reducing the impact of this pest, it is expected that not only will there be better corn yields but also better drought tolerance and fertilizer utilization due to the healthier root system. Research is also being done to genetically engineer tree crops to make them resistant to insects and herbicides and to increase their rate of growth. For example, a *Bt* gene has been inserted into hybrid poplars to protect them against defoliation by a leaf beetle. Acreage of hybrid poplars has increased because of their good wood pulp characteristics, but they have been susceptible to insect attack, which has prompted applications of insecticides. Many of these new traits for improving agricultural production on farm are ones that could have environmental impacts that are similar in kind to the present generation of transgenic crops. Their value accrues directly to the farmer and the seed company and only indirectly to other sectors of society. Their potential risks, however, are borne by a wider segment of society. Thus, risk analysis of this next generation of traits is likely to resemble present discussions and debates about biotechnology. The evaluation of these risks is likely to become more complicated and difficult as the range of transgenic crops expands from the major grain crops to the more wild and perennial plants, such as pines and poplar.

Over the long term, new knowledge regarding the physiology and development of plants and their interaction with microorganisms could eventually provide the foundation to modify plant structure and reproduction. It may become possible to genetically engineer crop plants that are more tolerant to drought, salinity, and other abiotic stresses (see below); that are able to grow more efficiently in the acidic, aluminum-containing soils found in tropical areas (Herrera-Estrella 1999); that can compete more effectively with weeds; that can reproduce in a shorter time; and that can potentially fix their own nitrogen.

Improved Postharvest Processing

Transgenic technology is also being applied to several commercially important tree species, including poplar, eucalyptus, aspen, sweet gum, white spruce, walnut, and apple (Kais 2001). The global demand for wood and wood products is growing along with the human population. To reduce pressure on existing forests, forest plantations that grow transgenic trees are expected to play an increasingly important role in meeting the demand for tree products (Tzfira et al. 1998). As mentioned above, the first traits being genetically engineered into trees are herbicide tolerance and insect resistance, which are useful for establishing and maintaining young trees. Several traits are under development to better adapt trees to postharvest processing, and these may become commercially available in the near future. For example, there is research under way to modify the lignin content of certain tree species, in order to improve pulping, the process by which wood fibers are separated to make paper. Reduced lignin may improve the efficiency of paper production and may reduce environmental pollution from the paper production process.

To restrict the transfer of transgenic traits to wild forest and orchard tree populations, it is generally considered essential to simultaneously genetically engineer reproductive sterility. Methods currently exist to do this in crop plants (Mariani et al. 1992, Williams 1995), so this technology is available. Besides restricting gene flow, sterility is expected to cause the trees to grow faster and produce more wood, since energy would not be wasted producing flowers or fruit. It is likely that this issue will be investigated by a future National Research Council (NRC) committee.

There is also interest in using genetic engineering technology to turn annual crop plants into factories that produce valuable chemicals (for a recent review, see Somerville and Bonetta 2001) and antibodies (Daniell et al. 2001). Plants have the capacity to synthesize a variety of complex molecules, given the simple inputs of a few minerals, carbon dioxide, water, and sunshine. It is widely thought that plants could provide a "green" renewable source of chemicals to replace those currently obtained from petroleum. This could also be a mechanism to create new markets for plant products as well as utilize excess production of agricultural commodities. The feasibility of producing a plastic precursor, polyhydroxybutyrate, in plants, was demonstrated several years ago (Poirier et al. 1992), but this was not found to be an economically viable process. Nevertheless, there is excellent potential for mass producing a variety of fatty acids in plants that serve as precursors for valuable polymers, such as nylon. The properties of plastics that incorporate starch could also be significantly improved as more knowledge is learned about the biochemistry of starch synthesis.

Research is also directed toward reducing pollution in postharvest production. For example, there is ongoing research to reduce the content of phytic acid in corn. Phytic acid stores phosphorus in the developing seed (Raboy 1997). Much of this mineral complex is not digested by livestock, so it ends up in the waste stream. Ultimately, it is released into ponds and lakes, where the phosphorus can create algal blooms (Tilman 1999b). Several molecular approaches, including genetic engineering, are being applied to reduce the phytate content of corn.

Some of these new plant products raise several risk issues that the first generation of products did not. If any of these applications displace crops for food and feed production, what are the marginal and aggregate effects on national and global food and feed supplies? If some applications lead to a more vertically integrated farm-to-product production system, what are the environmental impacts? These and other research questions may be important in the near future.

Improved Food Quality and Novel Products for Human Use

Corn and soybeans are two of the most important food and feed commodities in the United States and worldwide. Most (65 to 70%) of the 9 billion to 10 billion bushels of corn produced annually in this country are used for livestock feed; about 25% is exported, and the remaining 10%

is processed into food ingredients, nonfood coatings and adhesives, and ethanol (Corn Refiners Association 1999). In addition, approximately 20% of the dietary calories (of people) in the United States come from lipids obtained from plant seeds, with soybean oil accounting for about one-third of the total (*www.soygrowers.wegov2.com/file_depot/0-10000000/0-10000/735/folder/4944/2000SoyStats.pdf*). By altering the lipid, protein, and carbohydrate composition of these seeds, it may be possible to create more nutritious food and obtain byproducts with improved functional characteristics. Several of the transgenic products discussed below are in field trials or will soon be available for production.

Corn seed has a high caloric density because of its high starch and oil content, but the protein it contains is deficient in several amino acids (lysine, methionine, tryptophan) essential for swine, poultry, and human nutrition. Transgenic corn lines that contain higher than normal levels of these amino acids and/or that produce proteins with higher contents of these acids have been created, although they are not yet in commercial production. Varieties of high-oil corn have been developed through non-transgenic technology, but transgenic technology is also being used to increase the quantity and quality of corn oil. As is also true of soybean oil (see below), the stability and nutritional value of corn oil could be improved by increasing the proportion of monounsaturated fatty acid, and there are efforts under way to do so by genetic engineering.

Natural soybean oil contains a significant proportion of di- and tri-unsaturated fatty acids (linoleic and linolenic), and although these unsaturated fatty acids are generally considered healthier to eat than the saturated fatty acids found mainly in animal fat, they have a tendency to oxidize and become rancid. These unsaturated fatty acids are also liquid at room temperature, which limits their functional properties for making certain types of foods, such as margarine. The stability of soybean oil and its functional properties are improved by hydrogenating the oil. This reduces the double bonds in the unsaturated fatty acids, yielding mono-unsaturated trans-fatty acids. Although trans-unsaturated fatty acids have been consumed for many years, there is increasing evidence that they are unhealthy (Taubes 2001). To address this problem, soybean was genetically engineered to produce an oil that contains predominantly a cis-monounsaturated fatty acid (oleic acid; Mazur et al. 1999). This was achieved through genetic engineering by silencing the genes that produce linoleic and linolenic acid from oleic acid by a desaturation reaction. The new product is soybean oil with approximately 85% monounsaturated fatty acid, which has good stability, and reduced off-flavor and is healthier to consume.

The meal recovered after soybean oil has been extracted is rich in proteins that have excellent functional characteristics for creating a variety of foods. However, several of the most abundant soy proteins are deficient in sulfur-containing amino acids (methionine and cysteine), which are essential amino acids for humans and certain livestock. Other soy proteins are antinutritional factors, and one is a common food allergen (Ogawa et al. 1993). Thus, transgenic research has been done to remove the anti-nutritional factors and allergenic proteins from soybeans and to improve its protein quality (Mazur et al. 1999).

The deficiency in the sulfur-containing essential amino acids in soy protein cannot be addressed through conventional breeding because better genes simply do not exist in the natural gene pool. However, the problem can be approached through genetic engineering by altering the activity of the enzymes that synthesize methionine and cysteine or by overproducing a protein that contains them. A complementary approach is to block expression of the major soy proteins that lack methionine and cysteine, thereby increasing the percentage of these amino acids in the remaining proteins. Both of these strategies are being explored. The feasibility of producing a methionine-rich protein in transgenic soybeans and other pulse seeds has been demonstrated (Altenbach et al. 1989, Nordlee et al. 1996), but the trait has not been commercialized. Gene silencing has been shown to be an effective way to eliminate the major soybean allergen (Jung, 2001, personal communication), and other anti-nutritional proteins have been removed by mutagenesis (Mazur et al. 1999). The promoter used to silence the fatty acid desaturase genes in the high oleic acid transgeneic soybean effectively eliminates expression of one of the major classes of soy proteins that does not contain methionine and cysteine. Consequently, the transgenic seed that was modified for high oleic acid content also has an improved protein quality.

In addition to the macronutrients, starch, protein, and oil, plant foods provide many of the micronutrients essential in human diets. There are 17 minerals and 13 vitamins required at minimum levels to prevent nutritional disorders (DellaPenna 1999), and all of these have attracted biotechnology research. Clinical and epidemiological studies show an important role in health maintenance for several minerals (iron, calcium, selenium, and iodine) and vitamins (A, B_6, E, and folate), but these are typically not present in sufficient quantities in many diets throughout the world. Common reliance on rice, wheat, maize, and soybean for macronutrients limits the diets of many people to the micronutrients these seeds contain. In particular, these foods are deficient in iron, zinc, selenium, copper, riboflavin, and vitamins A and C.

More than two billion people face serious dietary problems due to inadequate quantities of micronutrients. For example, iron deficiency leads to anemia in 40% of all women and 50% of pregnant women and is thought to cause up to 40% of the half-million deaths at childbirth each year (Welch et al. 1997). Inadequate iron in children's diets impairs mental development. Vitamin A deficiency is considered a global epidemic (Ye et al. 2000). Annually, 250 million children suffer from vitamin A deficiency, which contributes to illness and death for some 10 million people annually. Vitamin A deficiency causes blindness in up to half a million children each year, half of whom die after losing their sight. Besides impairing vision, vitamin A deficiency can lead to protein malnutrition and poor immune system function. Folic acid deficiency increases the risk of birth defects, heart disease, and stroke.

Much remains to be learned about the uptake and accumulation of minerals and the synthesis of vitamins in plants, but significant progress is being made in some areas of research, such that transgenic plants producing increased levels of several micronutrients have been created. For example, mulled rice grains contain no beta-carotene, but it has been genetically engineered by the introduction of three genes, one from daffodil and two from bacteria, to produce significant levels of beta-carotene, which is made into vitamin A (Ye et al. 2000). Additional genetic engineering may be necessary to raise the beta-carotene level in transgenic rice such that a daily serving provides the recommended daily allowance (RDA) of vitamin A. The efficacy of this approach to reducing vitamin A deficiencies remains controversial (Nestle 2001).

Similar genetic approaches were used to increase the level of tocopherol, the lipid-soluble antioxidant known as vitamin E, in plant oils. The RDA for vitamin E is 10 to 13.4 international units (equal to about 7 to 9 mg of α-tocopherol), which is generally accessible through consumption of plant-derived dietary components, including soybean oil. However, an excess intake of vitamin E (100 to 1,000 IU/day) has been found to be associated with a reduced risk of cardiovascular disease and some cancers, improved immune function, and reduced progression of several human degenerative conditions (Traber and Sies 1996). Thus, there could be health benefits from increasing the level of vitamin E in commonly consumed foods, such as soybean (DellaPenna 1999). By overexpressing the gene responsible for the last step in vitamin E synthesis in the model plant *Arabidopsis thaliana*, the effective vitamin E level was increased nearly 10-fold (Shintani and DellaPenna 1998). A similar approach is now being used to create transgenic soybean and canola plants with enhanced levels of vitamin E.

Plants contribute a number of other health-promoting chemicals to our diet (DellaPenna 1999). The glucosinolates found in broccoli and related cruciferous vegetables are thought to help detoxify cancer-inducing carcinogens (Talalay and Zhang 1996). The isoflavones found in soybeans are phytoestrogens, and they appear to reduce the incidence of breast, prostate, and colon cancers; osteoporosis; and cardiovascular disease

(Kurzer and Xu 1997). Carotenoids, the red, orange, and yellow pigments in tomatoes and green leafy vegetables, appear to reduce the risk of certain cancers, cardiovascular disease, and blindness caused by macular degeneration (Charleux 1996). The genetic and metabolic factors influencing levels of glucosinolates, isoflavones, and carotenoids from various plant sources are poorly understood, as is the appropriate level at which they should be consumed for maximum health benefits. Western diets typically contain limited amounts of these chemicals, and there is growing interest in finding ways to increase their content in food and to determine their appropriate health-promoting levels.

Plant foods could also be used as edible vaccines (Langridge 2000, Walmsley and Arntzen 2000). Many seeds contain proteins that are allergenic in certain people. When these allergenic proteins are digested, small fragments derived from them are absorbed into patches of cells on the small intestine that are part of the immune system. Antibodies are produced against these proteins, and this leads to an immune response in the individual, with potentially severe consequences after subsequent exposure to the allergenic proteins. This same process can be used to create immunity against common viral and bacterial pathogens by producing antigenic proteins derived from them in edible plant parts. Of particular interest are a group of pathogens—Norwalk virus, *Vibrio cholerae* (the cause of cholera), and enterotoxigenic *Escherichia coli* (a source of "traveler's diarrhea")—that cause the deaths of several million children each year, mainly in developing countries. Preliminary studies indicate that uncooked plant foods, such as potatoes or bananas, can be used to produce pathogen-derived proteins (such as virus coat proteins). These foods might then be used to inoculate children and adults against a variety of common diseases. Although a great deal of research remains to be done to demonstrate the efficacy and economic viability of this approach, results from preliminary experiments are promising.

There is also interest in using plants to produce human monoclonal antibodies. Preliminary research has demonstrated that several types of plant tissues, including seeds and leaves, have the capacity to express genes encoding the protein subunits of monoclonal antibodies and assemble them into functional complexes (Daniell et al. 2001). It remains to be seen whether plants can produce these antibodies in sufficient quantities to meet therapeutic requirements. However, if it proves possible, the technology has tremendous potential because of the expense of producing monoclonal antibodies in mammalian tissue cultures. It would be essential to grow these plants in restricted locations, but the value of the products would easily be sufficient to offset the cost of growing the crop in isolation.

These various modifications to foods raise a number of issues, many that are beyond the scope of this study. One class of modifications addresses nutritional deficiencies by increasing the concentration of the deficient nutrient or precursor in the plant. As an analysis of vitamin A deficiencies suggests (Nestle 2001), alleviating such deficiencies may require more than merely increasing the nutrient in a food. A careful, considered approach to using biotechnology to address nutrient deficiencies might start by considering all alternatives and developing a systemic approach to alleviating the deficiency in the identified population. A second class of modifications aims to optimize food quality by incremental improvements. Again, it may be difficult to know that an anticipated improvement in food quality will actually lead to an improvement, such as changing the content of "health-producing" compounds in plants, rather than a decline in human well-being. Clearly there are many questions that other scientists, including nutritionists and public health specialists, need to engage to ensure that biotechnology will lead to improved human health.

All of these future products raise parallel questions related to environmental impact. There might be indirect human health risks mediated through the environment that would require new expert analysis. Nontarget risks associated with these plants with altered nutritional characteristics (both macronutrients and micronutrients), increased concentrations of "health-producing" compounds, or edible vaccines may be considerably more subtle than the direct mortality risks associated with plants producing insecticidal toxins, which are being evaluated presently.

Mitigation of Environmental Pollution

In the future, transgenic plants may be grown for reasons other than commercialization. For example, it has been proposed that transgenic plants could contribute to removing or detoxifying heavy metal pollutants in contaminated soils ("phytoremediation"). A particular problem in some locations is mercury. Plants have already been created that can accumulate mercury. Thus, growing such plants is a potential solution for cleaning up mercury pollution at despoiled sites (Pilon-Smits and Pilon 2000). This example is discussed in detail below.

Finding 7.1: For predicting environmental risks of future plant varieties and their novel traits, currently commercialized transgenic crops offer only limited experience and understanding.

Finding 7.2: The production of nonedible and potentially harmful compounds in crops such as cereals and legumes that have traditionally been used for food creates serious regulatory issues.

Finding 7.3: With few exceptions, the environmental risks that might accompany future novel plants cannot be predicted. Therefore, they should be evaluated on a case-by-case basis.

Finding 7.4: In the future many crops can be expected to include multiple transgenes.

Potential Environmental Impacts of Novel Traits

The Animal and Plant Health Inspection Service (APHIS) appropriately refrains from speculation on environmental risks associated with crops that may or may not reach the stage of commercialization. Full and fair discussion of such risks presupposes information that cannot be known until a functional phenotype has been developed and grown in limited field trials. However, a brief discussion of the general issues associated with some new crops currently being developed will help frame the context in which environmental risks from commercialization of the next generation of transgenic crops may be discussed.

Tolerance to Abiotic Stresses

Abiotic stresses significantly limit crop production worldwide. Cumulatively, these factors are estimated to be responsible for an average 70% reduction in agricultural production (Bresson 1999). Drought stress not only causes a reduction in the average yield for crops but also causes yield instability through high interannual variation in yield. Globally, about 35% of arable land can be classified as arid or semiarid. Of the remainder, approximately 25% consists of drought-sensitive soils. Even in nonarid regions where soils are nutrient-rich, drought stress occurs regularly for a short period or at moderate levels. Furthermore, it has been predicted that in the coming years rainfall patterns will shift and become more variable due to increased global temperatures. Thus, improved stress tolerance may improve agricultural production.

Research to create crop plants that are transformed to tolerate abiotic stresses, such as heat, drought, cold, salinity, and aluminum toxicity, is ongoing. Current research efforts in drought tolerance include the isolation of crop plant mutants to understand the molecular basis for salt responses (Borsani et al. 2001), studies on the transduction network for signaling guard cell responses and their subsequent control of carbon dioxide intake and water loss (Schroeder et al. 2001), and genetic activation and suppression screens that influence interrelationships among multiple signaling systems that control stress-adaptive responses in plants (Hasegawa et al. 2000). The actual development and field testing of novel crop plant varieties is likely to be 5 to 10 years in the future owing to the complex genetic mechanisms and multigene systems involved in abiotic stress tolerance in plants. Accurate assessments of the environmental risks posed by any of these stress-tolerant plants will not be possible until they are actually created because the genetic mechanism of stress tolerance will greatly determine the scope of potential risks.

Despite this uncertainty regarding the nature of stress-tolerant transgenic plants, they have raised concerns about environmental risks. Abiotic conditions, such as soil nutrient levels, water, cold, heat, salt, and metal toxicity, combined with their seasonal variations have strong determining effects on plant community structure worldwide, and the geographic distribution of many plant species is influenced strongly by these factors (Whittaker 1975). Thus, when plants are transformed to better tolerate these abiotic conditions, it raises risk questions about the possibility of impacts on plant community structure and expansion of the geographic range of a plant species. While these issues have been considered in discussions of the present generation of transgenic crops (see Chapter 4), most of these traits have not been expected to alter the invasiveness or weediness of the transformed plants (Crawley et al. 2001). The environmental risks associated with such stress-tolerant crops are both complicated and subtle. To clarify their analysis, the committee focuses here on drought-tolerant crops. Drought-tolerant phenotypes could be based on higher water use efficiency (WUE), which leads to greater biomass production per unit of water, or an increased ability to extract water from the soil. Obviously the environmental effects of these mechanisms will differ. Plants with improved water extraction will still require the same amount of water to grow, so potential environmental effects may often be related to competition for sunlight or nutrients in the soil due to the plant's metabolic needs associated with greater biomass. In contrast, ecological theory suggests that a plant with a higher WUE would be predicted to be a better competitor for water than a nontransformed plant. This hypothesized improved competitive ability is the source of some concerns about the environmental risks of drought-tolerant transgenic plants, whether it is the crop or a wild relative that might receive the transgenes by horizontal gene flow.

But increased competitive ability for water is not necessarily sufficient to cause a plant to expand its geographic range. A lupine plant in the oak savanna with better WUE may grow more luxuriantly than its conspecifics but might not expand into surrounding habitats because there is too much shade, water, or soil nitrogen, which might neutralize its advantage in the savanna. A maize variety with higher WUE may grow better in the dryland production systems of parts of Nebraska and Kansas but may still not displace spring wheat in the neighboring counties because it still needs water over a longer growing season than wheat. It is also possible that a farmer might clear droughty land and plant such a maize variety, leading to marginal increases in the area planted to maize and marginal decreases in the area in xeric prairie remnants. It is unlikely that maize can be transformed into a plant that could grow in arid or semiarid environments without irrigation. Thus, while there is clear potential for plants transformed to tolerate drought to expand their geographic range, there is also a limit to this potential due to the inherent characteristics of the plant, and seasonal limitations of other abiotic and biotic factors that restrict it from expanding. Assessment of these risks will require attention to the plant, trait, and environment. In a similar way, impacts on plant community structure and non-target species and interactions among traits (discussed below) might occur, but their assessment will be case specific. For example, certain transgenic salt-tolerant plants can also tolerate other stresses including chilling, freezing, heat, and drought (Zhu 2001).

Phytoremediation

In the future, transgenic plants may be grown to remove or detoxify pollutants from soils ("phytoremediation"). One persistent pollution problem is mercury contamination of soils. Plants have already been transformed that can mitigate this problem in various ways. One approach to pollution mitigation is to convert the highly toxic organic mercury in soil into less toxic elemental mercury, which is volatile (Pilon-Smits and Pilon 2000). Another approach is to accumulate mercury in plant tissues (Raskin 1996), where it can be harvested, extracted, or disposed of more safely. These "phytoextraction" plants are already capable of accumulating more than 1% of their biomass as mercury, and crop improvement techniques, including genetic engineering, may be able to increase that fraction.

The environmental impacts of these mitigation strategies are probably scale dependent. While phytoremediation by volatilization could have local benefits, the large-scale consequences are less clear. If the total amount of volatile mercury created is relatively small, it will probably be of little environmental consequence. Thus, small-scale use of the volatilization strategy could be beneficial for the environment. However, if the scale of volatilization is large, so that large amounts of mercury are volatilized, levels of atmospheric mercury may rise at regional or larger spatial scales. Atmospheric mercury is returned to terrestrial, aquatic, and marine ecosystems in precipitation as rain or snow. Once deposited, it is converted into more biologically toxic forms. The deposition of atmospheric mercury released by fossil fuel burning, garbage burning, and medical waste incineration has been identified as a potentially serious environmental problem (e.g., Jackson 1997, Macdonald et al. 2000). Thus, large-scale phytoremediation based on volatilization may exacerbate what is already a serious concern. Phytoremediation based on "phytoextraction" would not release elemental mercury into the environment. The environmental risks associated with this strategy may be less scale dependent and have the additional benefit of potential harvest for extraction of their heavy metal content.

Combinations of Transgenic Traits

Risk assessments of transgenic crops have focused on the impacts of single traits (e.g., herbicide tolerance) on the environment and human health. Already transgenic crops expressing multiple traits have been commercialized, such as cotton and corn, that have both pesticidal properties and herbicide tolerance. Many more cultivars with "stacked" genes (multiple transgenic traits) are expected to be commercialized soon, so it is important to assess if the environmental and health effects of multiple genes are simply the sum of the effects of each of the single genes or if interactions among the introduced genes quantitatively or qualitatively alter the impacts of these cultivars.

Focusing on environmental effects, there are several levels at which we can look for interactions between inserted genes. In this discussion the focus is on two levels—that of the individual plant phenotype and that of the whole-field or farming system level.

At the individual plant level, an answer to the following question is needed: If expression of each of the single genes engineered into a plant does not have unintended and/or unacceptable effects on the plant phenotype, could the interaction of these genes have such effects? A few illustrative examples may help in exploring this issue. Suppose an agency determines that adding a single gene for tolerance to herbicide A does not confer weedy characteristics to a plant and that adding a single gene for tolerance to herbicide B also does not confer weedy characteristics to a plant. Is it scientifically reasonable to assume that a plant with both genes will not have weedy characteristics? In APHIS's assessment of petitions for nonregulated status, one "weediness" management strategy hinges on whether "volunteer" plants with the herbicide tolerance gene can be controlled by alternative herbicides. While the answer to this may be *yes* for each of the single-gene insertions, the answer may be *no* for plants with multiple herbicide resistance gene insertions, if plants with the two genes are tolerant to the entire array of registered herbicides for the crop. Under current regulatory policies, APHIS does not regulate the plant with combined herbicide tolerance if it has already deregulated each of the traits separately.

In addition, suppose a plant has both a gene for drought tolerance and a gene for insect resistance (such as *Bt* toxin expression). The Environmental Protection Agency (EPA) requires most plants expressing *Bt* to produce a sufficiently high dose to enable effective resistance management. Is it scientifically reasonable to assume that the gene for drought tolerance, which alters plant physiology, will not also alter expression of the *Bt* gene? Under the statutory authority of Federal Insecticide, Fungicide and Rodenticide Act, the EPA can and may even be required to evaluate *Bt* toxin expression in such a drought-tolerant variety. Under the Federal Plant Pest Act and the Federal Plant Quarantine Act, however, APHIS may not be able to make such an evaluation.

At the field or farming systems level, interaction of traits in transgenic plants could involve complex indirect causal pathways. One example is how, in some cases, stacked transgenic traits might interfere with the practice of integrated pest management (IPM) and insecticide resistance management. One tenet of IPM is that pest suppression tools are only used when the level of the pest has increased above an economic threshold. The fact that *Bt* genes are typically expressed constitutively in itself raises concerns about their impacts on IPM.

A more serious concern was raised during the 2000 field season, when there was a shortage of transgenic cottonseed with only the single trait for herbicide tolerance to glyphosate (Roundup). The demand for glyphosate-tolerant cotton was much higher than the supply, so farmers in a number of southeastern states had to plant either a conventional nonherbicide-resistant cultivar or a transgenic cultivar with stacked *Bt* and herbicide tolerance genes. In 1999 approximately 22% of the cotton acreage in North Carolina was planted to *Bt*-expressing cultivars; in 2000, it was 54%. Cotton entomologists (Bradley, 2001, personal communication) have determined that about two-thirds of the increase was caused by farmers who wanted to use glyphosate-tolerant cotton and therefore needed to purchase a cultivar that contained the *Bt* gene. Similar increases in acreage planted to *Bt* cotton occurred in some areas of Texas and elsewhere. Such artificial increases in the area of *Bt* cotton can jeopardize resistance management in all target pests by creating geographic pockets where resistance could evolve rapidly. In addition, one important cotton pest, *Helicoverpa zea*, the cotton bollworm, is at high risk of evolving resistance, and unnecessarily high use of *Bt* cotton increases this risk. Stacking *Bt* and herbicide tolerance genes has actually caused an increase in the risk of resistance evolution.

Within the next three years, industry intends to commercialize corn cultivars with *Bt* genes that control corn rootworms. It is likely that these rootworm *Bt* genes will be stacked with the currently commercialized *Bt* genes that control the European corn borer in some of the new varieties. In 2000, approximately 20% of corn grown in the United States contained *Bt* genes that control the European corn borer (USDA-NASS 2000). The primary reason for the large proportion of nontransgenic corn is that losses caused by the European corn borer in many areas are not high enough to cover the additional costs associated with the *Bt*-expressing cultivars. There are geographic (micro and macro) regions where there will be positive economic incentives to plant corn cultivars with the new corn rootworm-active *Bt* toxins but where there is no need for European corn borer control. There will also be geographic regions and farm-specific situations where there will be a use for *Bt* toxins to suppress European corn borer but not corn rootworm. With the proliferation of traits, it will become increasingly difficult for all but the largest seed companies to maintain full inventories of all combinations of the transgenic traits and agronomic characteristics. If this results in situations where farmers must purchase corn cultivars with both rootworm and corn borer *Bt* toxins or cultivars without either toxin, we may see a repeat of the cotton situation. To date, neither the USDA nor the EPA have publicly addressed this issue.

In the future there is expected to be a proliferation of novel transgenic traits engineered into crop cultivars involving multiple genes. The engineered traits could be as diverse as salinity tolerance and fruit flavor. It is difficult to anticipate the nature of environmental interactions that may be associated with combinations of three or more transgenic traits. Based on past experience, such interactions must be examined on a case-by-case basis. Chapter 3 includes a review of the environmental risk determinations made in connection with the U.S. Department of Agriculture/APHIS petition (97-013-01p) for determination of nonregulated status 31807 and 31808 cotton, which illustrates the kind of issues that can arise with stacked genes (that is, multiple transgenic traits in one crop variety). As noted, APHIS's review did not consider the possibility of interaction with stacked genes, nor did it consider the impact on farming systems.

Finding 7.5: The current APHIS approach for deregulation does not assess the environmental effects of stacked genes for nonadditive or synergistic effects on the expression of individual genes, nor does it assess stacked genes for cumulative environmental effects at the field level.

Finding 7.6: There are at least two levels at which scientists and regulators must look for interactions between inserted genes with regard to environmental effects: (1) the individual plant phenotype and (2) the whole-field or farming system level.

Industrial Feedstocks and Agronomic Traits

There is significant potential to modify the biochemistry of plants to enable them to produce a number of novel biological chemicals and proteins that can be used in other industrial processes. This technology raises several risk issues. These transgenes may code for enzymes leading to nonedible products or potential mammalian toxins. There is a clear potential risk involved in environmentally mediated movement of these transgenes so that transgene products enter the human food system (see BOX 5.1). Segregation, isolation, or sterilization methods may be essential to manage these risks, but several uncertainties remain about the cost and efficacy of these methods (see Letourneau et al. 2001, 56–57).

FUTURE POLICY ISSUES

The policy questions raised in the future by new transgenic crops and the many other factors that influence shifts in policy perspectives worldwide will differ from those addressed in detail for the first generation of transgenic crops. Just as it is not possible to predict future environmental risk issues, it is also not possible to predict what the future policy issues will be. This section will identify and discuss several policy issues that the committee believes will become increasingly more prominent in the future as transgenic crops expand globally and new transgenic crops become commercialized. The committee concentrates on four issues, which should be interpreted as a selective slice of the many policy issues that will arise in the future. The committee does not necessarily believe that all of these issues will be crucial ones in the future, but it does believe that they will be prominent and important.

First, the committee discusses how agricultural biotechnology is influencing agricultural structure in the United States and the policy and environmental risk issues that this may raise. Next, the global context for the commercialization of transgenic crops from a societal perspective is considered, and how transgenic crops may or may not address issues related to global food supply is examined. Third, the need for involving the public to democratize decision making and the increasing role that communicating environmental risk will likely play in the future development of biotechnology are discussed. Finally, the focus is shifted back to regulatory policy issues, with a comment on the precautionary principle and drawing particular attention to some of the opportunities that the 2000 Plant Protection Act provides APHIS.

Finding 7.7: The types of new transgenic crops that are developed, as well as the rate at which they appear, will be affected by the interplay of complex factors, including public funding and private financial support for research, the regulatory environment, public acceptance of the foods and other products produced from them, and the resolution of debates over need- versus profit-driven rationales for the development of transgenic crops.

Agricultural Structure

Issues related to changing agricultural structure might have indirect environmental consequences. The socioeconomic considerations for evaluating transgenic crops have proven to be more volatile than was predicted only a few years ago. The desire of some consumers, both foreign and domestic, to avoid consumption of foods containing transgenes has led to the segmentation of markets for "genetically modified" ("GM") and "GM-free" crops. In the United States the creation of marketing standards for certified organic crops that do not contain transgenes or transgene products has established a market niche for producers. These forms of market segmentation create an opportunity for U.S. farmers to capture a price premium for nontransgenic crops. At the same time, the existence of a market for nontransgenic crops creates a new kind of environmental risk from transgenics: If they cannot be environmentally isolated, pollen and residues from transgenic crops can affect/influence the economic value of nontransgenic crop production. The development of market standards and international trade rules for nontransgenic crops will have economic implications for U.S. agriculture. In the worst-case scenario, U.S. producers may effectively be excluded from these emerging world markets, but economic incentives will be affected under any scenario. Changes in producer incentives generally lead to changes in agronomic practice, with attendant environmental effects.

Such changes may have already had both economic and environmental impacts. For example, widespread use of herbicide-tolerant crops has changed patterns of herbicide use, resulting in new types of damage claims from herbicide drift, forcing farmers to adjust their farming practices accordingly. Farmers may incur legal liability for technology fees as a consequence of neighboring fields inadvertently pollinating a crop, leading to transgenic seed production. This, in turn, may create new forms of environmental nuisance lawsuits, as farmers attempt to protect themselves from complaints lodged by the owners of transgenic technology. These issues need to be evaluated and should be given careful consideration by future study groups.

Biotechnology, World Food Supply, and Environmental Risk

Any attempt to mitigate environmental risk must be mindful of the fact that avoiding one risk can inadvertently cause another greater risk (Graham and Wiener 1995). This might occur if attempts to avoid risks from commercialization of transgenic crops resulted in the adoption or continuation of commercial agricultural practices with greater negative environmental impact. While at present this is a speculative issue, it is a key element in debates on the efficacy of biotechnology to address global food shortages. If, for example, a lower-yielding nontransgenic crop is chosen in lieu of a higher-yielding transgenic alternative, the result could be that fragile lands would need to be planted in order produce needed amounts of food. If, for example, a chemically intensive farming system is chosen over a reduced-chemical transgenic alternative, the result could be the continuation of an unnecessarily harmful chemical load on a local environment.

Unwanted environmental outcomes could also occur if regulatory oversight itself has unintended consequences. There may be some environmentally beneficial applications of recombinant DNA gene transfer that will never be attempted simply because the costs of obtaining regulatory approval exceed the profit potential. Because environmental benefits can display characteristics of public goods, private industry or individual farmers may derive little monetary benefit from environmentally beneficial farming practices. As such, transgenic crops with environmental benefit may be particularly vulnerable to real or perceived disincentives brought about by costly regulatory procedures. Again, this possibility is speculative, and only case-by-case evaluation of proposed transgenic technologies can tell whether the environmental costs of regulatory procedures outweigh their environmental benefits.

While these examples are speculative, the general point they illustrate is true. Avoiding one risk may inadvertently cause another greater risk. But its truth is more complex than it would appear initially.

The largest and most complex factor in making comparative judgments of agricultural technology is that the main goals of this technology are to provide more and better food for public consumption and to reduce the cost of production for growers. In circumstances where food supplies are inadequate, people may become more willing to accept perceived tradeoffs between environmental quality and food supply. Furthermore, changes in food production that lower prices are especially beneficial to the poor because, even in times of plenty, poor people spend a proportionally larger share of their income on food (Lipton and Longhurst 1989). The Nuffield Council on Bioethics report (1999) on ethical issues associated with biotechnology notes that there is an apparently persuasive argument for accepting environmental risks in order to increase the availability and reduce the cost of food for hungry people and poor farmers. However, this need-based argument itself must be evaluated in light of economic analyses suggesting that insufficient availability of food is seldom the cause of famine (Sen 1982) and that sweeping changes in agricultural production can have devastating effects on the poor (Dahlberg 1979). In other words, the perceived tradeoff between environmental quality and food supply may be illusory.

On the one hand, inadequate amounts of food and malnutrition have plagued humans from the beginning of civilization, environmental degradation is a widespread concern, and both arable land and water are limited. Indeed, as recently as the 1960s, there was starvation in China, resulting in millions of deaths (Brown 1995). The world's population has doubled in the past 40 years, and it is expected to continue to grow, reaching a maximum of 8 billion by 2020 and perhaps 10 to 11 billion by 2050 (Vasil 1998). The reasons for hunger and limits to food security in different places and at different times are clearly complex and dynamic. However, such population projections create cause for global concern. There would be less arable land to grow crops because cities would continue to expand and erosion would carry away valuable topsoil (Kishore and Shewmaker 1999, Tilman 1999b). It is also well documented that supplies of fresh water are declining worldwide (Johnson et al. 2001), and this will severely tax our ability to grow crops. Currently, 17% of the world's croplands are irrigated, and this area produces 40% of the world's food. It is projected that the requirement for food in the most rapidly growing areas of the world—Southeast Asia, South America, and Africa—might double by 2025 and nearly triple by 2050 (Vasil 1998). At the same time, improved economic conditions in Asia have been increasing demand for costlier food products, such as meat and poultry, which can only be obtained by producing larger amounts of feed grains (Vasil 1998, Kishore and Shewmaker 1999). Dramatic increases in population growth are not expected in many industrialized regions such as the United States and Europe, but per-capita consumption in industrialized nations is high and sustains a disproportionately high demand for diverse agricultural products. The increasing nutritional and economic demands of the less developed regions of the world are expected to impact the more economically secure countries. Some people believe that such problems will require increases in productivity that far exceed what agricultural technology has achieved in the past and that these increases cannot be achieved without the use of transgenic crop plants (McGloughlin 1999, Borlaug 2000).

Opinions regarding the severity of the problems resulting from projected rapid increases in human population and the requirements to feed and clothe the world's population differ (Alexandratos 1999, Johnson 1999, Borlaug 2000), but these issues and those of land and capital distribution are likely to impact agriculture worldwide. Finding ways to increase food and fiber production during the next 50 years would then become a major concern for societies around the world. Also, there is increasing evidence that food quality plays an important role in human health. According to the USDA, medical expenses and lost productivity resulting from chronic illnesses cost U.S. society some $250 billion annually, and $100 billion is related to poor nutrition (Welch et al. 1997). On the other hand, highly aggregated statistics on food production, population, and nutritional needs can present a misleading picture of the prospects for addressing these problems through agricultural production technologies, whether through transgenic crops or by other means. A wide array of national, regional and international policies have tremendous effects on incentives for food production and consumer exchange. Yield or nutrition-enhancing technological changes have variable and sometimes negative effects in a given policy context (Drèze and Sen 1989). Hence, the use of biotechnology to address growing needs for increased food production must, at a minimum, be accompanied by a call for research on economic and sociological impacts, and appropriate policy adjustments. One group of researchers has expressed concern that a concentration of economic power and intellectual property rights will skew the development of novel crops away from those that would be of most use to the neediest people (Bunders and Radder 1995, Buttel 1995). Still other researchers refute claims that research and development of novel crop traits through biotechnology are appropriate responses to present and future problems of hunger, even for less developed countries (Busch et al. 1990, Krimsky and Wrubel 1996, Altieri and Rosset 1999). They suggest instead that promotion and support of new rural development approaches and low-input technologies spearheaded by farmers and nongovernmental organizations will make a greater contribution to food security at the household, national, and regional levels in Africa, Asia, and Latin America (Pretty 1995).

The potential for commercially grown transgenic crops to increase total world food supplies and to reduce the need for utilizing marginal lands for food production is thus an appropriate but highly contested dimension of the context in which environmental risks for the next generation of crops will be debated (Kalaitzandonakes 1999). One driving factor in the debate is the suspicion that need-based arguments promoting transgenic crops are only a ruse for profit-seeking activities by the biotechnology industry. The argument for yield-enhancing agricultural research throughout the twentieth century stressed the economic viability of farmers (Rosenberg 1961, Danbom 1979). Yet analysis of the "technology treadmill" in agriculture has established that consumers and agricultural input companies are the primary economic beneficiaries of such technology (Cochrane 1979). Several early studies of biotechnology in agriculture placed a great deal of emphasis on this phenomenon while also noting that late adopters of biotechnology may experience economic losses (see Kalter 1985, Kenney 1986, Kloppenburg 1988). The result is that some farmers and farm groups take a jaundiced view of the claim that biotechnology responds to on-farm needs for greater efficiency while nonetheless moving quickly to adopt the first generation of transgenic crops (see Rundle 2000).

Thus, an argument that begins by noting that growing global food needs will create environmental stresses on soil and water resources becomes entangled with domestic social issues that have no obvious connection to the environmental risks of transgenic crops. While environmental risks and social impacts are logically distinct, they may have common causes that reside in what biotechnology firms do to make an adequate return on their research investments. Restricting the scope of an analysis of transgenic crops to environmental impact may seem justified in light of the way disciplinary scientific expertise tends to dissociate the causal factors that contribute to environmental impact from those that affect profitability and economic access to food. Such disciplinary divisions are also reflected in the organization of regulatory authority. Even well-meaning restrictions on the scope of an evaluation of transgenic crops may be perceived as strategically motivated attempts to limit debate on transgenic crops to topics on which they will be evaluated relatively favorably (Thompson 1997a, 2000).

The context for evaluation of environmental risks must take account of the need-based argument for expanding food capacity and production. Both human and environmental dimensions of the need for increased food production are critical to any balanced evaluation of the risks from transgenic crops. As the debate moves to consider the next generation of transgenic crops, it may be possible to formulate a way of framing the need for increased amounts of food in a manner that is more sensitive to the socioeconomic complexities of hunger and deprivation (see Conway 2000). It will be increasingly important for decision makers at all points in the global food system to be well versed in both sides of this debate.

As nonfood crops are added to the product mix of agricultural biotechnology, additional and more jarring points of tension, if not outright contradiction, will emerge. Some industrial products, such as pharmaceuticals, may be produced on such a limited spatial scale that they would have a negligible impact on total food production, but that is not necessarily the case for all transgenic crops aimed at producing industrial products. For example, interest in alternative energy sources has predictably been revived in the United States with recent news of diminishing domestic petroleum-based energy sources and the concomitant price increases. One commonly cited alternative is the production of gasohol from grain, especially corn. Corn and other grains are plentiful, renewable, and inexpensive; can be grown domestically; and might reduce U.S. dependence on foreign petroleum production (Woolsey 2000). Technically, generating gasohol from grains is feasible, but there are other factors to consider. To many the ethical implications of diverting substantial portions of foodstocks to produce an industrial chemical are themselves substantial. There may be arguments that only excess grains are used industrially and that the United States produces too much food anyway. Furthermore, gasohol could be made from the residual straw or chaff after the food grain is removed. Nevertheless, it is not difficult to see why the public might discern a contradictory message. On the one hand they are told that risks are offset because biotechnology is needed to meet increasing global food requirements. On the other hand they are told biotechnology is needed in order to allow farmers to grow something other than food.

Finding 7.8: In an era of globalization, applications of transgenic crops in developing countries will be an important component of the context in which environmental impacts from transgenic crops are evaluated.

Involving the Public and Communicating Environmental Risk

Members of the public express an interest in the environmental impacts of agriculture and a desire to be informed about the relative environmental risks and benefits of different production methods. The general public needs to know more about the environmental risks of transgenic crops in order to form opinions about whether to consume these products or to be concerned about the activities of those who promote and those who oppose biotechnology. One of the principal reasons for public concern about transgenic crops is the belief that environmental issues are not taken seriously by agricultural scientists, biotechnology companies, and farmers who utilize high-tech farming methods. The roots of this belief are not easy to trace but almost certainly include past experiences when agricultural chemicals were promoted as safe and without environmental impact, only to be regulated and withdrawn in subsequent years (Dunlap 1981, Perkins 1982). These issues linger. Some argue that the risks of chemicals have been overstated, and that scientific risk evaluations of past farming practices were neglected when chemicals were withdrawn (Fumento 1993). Others believe that substantive risks of agricultural chemicals continue to be neglected and alternatives continue to be ignored (Pimentel 1987). In either case, there is a need to do a better job of involving and informing the public about what is done to mitigate and manage environmental risks in agriculture. Issues arising in connection with the commercialization of transgenic crops are only one dimension of this general problem.

A 1996 NRC report, *Understanding Risk,* notes that the initial characterization of risk requires "an appropriately diverse participation or representation of interested and affected parties, of decision makers, and of specialists in risk analysis." The way that participants are identified and involved is critical to the success of any attempt to measure, manage, or mitigate risks associated with transgenic crops. There are different disciplinary and practical perspectives that have the potential to contribute information that will improve the modeling and measurement of hazards, exposure pathways, and the relative probability of unwanted outcomes. Also, because biotechnology is controversial, participant involvement is becoming increasingly critical to the role of risk analysis in forming the basis for authority, believability, and public confidence in regulatory decision making and in the subsequent commercialization and widespread adoption of transgenic crops.

As noted in Chapter 2, risk analysis should play at least two roles in the regulation of transgenic plants. In the past, the USDA emphasized the role of decision support and has limited participation in its environmental risk assessment procedures to those who, in the judgment of decision makers, could provide information pertinent to the anticipation and measurement of environmental hazards. However, as biotechnology has become controversial, the fact that risk assessments have been done has been cited to justify U.S. agriculture's rapid adoption of *Bt* and herbicide-tolerant crops. Such justifying citations involve a shift toward using risk assessment as the basis for claiming that the USDA has faithfully exercised its decision-making authority or that the public should have confidence in recombinant DNA technologies. However, USDA officials should not presume that risk analyses done purely to support internal decision making can bear the additional weight of supporting the public's confidence in technology or government.

Communicating the idea that risks are being taken seriously is a key element. One of the surest ways to do this is to ensure that the entire process of characterizing and measuring risk is open to an array of participants representing perspectives that reflect the various sectors of the public taking an interest in agriculture and the food system. Efforts to minimize the seriousness of risks may seem warranted in virtue of the improbability of an unwanted impact or in virtue of the relative risks posed by a transgenic technology when compared to its chemically based alternative. However, when risks are denigrated by industry, regulators, or their representatives, and people who were not present when key interpretive judgments were made are told that the risks are "not serious," the inadvertent message is that responsibilities are not being taken seriously. This situation is aggravated when citizens who express concern through intermediary organizations think they are being excluded from key processes of risk evaluation and management. The result can be a situation in which the belief of being at risk is amplified by the very activities that were intended to put the public's mind at ease (Sandman et al. 1987, Kasperson 1992).

The practice of sending messages that persistently minimize the significance of risks can also have the unintended effect of making those who should in fact be attentive to environmental impacts overconfident. In some cases the individuals needed to mitigate an environmental hazard may be farmers or nonscientific employees of agricultural companies. Such individuals need to be advised about environmental risks and the methods to contain them in unvarnished language (Otway 1992). Although there were no enduring food or environmental safety issues associated with the contamination of corn stocks with unapproved Cry9c protein in the autumn of 2000, one lesson is that, when key individuals—in this case farmers and grain handlers—are persistently told that risks are minimal, a general culture of laxity may come to prevail. Adequate risk management for the coming products of biotechnology depends on a culture that reinforces the seriousness with which environmental risks must be addressed.

A more inclusive approach is especially important in environmental risk assessment, as compared to human health risk assessments, because classifying (or failing to classify) a possible event as a hazard involves value judgments that are more likely to be controversial. Describing some possible event or set of events in terms of risk always implies that these events are regarded as adverse. Although we talk about the risk of dying or losing money, we do think of our chances of recovering from disease or winning the lottery as forms of risk. The tendency to overlook the value judgments made in risk assessments that emphasize adverse effects on human health arises simply because these value judgments are not controversial. In comparison to human health, the adversity of events in the environmental arena may not be at all obvious, especially to individuals lacking key background knowledge of agriculture and ecology. In addition, conflicting values and attitudes about which nature is wild, which needs protecting, and about the relative value of wild and agricultural land use will affect the way in which environmental risks from agricultural production are defined (Cranor 1997, Thompson 1997b).

The question of whether genetically engineered crops involve environmental risks is thus not wholly a scientific one. An answer depends partially on how one understands for classifying events associated with crop production as adverse. One interpretation of adversity might stress the impact on such factors as insect or disease resistance, soil fertility, and prevalence of pollinators or noxious weeds. Such factors affect the chances of successful farming. Other effects related to successful farming are indirect: herbicide-tolerant crops affect patterns of herbicide use and with that comes sources of risk associated with spray drift, for example, and pollen from transgenic crops can substantially damage the value of crops grown for organic markets. Such events might be regarded as adverse from one farmer's perspective but not from another. An altogether different interpretation of environmental risk might stress the impact on habitats for nonagricultural plants and animals or on genetic and species diversity.

Some of the parameters for interpreting what counts as adverse have been reasonably well articulated in the statutes on which the USDA exercises its regulatory authority. However, it is not likely that risk assessment will succeed in building the public's confidence in biotechnology unless there is a more systematic and more public effort to acknowledge the diverse values that influence judgments on adversity. Each of the actors in the agri-food system must find ways to involve a broader array of value perspectives throughout all phases of research planning, product development, and regulatory oversight.

Finding 7.9: Adequate risk management for future biotechnology products will depend on a regulatory culture that reinforces the seriousness with which environmental risks are addressed.

Finding 7.10: Public confidence in biotechnology will require that socioeconomic impacts are evaluated along with environmental risks and that people representing diverse values have an opportunity to participate in judgments about the impact of the technology.

Regulatory Issues

The preceding discussion presents the general context in which the risks of coming generations of transgenic plants should be evaluated. The types of crops developed in the second generation and the rate at which new transgenic crops become available will depend in part on specific regulatory policies, on how need-based versus profit-based technology debates play out, on continued public and private financial support for the research needed to create new transgenic crops, and finally on public acceptance of the foods and other products produced from them. Although these factors can be conceptually and logically distinguished from one another, they impact one another in complex ways. Regulatory policies affect financing and profitability of research ventures. A lack of public acceptance shapes markets for transgenic crops and creates political pressure for regulatory responses that may not be strictly justifiable based on scientific measures of health and environmental risk.

The probability of environmental damage from the next generation of transgenic crops will be determined by the specific phenotypic traits of each crop as well as by agronomic practices and land uses specific to each particular crop that is commercialized. Throughout this report, discussion of risk characterization and risk management has focused on procedures that can and should be used to measure the likelihood and extent of such damage and to mitigate environmental risks. It is appropriate that regulators focus intently on such specifics when evaluating any particular transgenic crop. In looking toward the future, however, it is also important to ensure that the details of risk analysis do not obscure the larger context in which discussions about agriculture and the environment are situated. In this section, the discussion is focused on three related regulatory concerns from this larger perspective. The committee considers how the future of commercialized transgenic crops may challenge both U.S. and global capacity for environmental risk analysis and the ambiguities surrounding the precautionary principle and the development of its use worldwide. Then, the committee turns to the changing domestic standards for environmental regulation and some implications for regulation of any agricultural practice.

U.S. and Global Capacity for Environmental Regulation

The overall adequacy of the U.S. Coordinated Framework for the Regulation of Biotechnology has been a frequent subject of discussion throughout the brief history of transgenic crops. Krimsky (2000) finds the coordinated framework to have failed to adequately manage environmental risk, while Malinowski (2000) finds it to have been remarkably successful. It is beyond the scope of this report to evaluate the entire framework, and so instead several points are provided for consideration that may challenge the regulatory system in the near future.

As noted above, some of the coming applications of biotechnology may involve the use of plants to produce pharmaceutical products, biologics, fuels, and other substances not intended for human food use. The introduction of such transgenes poses the potential for environmentally associated risks of a wholly different order than those associated with existing transgenic crops. If such a transgene moves into food crops, either through pollen transfer or physical contamination, there could be serious human safety risks. If such a transgene moves into a wild relative, there could be widespread environmental dissemination of the pharmaceutical substance or other nonfood substances that could have impacts on wildlife as well as microbial populations.

Provisions for regulation of such environmental risks are identified under the existing coordinated framework. Environmental risks associated with pharmaceutical-producing plants, for example, could be regulated by the Food and Drug Administration under the authority it currently exercises to regulate drug production. It is not too soon to raise questions about the interagency technical capacity to review and manage such risks, the potential for duplicative regulation and new regulatory gaps, and the need to plan research to understand and adequately assess such risks. In addition, there may be a need for changes or modifications in the current regulatory approach to commercialization of transgenic plants.

A broader issue associated with global regulatory capacity relates to a potential unintended consequence of U.S. regulatory decision making. The United States has conducted environmental reviews of more transgenic crops than any other country. Many countries lack sufficient regulatory oversight of environmental impacts, especially the scientific capacity to conduct environmental risk analyses and the administrative capacity to enforce environmental decisions that are made. In these conditions there becomes a tendency to rely on previous regulatory reviews conducted in other countries, whether indirectly when an applicant submits the identical data package for risk analysis or overtly when regulatory personnel parrot the previous review. As discussed in Chapter 3, while APHIS notes some non-U.S. environmental risks of transgenic crops in some of its reviews, it does not and cannot do so in every case. Even when some global environmental impacts are noted, U.S. environmental reviews are keyed to environmental conditions and farming practices prevalent in the United States. Transgenic crop varieties that may be environmentally benign in this country might have a high risk of ecological disruption were they to be grown in some developing countries. For example, corn grown in southern Mexico and Guatemala and potatoes grown in the highlands of Peru will cross readily with wild relatives, which could lead to environmental risks. The global scientific capacity to conduct locally relevant environmental risk analyses is sorely deficient, and this needs immediate attention. The relationships among nationally-based environmental risk assessment and regulation, multinational agreements, and global environmental hazards must also be included in the broader context for future discussions of environmental risk from transgenic crops.

The Precautionary Principle

At the same time that new products and applications of gene transfer are being applied in crop production, there has been discussion of new regulatory approaches. One important component of this discussion has been the "precautionary principle." A review of this discussion, however, reveals that there are actually several principles being proposed rather than only one. Language endorsing a precautionary principle and taking a precautionary approach in regulatory decision making is in the European Union's (EU) Directive 90/220 on biotechnology and its revision in 2001, the Cartagena Protocol on Biosafety, which was finalized in Montreal during 2000, under the Convention on Biological Diversity (CBD). The CBD has been ratified by many countries but not the United States. It is an international agreement to conserve and use sustainably biological diversity for the benefit of present and future generations.

The Rio Declaration of 1992, which is a part of the CBD, includes the following language in Principle 15:

> In order to protect the environment, the precautionary approach shall be widely applied by States according to their capabilities. Where there are threats of serious or irreversible damage, lack of full scientific certainty shall not be used as a reason for postponing cost-effective measures to prevent environmental degradation.

On January 20, 2000, the Cartagena Protocol on Biosafety (Biosafety Protocol) was adopted by consensus of the parties. It applies to transgenic organisms that will be intentionally introduced into the environment. Food and feed commodities, which might lead to accidental introductions, are excluded. The protocol reaffirms Principle 15 and expands it in its Articles 10.7 and 11.8.

> Lack of scientific certainty due to insufficient relevant scientific information and knowledge regarding the extent of potential adverse effects of a living modified organism [some transgenic organisms] on the conservation and sustainable use of biological diversity in the Party of import, taking also into account risks to human health, shall not prevent that Party from taking a decision, as appropriate, with regard to the import of the living modified organisms in question in order to avoid or minimize such potential adverse effects.

This statement of the precautionary principle enables regulatory decisions to err on the side of precaution when there is scientific uncertainty. Most of the details on implementation are left to the discretion of the parties, and it is within these details that the scope of the precautionary principle will be established. Consequently, unless there are additional negotiations, the regulatory impact of the precautionary principle will be determined by practice and only after its use will it be possible to articulate what it actually is.

The European discussion of the precautionary principle has occurred in the context of attempts to harmonize grades, standards, and regulatory approaches among member states. These discussions have occurred during a spate of food scares and failings in government oversight, most notably associated with transmissible bovine spongiform encephalopathy (Mad Cow Disease), that have rocked public confidence in government regulation of the food system. Although debate over the food safety and environmental risks of transgenic crops has figured prominently in Europe, European pronouncements on the precautionary principle neither offer a specific evaluation of transgenic crops nor provide language that would give regulators specific instructions on how to take a precautionary approach with respect to transgenic crops. The March 12, 2001, revision to EU Directive 90/220 states that the requirements of the Biosafety Protocol, including the precautionary principle, should be respected. The 2001 directive indicates that the precautionary principle was taken into account in the drafting of the directive and must be taken into account when implementing it. Consequently, the directive does not provide a statement of the precautionary principle but identifies regulatory processes that are precautionary. It is beyond the scope of this report to provide a detailed analysis of this new directive, but suffice it to say that the precautionary principle has been incorporated into at least the labeling and traceability standards, the monitoring standards, and the EU approval process. Much the same as in the Biosafety Protocol, the precautionary principle will be defined in Europe through its application to specific cases.

U.S. discussion of the precautionary principle has been sparked by both these European developments and the publication of papers from a 1998 working group conference on the meaning and applicability of the precautionary principle in a number of legal and public health settings (Raffensperger and Tickner 1998). This collection of essays on the precautionary principle includes a number of interpretations and applications, all of which should be regarded as supportive of the precautionary principle and the precautionary approach. There is considerable variation even among contributors to Raffensperger's and Tickner's book advocating implementation of the precautionary principle. Cranor (1998) focuses on the need to emphasize the minimization of type II statistical errors in science intended for regulatory application. Ozonoff (1998) argues that the precautionary principle should be interpreted as a screening device to apply "scientific evaluations to situations where the proportion of cases that are hazards is high and/or use methods with high specificity, that is, that correctly identify 'no-hazard' situations" (104). Arguably, these approaches simply reflect the current regulatory philosophy in place for transgenic plants under the coordinated framework. However, other contributors explicitly advocate the use of nonscientific criteria, such as political solidarity or respect for life, to evaluate the possibility and potential for environmental hazards (Bernstein 1998, M'Gonigle 1998).

In summary, a number of different formulations are given for the precautionary principle. Advocacy of a precautionary approach can be interpreted to convey a need to reconsider and possibly strengthen regulatory oversight of technology. One key is the advocacy of statistical and scientific evidentiary decision standards that err on the side of preventing serious and irreversible health and environmental effects, even in the absence of definitive demonstration of harm to humans or non-target organisms. Given the ambiguity in its various formulations as well as opportunities to interpret the seriousness of environmental hazards in different ways, the precautionary principle does not provide a single alternative to existing regulatory policies for the U.S. government. Indeed, some interpretations of the precautionary principle would be consistent with current approaches. Nor does the principle provide unambiguous guidance for the evaluation of transgenic crops. While both advocates and critics of the precautionary approach appear to have assumed that this would result in an adverse evaluation of transgenic crops (Carr and Levidow 2000, Miller and Conko 2001), it is not necessary that such a result would follow. A more definitive evaluation of the precautionary principle must await more specific criteria for its application in agriculture (Soule 2000). For the meantime, the "precautionary principle" should be regarded as a number of potential regulatory approaches pertaining to the use of science to formulate public health and environmental policy, rather than a specific proposal for change. However, as it is applied in Europe and under the Biosafety Protocol, its meaning will become clear. As it gains clarity, there may be significant implications of its use with respect to U.S. regulatory policy.

Finding 7.11: The precautionary principle forms an important but imprecise context for the regulation of future transgenic organisms.

Raising the Regulatory Bar

It will soon become necessary to reconsider the general philosophy of regulation for environmental impact that has been applied to all forms of agricultural technology since World War II. Plant scientists have developed an array of techniques for introducing genetic novelty into crop varieties through conventional breeding methods such as mutagenesis, wide crosses, and embryo rescue. These techniques have the potential to introduce genes and genetic variations into crops that in some cases equal the novelty associated with recombinant DNA techniques. It is entirely possible that crops developed with either of these techniques could possess phenotypic traits associated with elevated levels of environmental risk.

As BOX 1.1 notes, Green Revolution varieties that increased commercial crop capacity to utilize nitrogen fertilizer shifted land-use patterns and sparked a worldwide increase in the adoption of chemical inputs. A comparison of the environmental risks from Green Revolution varieties with those of the first, second, or third generations of transgenic crops would be a highly speculative enterprise. Yet in retrospect it is difficult to argue that Green Revolution impacts on the environment were automatically safe, benign, or even acceptable simply because they were developed using conventional breeding. With the potential for introducing additional novel traits through nontransgenic methods, it becomes increasingly difficult to defend the idea that conventional crops should automatically be excluded from scrutiny for environmental impact.

As noted earlier, NRC reports have consistently found that use of recombinant DNA technology in the development of an agricultural crop does not in itself create a new class of risks. As with conventionally bred crops, it is the phenotypic characteristics of the plant that are the source of environmental risks. This report and the 2000 NRC report (2000c) on pest-protected plants cite a number of environmental risks that should be accounted for in the regulation of both current and future crop varieties. When these observations are combined, the possibility that nontransgenic crops may also pose environmental risks requiring a regulatory response becomes logically inescapable. Yet new crop varieties posing potential hazard fail to require regulatory scrutiny under APHIS while potentially benign transgenic crops do because APHIS oversight excludes conventional crops. Those plants produced via recombinant DNA are regulated, and those produced by other methods are exempt, even if the final product has an identical phenotype and therefore presents similar potential risks.

Moving beyond the realm of crop modification, it is also clear that the bar has been raised substantially for acceptable environmental effects for novel pesticides and new agricultural practices (e.g., changes in crop rotations and changes in cultivation practices). As our perspective on the ecological interactions and interchange between agricultural and nonagricultural lands evolves (see Chapter 1), the environmental standard being set for transgenic plants may be a better overall environmental effects model for agriculture than the model developed in the early 1900s for assessing the acceptability of conventional crop varieties and agricultural practices.

Although government regulation of conventionally modified plants has been virtually nonexistent, the agricultural research establishment has not ignored the potential of genetic modification of crops to result in environmental change. Indeed, some of the work begun in the 1960s to assess the long-term impacts of Green Revolution cultivars and cropping practices demonstrated insight into the needs for examining environmental effects on long and large scales. As discussed in Chapter 1, it was only because of long-term 30-year experiments on the impact of cropping intensification that researchers were able to clearly document the effects of the new cropping practices on soil characteristics. These long-term experiments were sponsored by nonprofit groups and local governments. Today, most efforts at examining environmental impacts of transgenic crops emphasize short-term laboratory and field plot experiments. While such experiments are useful, there is a need to examine effects that cannot be seen at such small scales. For example, short-term and small-scale experiments would not have predicted the effects of Green Revolution practices on lowering the water table in semiarid regions of India or the rise in water tables in other areas that has been accompanied by salinization of the soils in the root zones of crops.

Finding 7.12: The environmental impacts of transgenic plants and other new agricultural practices can be studied at a number of ecological scales ranging from the specific toxicological effects of a newly produced compound to the large-scale and long-term spatial and temporal effects of changes in agricultural practices induced by the introduction of a novel crop variety.

Finding 7.13: Currently APHIS environmental assessments focus on the simplest ecological scales, even though the history of environmental impacts associated with conventional breeding points to the importance of large-scale effects, as seen in the impacts of Green Revolution cultivars.

Recommendation 7.1: APHIS should include any impact on regional farming practices or systems in its deregulation assessments.

Society demands that commercial products are deemed safe for health and the environment. The public trusts government regulators to assess risks adequately, to exclude from commerce products posing unacceptable risks, and to impose appropriate risk management strategies to reduce risks. Some people want regulators to test "everything for everything" prior to commercial release. All regulatory agencies, including APHIS, must work with limited financial, temporal, and human resources, so testing everything is not feasible. Instead, prudence and fiscal reality dictate that regulators identify those products most likely to be hazardous and concentrate scrutiny on them, applying a science-based approach to risk assessment.

Currently, one of the risk-based triggers for assessment of a crop by APHIS is that it is a transgenic crop variety (see Chapter 2). This trigger captures all products of genetic engineering but excludes potentially hazardous products derived from conventional methods. This regulatory trigger is imperfect because it does not provide regulatory scrutiny for certain conventional crop plants that may have environmental risks. As discussed more thoroughly in Chapter 2, it fails to capture potentially hazardous conventional products, such as stress-tolerant canola. At the same time, it brings under regulatory scrutiny some transgenic crop plants that it later determines to have no significant environmental risks. Indeed, an intraspecific recombinant DNA variety would be regulated even if the gene were removed and replaced back in its original location with no extraneous DNA. Thus, there are reasons for considering other or additional potential triggers for the regulation of crop varieties.

APHIS has the authority to regulate all plants that pose a plant pest risk under its present statutory authority. On May 25, 2000, the U.S. Senate and House of Representatives agreed on a conference report on a new Plant Protection Act (PPA), which was developed to improve the government's ability to prevent and mitigate the effects of organisms that might harm agriculture and the environment. This act, signed by President Clinton, on June 20, 2000, includes products of genetic engineering but is not limited to them. The PPA offers an opportunity for APHIS to refocus regulation on those plant products that pose risk, regardless of the method of derivation. As developed in more detail in Chapter 2, however, scientifically defensible ex ante risk assessment is not yet possible, so it will not be possible to develop a science-based trigger based solely on predictions of the risks associated with particular crop varieties. In addition, it would be imprudent to bring all conventional crop varieties under regulatory oversight. Thus, careful scientific thought must be applied to this problem to avoid extraneous regulation.

Finding 7.14: The committee finds that the Plant Protection Act of 2000 can be viewed as an opportunity to clearly define the types of novel plants, regardless of method of breeding, that trigger regulatory scrutiny.

As explained in previous chapters, APHIS regulates transgenic plants under the authority of the Federal Plant Pest Act (FPPA) and the Federal Plant Quarantine Act (FPQA). While the scope of these acts is quite broad, there are some taxonomic and functional limits that make regulation of some potentially hazardous organisms problematic. For example, the FPPA does not recognize vertebrates as pests. Furthermore, transgenic plants that have been modified using *Agrobacterium* DNA or the DNA from certain viral species fall clearly within the regulatory authority of the FPPA because these organisms are plant pests. Although APHIS can regulate plants that have been genetically transformed without the use of DNA from a plant pest, determining which plants will be regulated is not simple. As identified early in this report, the regulatory options available to APHIS through the FPPA are limited. For example, once a plant is deregulated, APHIS has no authority to restrict its use or to even monitor it.

The broad language of the new Plant Protection Act defines "plant pest" as including all vertebrate and invertebrate animals except humans. The PPA repeals the FPPA and the FPQA, however, it provides for regulations issued under the repealed acts to be enforced until new regulations are developed. The PPA also provides the potential for developing new procedures for regulating plant products that pose risk. For example, regulations could presumably be written giving directives to APHIS on how to involve the public and external scientific experts in its review process. In addition, regulations under the PPA could be used to provide the flexibility lacking in present regulations. For example, it would be useful to allow APHIS to recall or rescind deregulated plants under appropriate circumstances. Current federal policymakers will determine to what extent the new PPA will be used as a vehicle for change.

Recommendation 7.2: The committee recommends that the Plant Protection Act of 2000 be viewed as an opportunity to increase the flexibility, transparency, and rigor of the APHIS decision-making process.

Finding 7.15: Nontransgenic crop breeding techniques have the potential to introduce genes and genetic variation into crops that equal or surpass the novelty associated with recombinant DNA techniques.

It is entirely possible that crops developed using these techniques could possess phenotypic traits associated with elevated levels of environmental risk. When evaluating transgenic crops for deregulation, it becomes increasingly difficult to defend the idea that these new nontransgenic crops should automatically be excluded from scrutiny for environmental impact.

THE NEED FOR STRATEGIC PUBLIC INVESTMENT IN RESEARCH

Perhaps more than anything else, the experience with commercialization of transgenic crops has revealed gaps in the knowledge base for understanding and measuring the environmental risks of crop production, irrespective of whether recombinant DNA technologies have been applied. Crops developed through nontransgenic methods can and have resulted in avoidable environmental damage. Managing the risks of crop production in the future will depend on improving basic scientific knowledge and research capacity with respect to five key areas discussed below.

Improved Risk Analysis Methodologies and Protocols

Formal research support in the United States for the study of environmental impacts of transgenic plants has been sparse. The longest continuous funding, now almost a decade, has come from the USDA's Biotechnology Risk Assessment Research Grants Program (BRARGP) to assist federal regulatory agencies in making science-based decisions about the safety of introducing genetically modified organisms into the environment. The program accomplishes its purpose by funding scientific research in priority areas determined in part by the input of the regulatory agencies. Research proposals submitted to this competitive grants program must address risk assessment, not risk management, and are evaluated by a peer panel of scientists. The program has allocated no more than a few million dollars for research each year. Recently, the USDA's Initiative for Future Agriculture and Food Systems (IFAFS) program has included a competition for funding research, education, and extension on the management of environmental risks of agricultural biotechnology. Both funding programs have substantial limitations—BRARGP because its focus is only on assessment and because the total amount of funding is so low; IFAFS because the focus is only for risk management and the funding program itself is anticipated to have a short life. Neither program funds monitoring or research related to monitoring.

Research on the environmental impacts of transgenic plants can be accomplished through other funding sources if the research questions asked have general significance. For example, issues directly associated with the impacts of transgenic plants may often be associated with critical, but largely unanswered, questions in other fields. For example, whether or not the introgression of pest resistance transgenes into wild populations will result in the evolution of weediness or invasiveness is directly associated with important questions in population biology regarding the genetic and ecological causes and correlates of invasiveness (Traynor and Westwood 1999).

There are several critical areas of research related to risk analysis that would benefit from increased funding.

(1) In the area of hazard identification and risk assessment, there is a need for improved, scientifically sound protocols to detect effects of transgenes and transgene products on non-target organisms. Present protocols are better adapted for screening for effects of toxic chemicals with broad ecological effects, rather than biologically released materials with more targeted specificity. This includes better understanding of effects on pollinators, natural enemies, species of conservation concern and soil organisms. There is also a need for improving understanding of the environmental effects of transgene

movement. Present research has concentrated on determining the conditions under which transgene movement would likely occur. It will be essential to develop a deeper theoretical and empirical understanding of the kinds of environmental effects that could result from transgene movement and the conditions under which such effects are likely to occur. This needs to be understood for transgene movement associated with pollen, viruses, and bacteria. In addition, there is a need to develop assessment protocols that can identify the circumstances under which a plant is likely to become invasive. Because deliberate introduction of plants is the most common route of establishment of invasive plants, it will be critical to understand how genetic modification affects invasiveness.

(2) Research on the effectiveness and efficiency of present regulatory systems is needed to develop approaches to enable the regulatory systems to improve based on scientific principles and data as they acquire information and experience. This includes analysis of alternative methodologies as discussed in Chapter 2.

(3) Research is needed to evaluate the environmental risks associated with conventionally produced crop plant varieties. This will involve understanding the kinds of risk and the circumstances under which these risks might occur. Because conventional crop breeding methods are constantly changing, this evaluation should include a dynamic assessment.

(4) Because many novel transgenic organisms are likely to be developed in the future, it will be useful to fund research to identify and investigate possible environmental hazards associated with these new types of transgenic organisms. These efforts might best be initiated several years before any of these plants become likely to be commercialized. This research effort could be coordinated with research on transgenic methods to minimize risk, as discussed below.

Postcommercialization Validation and Monitoring

As discussed in Chapter 6, there is a need for formal postcommercialization validation testing to determine if the results of small-scale precommercialization tests are relevant at larger spatial and temporal scales. The infrastructure for such testing exists but resources must be targeted to carry out such testing in a rigorous manner. Development of long-term monitoring systems, and development and coordination of trained observer monitoring will require a major commitment to building infrastructure.

Long-term monitoring is based on the use of ecological indicators (NRC 2000a), which are intended to provide valid, reliable, and cost-effective information about the status of important ecological systems. Repeated observations of these indicators can enable the identification of associations between changes in indicator values and the use of transgenic crops. To improve long-term monitoring of transgenic crops, research is needed to fill knowledge gaps related to ecological indicators. Important research areas include temporal behavior of ecological indicators, and the need for improvements in mechanistic understanding of the links between ecological dynamics at population and other levels and the ecosystem process measures frequently used as indicators. The latter area is particularly important to ensure that ecological indicators will be able to resolve the ecological effects of transgenics from the effects of other factors that may be associated with the use of transgenic crops. There also are important research questions about the structure and organization of scientific efforts to accomplish long-term monitoring. For example, how should feedback loops be established between validation efforts and long-term monitoring efforts so as to clarify the adequacy of available ecological indicators for monitoring of ecological effects of transgenic crops? Finally, research is needed on processes for establishing long-term monitoring indicators in relation to transgenic crops as well as research to evaluate the efficacy and effectiveness of potential long-term indicators.

Trained observer monitoring attempts to detect effects the transgenic crops that are so poorly understood—or simply unforeseen—that they cannot yet be the target of more specific monitoring efforts. The goal is to develop a network of trained observers that could detect effects of this sort. The outstanding research needs related to this kind of monitoring concern questions about what sort of practical approaches could serve this purpose, as well as the broader challenge of "posting" observers that can monitor environmental effects of technological change in agriculture generally, since it is unlikely that monitoring aimed solely at transgenic effects will be cost effective.

Specific research needs include improving understanding of the organization and facilitation of observer activities that are or could be performed by existing organizations that have a current or potential interest in performing relevant ecological monitoring (e.g., the Christmas Bird Count of the National Audubon Society). Another important issue is the structure and functioning that would provide integration, organization, and development for the trained observer network. Many questions remain about how such a network would perform its functions—for example, proactively conducting two-way communication about monitoring issues with interested parties in organizations that do monitoring, developing curricula for training sessions for observers, providing a clearinghouse for reports of notable observations from field monitoring efforts, and developing quality assurance standards.

Improved Transgenic Methods to Reduce Risks and Improve Benefits to the Environment

Some of the concerns raised about early biotechnology-derived crop varieties can be avoided in the future with our expanding knowledge base. For example, pollen flow issues could be reduced if there are advances in plastid transformation, if sterility systems are engineered into the varieties, or if gametophytic factors (i.e., genes expressed in the pollen or egg) can be used to restrict fertilization of the egg. As nontissue culture methods are developed for gene transfer, such as cocultivation of developing flowers with *A. tumefaciens* (see Chapter 1), gene transformation in a number of plant species might become much less genotype dependent, perhaps even to the point of using an elite cultivar for the initial transformation event, thus avoiding linkage drag as the result of subsequent breeding procedures. In addition, tissue culture itself often introduces unwanted and unpredictable genetic and epigenetic variation to transgenic plants that such methods may avoid.

Alternatively, different promoters could allow greater control of where in the plant the gene product is produced; tissue-specific promoters could preclude expression of pesticidal proteins, for example, in pollen and other tissues. Temporal and spatial regulation of gene expression also can be controlled sometimes via an exogenous inducer. The ability to identify and transfer genes from elite germplasm collections of the same species could dramatically improve as a result of knowledge gained from genomics and proteomics research. Antisense technology will improve our ability to knock out the function of expressed genes. Targeted recombination and gene replacement technologies can be expected to improve, which will allow greater precision in gene insertion following transformation. Our ability to produce primary transformation events with selectable markers that can be removed through genetic crosses or that make use of benign (more acceptable) selectable markers will be enhanced. Improvements in DNA sequencing already allow the full sequence analysis of primary transformation events, such that optimal ones with simple single-gene insertions are selected early in the process of engineering transgenic traits.

Value-Oriented Research

When new technologies are involved in large-scale social changes, public debate and opinion formation on value-based issues can be facilitated by research that articulates and analyzes ethical, legal, and cultural traditions as they might bear on novel questions, as well as more traditional economic research on costs, benefits, and likely effects on the prices and availability of food. Such research helps both scientists and the public understand what is at stake and how the issues might be viewed from differing perspectives. The potential for economic and sociological research has been well established in agriculture, and there is clearly a need for more studies that focus on biotechnology. The potential for philosophical, legal, and culturally-oriented research on issues in involving genetics has been demonstrated by the Ethical, Legal, and Social Issues (ELSI) program associated with the Human Genome Initiative. This program has produced both scholarly studies and educational materials that help people go beyond initial reactions of enthusiasm *or* repugnance. The USDA has funded little research of this kind, and few U.S. colleges of agriculture offer training or coursework to prepare professionals for ethical decision making—coursework that is now routinely offered in medical schools. The possibility cannot be dismissed that at least some of the controversy and turmoil that have greeted genetic engineering in agriculture is due to this omission. The USDA should encourage the development of a systematic program of research and teaching on ethical, legal, and cultural dimensions of agriculture and food issues, especially as they involve genetic techniques.

Recommendation 7.3: Significant public-sector investment is called for in the following research areas: improvement in risk analysis methodologies and protocols; improvement in transgenic methods that will reduce risks and improve benefits to the environment; research to develop and improve monitoring for effects in the environment; and research on the social, economic, and value-based issues affecting environmental impacts of transgenic crops.

References

Abbott, R. 1992. Plant invasions, interspecific hybridization and the evolution of new plant taxa. Trends in Ecological Evolution 7:401–405.

Abelson, P.H., and P.J. Hines. 1999. The plant revolution. Science 285:367–368.

Adams, K.L., M.J. Clements, and J.C. Vaughn. 1998. The Peperomia cox I group I intron: Timing of horizontal transfer and subsequent evolution of the intron. Journal of Molecular Evolution 46:689–696.

Addison, J.A. 1993. Persistence and nontarget effects of *Bacillus thuringiensis* in soil: A review. Canadian Journal for Restoration. Ottawa 23(11):2329–2342.

Aerts, R., and F. Berendse. 1988. The effect of increased nutrient availability on vegetation dynamics in wet heathlands. Vegetation 76:63–69.

Alexandratos, N. 1999. World food and agriculture: Outlook for the medium and long term. Proceedings of the National Academy of Sciences of the United States of America 96:5908–5914.

Allen, W. 2000. Restoring Hawaii's dry forests: Research on Kona slope shows promise for native ecosystem recovery. BioScience 50(12):1037–1041.

Alstad, D.N., and D.A. Andow. 1995. Managing the evolution of insect resistance to transgenic plants. Science 268:1894–1896.

Altenbach, S.B., K.W. Pearson, G. Meeker, L.C. Staraci, and S.S.M. Sun. 1989. Enhancement of the methionine content of seed proteins by the expression of a chimeric gene encoding a methionine-rich protein in transgenic plants. Plant Molecular Biology 13:513–522.

Altieri, M.A., and P. Rosset. 1999. Ten reasons why biotechnology will not ensure food security, protect the environment, and reduce poverty in the developing world. AgBioForum, 2(3&4):155–162. Available online at *www.agbioforum.org*.

Anderson, D.R., and C.J. Henny. 1972. Population ecology of the mallard. I. A review of previous studies and the distribution and migration from breeding areas. Publication 105. Washington, DC: Fish and Wildlife Service, U.S. Department of the Interior.

Andow, D.A. 1994. Community response to transgenic plant release: Using mathematical theory to predict effects of transgenic plants. Molecular Ecology 3:65–70.

Andow, D.A., and D.N. Alstad. 1998. The F2 screen for rare resistance alleles. Journal of Economic Entomology 91:572–578.

Andow, D.A., and O. Imura. 1994. Specialization of phytophagous communities on introduced plants. Ecology 75:296–300.

Andow, D.A., and K.R. Ostlie. 1990. First-generation European corn borer (Lepidoptera: Pyralidae) response to three conservation tillage systems in Minnesota. Journal of Economic Entomology 86:2455–2461.

Andow, D.A., Lane, C.P., and Olson, D.M. 1995. Use of Trichogramma in maize-estimating environmental risks. Pp. 101–118 in Benefits and Risks of Introducing Biocontrol Agents, J.M. Lynch and H.H. Hokkanen, eds. Blackwell: London.

APHIS (Animal and Plant Health Inspection Service). 1987. 7 CFR Parts 330 and 340, Plant pests; introduction of genetically engineered organisms or products; final rule. Federal Register 52:22892–22915.

APHIS (Animal and Plant Health Inspection Service). 1988. Genetically engineered organisms and products; exemption for interstate movement of certain microorganisms under specified conditions. Federal Register 53:12910–12913.

APHIS (Animal and Plant Health Inspection Service). 1990. Genetically engineered organisms and products; exemption for interstate movement of the genetically-engineered plant ("Arabidopsis thaliana") under specified conditions. Federal Register 55:53275–53276.

APHIS (Animal and Plant Health Inspection Service). 1993. 7 CFR Part 340, Genetically engineered organisms and products; notification procedures for the introduction of certain regulated articles; and petition for nonregulated status; final rule. Federal Register 58:17044–17059.

APHIS (Animal and Plant Health Inspection Service). 1997a. 7 CFR Part 340, Genetically Engineered Organisms and Products; Simplification of Requirements and Procedures for Genetically Engineered Organisms; final rule. Federal Register 62:19903–19917.

APHIS (Animal and Plant Health Inspection Service). 1997b. User's Guide for Introducing Genetically Engineered Plants and Microorganisms. Technical Bulletin 1783. Washington, DC: USDA-Animal Plant and Health Inspection Service.

Asano, Y., Y. Otsuki, and M. Ugaki. 1991. Electroporation-mediated and silicon carbide whisker-mediated DNA delivery in *Agrostis alba* L. (Redtop). Plant Science 79:247–252.

Attayde, J.L., and L.A. Hansson. 2001. Press perturbation experiments and the indeterminacy of ecological interactions: Effects of taxonomic resolution and experimental duration. Oikos 92 235–244.

Azarovitz, T.R., C.J. Byrne, M.J. Silverman, B.L. Freeman, W.G. Smith, and S.C. Turner. 1979. In oxygen depletion and associated benthic moralities in New York bight, 1976. Pp. 295–314 in R.L. Swanson and C.J. Sindermann, eds. NOAA Professional Paper No. 11. Washington, DC: National Oceanic and Atmospheric Administration.

Azpironz-Leehan, R., and K.A. Feldmann. 1997. T-DNA insertion mutagenesis in *Arabidopsis*: Going back and forth. Trends in Genetics 13:152–156.

Bacheler, J. 2000. North Carolina State University Extension. Personal communication.

Baez-Saldana, A., G. Diaz, B. Espinoza, and E. Ortega. 1998. Biotin deficiency induces changes in subpopulations of spleen lymphocytes in mice. American Journal of Clinical Nutrition 67:431–437.

Baker, H.G. 1965. Characteristics and modes of origin of weeds. Pp. 147–142 in The Genetics of Colonizing Species, H.G. Baker and G.L. Stebbins, eds. New York: Academic Press.

Ball, D.A., D. Cudney, S.A. Dewey, C.L. Elmore, R.G. Lym, D.W. Morishita, R. Parker, D. G. Swan, T.D. Whitson, and R.K. Zollinger. 2000. Weeds of the West, 9th ed. Newark, CA: Western Society of Weed Science.

Bandara, C.M.M. 1977. Hydrological consequences of agrarian change, Pp. 323–339 in Green Revolution? Technology and Change in Rice-Growing Areas of Tamil Nadu and Sri Lanka, B.H. Farmer, ed. Boulder, CO: Westview Press.

Barrett, S.C.H. 1983. Crop mimicry in weeds. Economic Botany 37:255–282.

Barry, D., D. Alfaro, and L.L. Darrah. 1994. Relation of European corn borer (Lepidoptera: Pyralidae) leaf-feeding resistance and DIMBOA content in maize. Environmental Entomology 23(1):177–182.

Bentur, J.S., M.B. Cohen, and F. Gould. 2000. Insecticide resistance and resistance management. Journal of Economic Entomology 93(6):1773–1778.

Berlow, E.L. 1999. Strong effects of weak interactions in ecological communities. Nature 398:330–334.

Bernstein, A. 1998. Precaution and respect. Pp. 148–157 in Protecting Public Health & the Environment: Implementing the Precautionary Principle, C. Raffensperger and J. Tickner, eds. Washington, DC: Island Press.

Biotech Basics. 2001. A Brief Biotech Timeline. Monsanto Company. Available online at *www.biotechbasics.com/timeline.html.*

Birch, R.G. 1997. Plant transformation: Problems and strategies for practical application. Annual Review of Plant Physiology and Plant Molecular Biology 48:297–326.

Blossey, B., and R. Nötzold. 1995. Evolution of increased competitive ability in invasive non-indigenous plants: A hypothesis. Journal of Ecology 83:887–889.

Borlaug, N.E. 2000. Ending world hunger: The promise of biotechnology and the threat of antiscience zealotry. Plant Physiology 124: 487–490.

Borsani, O., J. Cuartero, J.A. Fernandez, V. Valpuesta, and M.A. Botella. 2001. Identification of two loci in tomato reveals distinct mechanisms for salt tolerance. Plant Cell 4:873–887.

Boudry, P., M. Mörchen, P. Saumitou-Laprade, P. Vernet, and H. Van Dijk. 1993. The origin and evolution of weed beets: Consequences for the breeding and release of herbicide-resistant transgenic sugar beets. Theoretical and Applied Genetics 87:471–478.

Boudry, P., K. Broomberg, P. Saumitou-Laprade, M. Murchen, J. Cuguen, and H. Van Dijk. 1994. Gene escape in transgenic sugar beet: What can be learned from molecular studies of weed beet populations? Pp. 75–87 in Proceedings of the 3rd International Symposium on the Biosafety Results of Field Tests of Genetically Modified Plants and Microorganisms, D.D. Jones, ed. Oakland, CA: University of California, Division of Agriculture and Natural Resources.

Bradley, J.R. 2001. North Carolina State University, Department of Entomology. Personal communication.

Bresson, R. 1999. Testimony before the Committee on Agriculture, Nutrition and Forestry, U.S. Senate regarding Development of Biotechnology and Its Potential Applications in the Agriculture and Food Sectors.

Brossman, G.D., and J.R. Wilcox. 1984. Induction of genetic variability for oil properties and agronomic characteristics of soybean. Crop Science 24:783–787.

Brown, L. 1995. Who Will Feed China?: Wake-up Call for a Small Planet. Washington, DC: Worldwatch Institute.

Bruins, B.G., W. Scharloo, and G.E.W. Thorig. 1991. The harmful effect of light on *Drosophila* is diet dependent. Insect Biochemistry 21:535–539.

Bunders, J., and H. Radder. 1995. The appropriate realization of agricultural biotechnology. Pp. 177–204 in Issues in Agricultural Bioethics, T.B. Mepham, G.A. Tucker, and J. Wisman, eds. Nottingham: University of Nottingham Press.

Burdon, J.J., and G.A. Chilvers. 1994. Demographic changes and the development of competition in a native Australian eucalypt forest invaded by exotic pines. Oecologia 97:419–423.

Busch, L., Lacey, W.B., J. Burkhardt, and L. Lacey. 1990. Plants, Power and Profit. Oxford: Basil Blackwell.

Buttel, F.H. 1995. The global impacts of agricultural biotechnology: A post-green revolution perspective. Pp. 345–360 in Issues in Agricultural Bioethics, T.B. Mepham, G.A. Tucker, and J. Wisman, eds. Nottingham: University of Nottingham Press.

Byers, J.E., and L. Goldwasser. 2001. Exposing the mechanisms and timing of impact of non-indigenous species on native species. Ecology 85:1330–1343.

Camazine, S., and R.A. Morse. 1988. The Africanized honeybee. American Science 76:465–471.

Campbell, C.L., and L.V. Madden. 1990. Introduction to Plant Disease Epidemiology. New York: John Wiley.

Campbell, N.A, L.G. Mitchell, and J.B. Reece. 1997. Biology: Concepts and Connections, 2nd ed. Menlo Park, CA: Benjamin Cummings.

Carpenter, S.R., and K.L. Cottingham. 1997. Resilience and restoration of lakes. Conservation Ecology 1(1):2. Available online at *www.consecol.org/vol1/iss1/art2*.

Carr, S., and L. Levidow. 2000. Exploring the links between science, risk, uncertainty, and ethics in regulatory controversies about genetically modified crops. Journal of Agricultural and Environmental Ethics 12:29–39.

Carroll, C.R. 1990. The interface between natural areas and agroecosystems. Pp. 362–384 in Agroecology, C.R. Carroll, J.H. Vandermeer, and P.M. Rosset, eds. New York: McGraw-Hill.

Carson, R. 1962. Silent Spring. Boston, MA: Houghton-Mifflin.

Cassman, K.G., S.K. De Datta, D.C. Olk, J. Alcantara, M. Samson, J.P. Descalosota, and M. Dizon. 1995. Yield decline and the nitrogen economy of long-term experiments on continuous, irrigated rice systems in the tropics. Pp. 181–222 in Soil Management: Experimental Basis for Sustainability and Environmental Quality, R. Lal and B.A. Stewart, eds. Boca Raton, FL: CRC Press.

CBD (Secretariat of the Convention on Biological Diversity). 2000. Cartagena Protocol on Biosafety to the Convention on Biological Diversity: Text and annexes. Montreal: Secretariat of the Convention on Biological Diversity. Available online at *www.biodiv.org/doc/legal/cartagena-protocol-en.pdf*.

Chan, M-T., H-H. Chang, and T-M. Lee. 1992. Transformation of indica rice (*Oryza sativa* L.) mediated by *Agrobacterium tumefaciens*. Plant Cell Physiology 33:577–583.

Chandler V.L., and H. Vaucheret. 2001. Gene activation and gene silencing. Plant Physiology 125:145–148.

Charleux, J-L. 1996. Beta-carotene, vitamin C, and vitamin E: The protective micronutrients. Nutrition Review 54:S109–S114.

Chazallon, G. 2000. Mesures complémentaires à prendre an vue de la production de semences tolérantes à un herbicide non sélectif. Pp. 221–229 in Proceedings of the 63rd IIRB congress, February 2000. Interlaken (CH).

Cisar, J.L., K.E. Williams, H.E. Vivas, and J.J. Haydu. 2000. The occurrence and alleviation surfactants of soil-water repellency on sand-based turfgrass systems. Journal of Hydrology 231–232:352–358.

Clegg, M.T., and M.L. Durbin. 2000. Flower color variation: A model for the experimental study of evolution. Proceedings of the National Academy of Sciences of the United States of America 97:7016–7023.

Cochrane, W.W. 1979. The Development of American Agriculture: A Historical Analysis. Minneapolis: University of Minnesota Press.

Cohen, M.B., A.M. Romena, R.M. Aguda, A. Dirie, and F.L. Gould. 1996. Evaluation of resistance management strategies for *Bt* rice. Pp. 496–505 in Proceedings, Second Pacific Rim conference on biotechnology of *Bacillus thuringiensis* and its impact to the environment. Entomology and Zoology Association of Thailand, Bangkok.

Colwell, R.E., E.A. Norse, D. Pimentel, F.E. Sharples, and D. Simberloff. 1985. Genetic Engineering in Agriculture. Science 229:111–112.

Conway, G. 1998. The Doubly Green Revolution: Food for All in the 21st Century. Ithaca, NY: Comstock.

Conway, G. 2000. Genetically modified crops: Risks and promise. Conservation Ecology 4(1):2. Available online at *www.consecol.org/vol4/iss1/art2*.

Cooke, B.J., and J. Roland. 2000. Spatial analysis of large-scale patterns of forest tent caterpillar outbreaks. Ecoscience 7:410–422.

Corn Refiners Association. 1999. Corn Annual. Washington, DC: Corn Refiners Association, Inc.

Cranor, C.F. 1997. The normative nature of risk assessment: Features and possibilities. Risk: Health, Safety, and Environment 8:123–136.

Cranor, C.F. 1998. Asymmetric information: The precautionary principle and burdens of proof. Pp. 74–99 in Protecting Public Health & the Environment: Implementing the Precautionary Principle, C. Raffensperger and J. Tickner, eds. Washington, DC: Island Press.

Crawley, M.J., and S.L. Brown. 1995. Seed limitation and the dynamics of oilseed rape on the M25 motorway. Proceedings of the Royal Society London Series B Biological Sciences 259:49–54.

Crawley, M.J., R.S. Hails, M. Rees, D. Kohn, and J. Buxton. 1993. Ecology of transgenic oilseed rape in natural habitats. Nature 363:620–623.

Crawley, M.J., S.L. Brown, R.S. Hails, D.D. Kohn, and M. Rees. 2001. Biotechnology: Transgenic crops in natural habitats. Nature 409(6821):682–683.

Crecchio, C., and G. Stotzky. 2001. Biodegradation and insecticidal activity of the toxin from *Bacillus thuringiensis* subsp kurstaki bound on complexes of montmorillonite-humic acids-Al hydroxypolymers. Soil Biology and Biochemistry 33(4–5):573–581.

Crooks, J.A., and M.E. Soulé. 1999. Lag times in population explosions of invasive species: Causes and implications. Pp. 103–125 in Invasive species and biodiversity management, O.T. Sandlund, P.J. Schrei, and A. Viken, eds. Dordrecht: Kluwer Academic Publishers.

Dahlberg, K. 1979. Beyond the Green Revolution: The Ecology and Politics of Global Agricultural Development. New York: Plenum Press.

Daily, G.C. 1999. Developing a scientific basis for managing Earth's life support systems. Conservation Ecology 3(2):14. Available online at *www.consecol.org/vol3/iss2/art14*.

Dale, V.H., S. Brown, R.A Haeuber, N.T. Hobbs, N. Huntly, R.J. Naiman, W.E. Riebsame, M.G. Turner, and T.J. Valone. 2000. Ecological principles and guidelines for managing the use of land. Ecological Applications 10:639–670.

Dalrymple, D.G. 1986. Development and spread of high-yielding rice varieties in developing countries. Washington, DC: Agency for International Development.

Danbom, D. 1979. The Resisted Revolution: Urban America and the Industrialization of Agriculture, 1900–1930. Ames: Iowa State University Press.

Daniell, H., S.J. Streatfield, and K. Wycoff. 2001. Medical molecular farming: Production of antibodies, biopharmaceuticals and edible vaccines in plants. Trends in Plant Science 6:219–226.

D'Antonio, C.M., and P.M. Vitousek. 1992. Biological invasions by exotic grasses: The grass/fire cycle, and global change. Annual Review of Ecology and Systematics 23:63–87.

Darmency, H. 1994. The impact of hybrids between genetically modified crop plants and their related species: Introgression and weediness. Molecular Ecology 3:37–40.

Davidson, R.H., and W.F. Lyon. 1987. Insect Pests of Farm, Garden, and Orchard, 8th ed. New York: John Wiley & Sons.

Dawe, D., A. Dobermann, P. Moya, S. Abdulrachman, B. Singh, P. Lai, S.Y. Li, G. Panaullah, O. Sariam, Y. Singh, A. Swarup, P.S. Tan, and Q.X. Zhen. 2000. How widespread are yield declines in long-term rice experiments in Asia? Field Crops Research 66: 175–193.

De Leo, G.A., and S. Levin. 1997. The multifaceted aspects of ecosystem integrity. Conservation Ecology 1(1):3. Available online at *www.consecol.org/vol1/iss1/art3.*

DellaPenna, D. 1999. Nutritional genomics: Manipulating plant micronutrients to improve human health. 1999. Science 285:375–379.

Delp, C.J. 1988. Fungicide resistance problems in perspective. Pp. 4–5 in Fungicide Resistance in North America, C.J. Delp, ed. St. Paul, Minn.: APS Press.

Desbiez, C., and H. Lecoq. 1997. Zucchini yellow mosaic virus. Plant Pathology 46:809–829.

De Wet, J.M.J., and J.R. Harlan. 1975. Weeds and domesticates: Evolution in the man-made habitat. Economic Botany 29:99–107.

Dewey, R.E., C.S. Levings, and D.H. Timothy. 1986. Novel recombinations in the maize mitochondrial genome produces a unique transcriptional unit in the Texas male-sterile cytoplasm. Cell 44:439–449.

Diaz, R.J., and A. Solow. 1999. Ecological and economic consequences of hypoxia: Topic 2 report for the Integrated Assessment on Hypoxia in the Gulf of Mexico. NOAA Coastal Ocean Program Decision Analysis Series No. 16. Silver Spring, MD: NOAA Coastal Ocean Program.

Dobermann, A., D. Dawe, R.P. Roetter, and K.G. Cassman. 2000. Reversal of rice yield decline in a long-term continuous cropping experiment. Agronomy Journal 92:633–643.

Doebley, J. 1990. Molecular evidence for gene flow among *Zea* species. BioScience 40:443–448.

Doetschaman, T., R.G. Gregg, N. Maeda, M.L. Hooper, D.W. Melton, S. Thompson, and O. Smithies. 1987. Nature 330:576–578. Targeted correction of a mutant HPRT gene in mouse embryonic stem cells.

Donagan, K.K., and R.J. Seidler. 1999. Effects of transgenic plants on soil and plant microorganisms. Recent Research Developments in Microbiology 3:415–424.

Donagan K.K., C.J. Palm, V.J. Fieland, L.A. Porteous, L.M. Ganio, D.L. Schaller, L.Q. Bucao, R.J. Seidler. 1995. Changes in levels, species, and DNA fingerprints of soil microorganisms associated with cotton expressing the *Bacillus thuringiensis* var. *kurstaki* endotoxin. Applied Soil Ecology 2:111–114.

Donagan K.K., D.L. Schaller, J.K. Stone, L.M. Ganio, G. Reed, P.B. Hamm, R.J. Seidler. 1996. Microbial populations, fungal species diversity and plant pathogen levels in field plots of potato plants expressing the *Bacillus thuringiensis* var. *tenebrionis* endotoxin. Transgenic Restoration 5:25–35.

Draaijers, G.P., et al. 1989. The contribution of ammonia emissions from agriculture to the deposition of acidifying and eutrophying compounds onto forests. Environmental Pollution 60:55–66.

Drèze, J., and A.K. Sen. 1989. Hunger and Public Action. Oxford: Oxford University Press.

Dunlap, T.R. 1981. DDT: Scientists, Citizens, and Public Policy. Princeton, NJ: Princeton University Press.

Durant, J., and G. Gaskell. In Press. Biotechnology 1996–1999: The Years of Controversy. London: The Science Museum.

Durant, J., M.W. Bauer, and G. Gaskell. 1998. Biotechnology in the Public Sphere. London: The Science Museum.

Ellstrand, N.C. 1988. Pollen as a vehicle for the escape of engineered genes? Pp. S30–S32 in Planned Release of Genetically Engineered Organisms, J. Hodgson and A.M. Sugden, eds. Cambridge: Elsevier Publications.

Ellstrand, N.C. 2001. When transgenes wander, should we worry? Plant Physiology 125:1543–1545.

Ellstrand, N.C., and D.R. Elam. 1993. Population genetic consequences of small population size: Implications for plant conservation. Annual Review of Ecology and Systematics 24:217–242.

Ellstrand, N.C., and K. Schierenbeck. 2000. Hybridization as a stimulus for the evolution of invasiveness in plants? Proceedings of the National Academy of Sciences of the United States of America 97:7043–7050.

Ellstrand, N.C., H.C. Prentice, and J.F. Hancock. 1999. Gene flow and introgression from domesticated plants into their wild relatives. Annual Review of Ecological Systems 30:539–563.

EPA (Environmental Protection Agency). 1995. Pesticide Fact Sheet, Issued 8/10/1995 for *Bacillus thuringiensis* CryIA(b) δ-endotoxin and the genetic material necessary for its production (Plasmid vector pCIB4431) in corn. Washington, DC: United States Environmental Protection Agency, Office of Prevention, Pesticides and Toxic Substances (7501C).

EPA (Environmental Protection Agency). 2000a. *Bt* Plant-Pesticides Biopesticides Registration Action Document. Available online at *www.epa.gov/scipoly/sap/2000/october/brad3_enviroassessment.pdf.*

EPA (Environmental Protection Agency). 2000b. FIFRA Scientific Advisory Panel Meeting, October 18–20, 2000 EPA Draft report on *Bt* plant-pesticides risk and benefit assessments. Washington, DC: United States Environmental Protection Agency, Office of Prevention, Pesticides and Toxic Substances.

Ervin, D., S. Batie, R. Welsh, C. Carpentier, J. Fern, N. Richman, and M. Schulz. 2000. Transgenic crops: An environmental assessment. Policy Studies Report, Wallace Center for Agricultural and Environmental Policy, Winrock International.

Esquinas-Alcazar, J.T., and P.J. Gulick. 1983. Genetic resources of Cucurbitaceae: A global report. Rome: International Board of Plant Genetic Research.

European Commission. 2000. Communication from the Commission on the Precautionary Principle COM 1. Brussels.

European Commission. 2001. Directive 2001/18/EC of the European Parliament and of the Council of 12 March 2001 on the deliberate release into the environment of genetically modified organisms and repealing Council Directive 90/220/EEC. Official Journal of the European Communities 17.4.2001: L 106:1–38.

Evenson, R., and C. David. 1993. Adjustment and technology. Paris: Organization for Economic Cooperation and Development.

Ewel, J.J., D.J. O'Dowd, J., Bergelson, C.C. Daehler, C.M. D'Antonio, L.D. Gomez, D.R. Gordon, R.J. Hobbs, A. Holt, K.R. Hopper, C.E. Hughes, M. LaHart, R.R.B. Leakey, W. G. Lee, L.L. Loope, D.H. Lorence, S.M. Louda, A.E. Lugo, P.B. McEvoy, D.M. Richardson, and P.M. Vitousek. 1999. Deliberate introductions of species: Research needs—Benefits can be reaped, but risks are high. BioScience 49:619–630.

Fahrig, L. 1997. Relative effects of habitat loss and fragmentation on population extinction. Journal of Wildlife Management 61:603–610.

FDA (Food and Drug Administration). 1992. Statement of policy: Foods derived from new plant varieties; Notice. Federal Register 57:22984–23005.

Federal Register. 1992. Genetically Engineered Organisms and Products; Notification Procedures for the Introduction of Certain Regulated Articles; and Petition for Nonregulated Status. Final Rule. Animal and Plant Health Inspection Service, USDA. Docket No. 92-156-02. Washington, DC: U.S. Government Printing Office.

Fehr, W.R. 1989. Soybean. Pp. 283–300 in Oil Crops of the World, G. Robbelen, R.K. Downey, and A. Ashri, eds. New York: McGraw-Hill.

Feitelson, J.S., J. Payne, and L. Kim. 1992. *Bacillus thuringiensis*: Insects and beyond. Bio/Technology 10(3):271–275.

Fernandez-Cornejo, J., and W.D. McBride. 2000. Genetically engineered crops for pest management in US agriculture: Farm level effects. Agriculture Economic Report No. 786. Washington, DC: Resource Economics Division, Economic Research Service, U.S. Department of Agriculture.

Ferro, D.N., Q.C. Yuan, A. Slocombe, and A.F. Tuttle. 1993. Residual activity of insecticides under field conditions for controlling the Colorado potato beetle (coleoptera, chrysomelidae). Journal of Economic Entomology 86(2):511–516.

Ford-Lloyd, B. 1995. Sugarbeet, and other cultivated beets. Pp. 35–40 in Evolution of Crop Plants, 2nd ed., J. Smartt and N.W. Simmonds, eds. Harlow, UK: Longman.

Forman, R. 1997. Land Mosaics—The Ecology of Landscapes and Regions. Cambridge: Cambridge University Press.

Frankel, R., and E. Galun. 1977. Pollination Mechanisms, Reproduction and Plant Breeding. Berlin: Springer-Verlag.

Frewer, L. J., C. Howard, and R. Shepherd. 1997. Public concerns in the United Kingdom about general and specific applications of genetic engineering: Risk, benefit and ethics. Science, Technology, and Human Values 22:98–124.

Frey, M, P. Chomet, E. Glawischnig, C. Stettner, S. Grün, A. Winklmair, W. Eisenreich, A. Bacher, R.B. Meeley, S.P. Briggs, K. Simcox, and A. Gierl. 1997. Analysis of a chemical plant defense mechanism in grasses. Science 277(5326):696–699.

Fukami, T. 2001. Sequence effects of disturbance on community structure. Oikos 92:215–224.

Fumento, M. 1993. Science Under Siege: Balancing Technology and the Environment. New York: William Morrow.

Funtowicz, S.O., and J.R. Ravetz. 1990. Uncertainty and Quality in Science for Policy. Amsterdam: Kluwer.

Galinat, W.C. 1988. The origin of corn. Pp. 1–31 in Corn and Corn Improvement, 3rd ed, G.F. Sprague and J.W. Dudley, eds. Madison, WI: American Society of Agronomy.

Gallagher, K.D., P.E. Kenmore, and K. Sogawa. 1994. Judicial use of insecticides deter planthopper outbreaks and extend the life of resistant varieties in southeast Asian rice. Pp. 599–614 in Planthoppers: Their Ecology and Management, R.F. Denno and J.T. Perfect, eds. New York: Chapman & Hall.

Gaskell, G., J. Durant, and M. Bauer, eds. Biotechnology: The Years of Controversy. Cambridge: Cambridge University Press. In press.

Gates, C.A., and M.M. Cox. 1988. FLP recombinase is an enzyme. Proceedings of the National Academy of Sciences of the United States of America 85:4628–4632.

Gazey, W.J., B.J. Galloway, R.C. Fechhelm, L.R. Martin, and L.A. Reitsema. 1982. Shrimp mark and release and port interview sampling survey of shrimp catch and effort with recovery of captured tagged shrimp. Shrimp Population Studies: West Hackberry and Big Hill Brine Disposal Sites Off Southwest Louisiana and Upper Texas Coasts, 1980–1982, Vol. 3. W.B. Jackson, ed. Washington, DC: U.S. Department of Commerce.

Georghiou, G.P. 1986. The magnitude of the resistance problem. Pp. 14–34 in Pesticide Resistance: Strategies and Tactics for Management. National Research Council. Washington, DC: National Academy Press.

Georghiou, G. P., and A. Lagunes. 1988. The Occurrence of Resistance to Pesticides: Cases of Resistance Reported Worldwide Through 1988. Rome: Food and Agriculture Organization of the United Nations.

Ghosh, B.C., and R. Bhat. 1998. Environmental hazards of nitrogen loading in wetland rice fields. Environmental. Pollution 102:S123–S126.

Gianessi, L.P., and J.E. Carpenter. 1999. Agricultural Biotechnology: Insect Control Benefits. Washington, DC: National Center for Food and Agricultural Policy. Available online at *www.ncfap.org/pup/biotech/insectcontrolbenefits.pdf*.

Goebel, J.J. 1998. The National Resources Inventory and Its Role in U.S. Agriculture. Agricultural Statistics 2000. Proceedings of the Conference on Agricultural Statistics Organized by the National Agricultural Statistics Service of the U.S. Department of Agriculture, under the auspices of the International Statistical Institute. Available online at *www.statlab.iastate.edu/survey/nri/#Goebel*.

Goodman, R.M., and N. Newell. 1985. Genetic engineering of plants for herbicide resistance: Status and prospects. Pp. 47–53 in Engineered Organisms in the Environment: Scientific issues, H. Halvorson, D. Pramer, and M. Rogul, eds. Washington, DC: American Society for Microbiology.

Goolsby, D.A., W.A. Battaglin, G.B. Lawrence, R.S. Artz, B.T. Aulenbach, R.P. Hooper, D.R. Keeney, and G.J. Stensland. 1999. Flux and Sources of Nutrients In the Mississippi-Atchafalaya River Basin. Topic 3 Report for the Integrated Assessment of Hypoxia in the Gulf of Mexico. NOAA Coastal Ocean Program Decision Analysis Series No. 17. NOAA Coastal Ocean Program, Silver Spring, MD.

Gould, F. 1998. Sustainability of transgenic insecticidal cultivars: integrating pest genetics and ecology. Annual Review of Entomology 43:701–726.

Gould, F.W. 1968. Grass Systematics. New York: McGraw-Hill.

Graham, J.D., and J.B. Wiener. 1995. Risk Versus Risk: Tradeoffs in Protecting Health and the Environment. Cambridge, MA: Harvard University Press.

Gray, A.J., D.F. Marshall, and A.F. Raybould. 1991. A century of evolution in *Spartina anglica*. Advances in Ecological Research 211–261.

Green, A.G., 1986. A mutant genotype of flax (*Linum usitatissimum* L.) containing very low levels of linolenic acid in its seed oil. Canadian Journal of Plant Science 66: 99–503.

Green, M.B., H.M. LeBaron, and W.K. Moberg, eds. 1990. Managing Resistance to Agrochemicals: From Fundamental Research to Practical Strategies. Washington, DC: American Chemical Society.

Greenland, D.J. 1997. The Sustainability of Rice Farming. Oxford, UK: CAB International.

Greenwalt, L.A. 1992. Global indicators: What the people expect. Pp. 109–114 in Ecological Indicators, vol. 2, D.H. McKenzie, D.E. Hyatt, and V.J. McDonald, eds. London: Elsevier Applied Science.

Griffitts, J.S., J.L. Whitacre, D.E. Stevens, R.V. Aroian. 2001. *Bt* toxin resistance from loss of a putative carbohydrate-modifying enzyme. Science 293(5531):860–864.

Grumbine, R.E. 1994. What is ecosystem management? Conservation Biology 8:27–38.

Grumet, R. 1995. Genetic engineering for crop virus resistance. Hortscience 30: 449–456.

Gunderson, L.H. 2000. Ecological resilience in theory and application. Annual Review of Ecological Systems 31:425–439.

Hadidi, A., R.K. Khetarpal, and H. Koganezawa. 1998. Plant Virus Disease Control. St. Paul, MN: APS Press.

Hails, R.S. 2000. Genetically modified plants—the debate continues. Trends in Ecological Evolution 15:15–18.

Haimes, Y.Y. 1998. Risk Modeling, Assessment and Management. New York: John Wiley.

Hall, L., K. Topinka, J. Huffman, L. Davis, and A. Good. 2000. Pollen flow between herbicide-resistant *Brassica napus* is the cause of multiple-resistant *B. napus* volunteers. Weed Science 48(6):688–694.

Hallauer, A.R. 1990. Methods used in developing maize inbreds. Maydica 35:1–16.

Handel, S.N. 1983. Pollination ecology, plant population structure, and gene flow. Pp. 163–212 in Pollination Biology, L. Real, ed. New York: Academic Press.

Hanna, W.W., and J.E. Elsner. 1999. Registration of "TifEagle" bermudagrass. Crop Science 39:1258.

Hannah, L.C. 1997. Starch synthesis in the maize seed. Pp. 375–405 in Cellular and Molecular Biology of Plant Seed Development, B.A. Larkins and I.K. Vasil, eds. Kluwer: Dordrecht.

Hanson, H., N.E. Borlaug, and R.G. Anderson. 1982. Wheat in the Third World. Boulder, CO: Westview Press.

Harries, H.C. 1995. Coconut. Pp. 57–62 in Evolution of Crop Plants, 2nd ed., J. Smartt and N.W. Simmonds, eds. Harlow: Longman.

Hasegawa, P.M., R.A. Bressan, J.K. Zhu, and H.J. Bohnert. 2000. Plant cellular and molecular responses to high salinity. Annual Review of Plant Physiology and Plant Molecular Biology 51:463–499.

Haycock, N.E., G. Pinay, and C. Walker. 1993. Nitrogen retention in river corridors: European Perspective. Ambio 22:340–346.

Heap, I. 2000. International Survey of Herbicide Resistant Weeds. March 23. Available at *www.weedscience.com*.

Hebert, C., R. Berthiaume, A. Dupont, and M. Auger. 2001. Population collapses in a forecasted outbreak of Lambdina fiscellaria (Lepidoptera: Geometridae) caused by spring egg parasitism by Telenomus spp. (Hymenoptera: Scelionidae). Environmental Entomology 30:37–43.

The Heinz Center. 1999. Designing a Report on the State of the Nation's Ecosystems: Selected Measurements for Croplands, Forests, and Coasts & Oceans. Washington, DC: The H. John Heinz III Center.

Hellenas, K.E., C. Branzell, H. Johnsson, and P. Slanina. 1995. High levels of glycoalkaloids in the established Swedish potato variety Magnum Bonum. Journal of Science and Food Agriculture 23:520–523.

Hellmich, R.L., B.D. Siegfried, M.K. Sears, D.E. Stanley-Horn, M.J. Daniels, H.R. Mattila, T. Spencer, K.G. Bidne, and L.C. Lewis. 2001. Monarch larvae sensitivity to *Bacillus thuringiensis*-purified proteins and pollen. Proceedings of the National Academy of Sciences of the United States of America 98(19).

Hengeveld, R. 1999. Modelling the impact of biological invasions. Pp. 127–138 in Invasive Species and Biodiversity Management, O.T. Sandlund, P.J. Schrei, and A. Viken, eds. Dordrecht: Kluwer Academic Publishers.

Heong, K.L., and K.G. Schoenly. 1998. Impact of insecticides on behaviore-natural enemy communities in tropical rice ecosystems. Pp. 381–403 in Ecotoxicology, Pesticides and Beneficial Organisms, P.T. Haskell and P. McEwen, eds. London: Chapman & Hall.

Herrera-Estrella, L. 1999. Transgenic plants for tropical regions: Some considerations about their development and their transfer to the small farmer. Proceedings of the National Academy of Sciences of the United States of America 96:5978–5981.

Hickman, J.C. 1993. The Jepson Manual: Higher Plants of California. University of California Press: Berkeley.

Hilbeck, A. 2001. Transgenic host plant resistance and non-target effects. Pp. 167–185 in Genetically engineered organisms: Assessing environmental and human health risks, D.K. Letourneau, and B.E. Burrows, eds. Boca Raton: CRC Press.

Hilbeck, A., M.S. Meier, and A. Raps. 2000. Review on Nontarget Organisms and *Bt* Plants. Zurich: EcoStrat GmbH.

Hilbeck, A., M. Baumgartner, P.M. Fried, and F. Bigler. 1998a. Effects of transgenic *Bacillus thuringiensis* corn-fed prey on mortality and development time of immature *Chrysoperla carnea* (Neuroptera: Chrysopidae). Environmental Entomology 27:480–487.

Hillbeck, A., W.J. Moar, M. Pusztai-Carey, A. Filippini, and F. Bigler. 1998b. Toxicity of *Bacillus thuringiensis* Cr1Ab toxin to the predator *Chrysoperla carnea* (Neuroptera: Chrysopidae). Environmental Entomology 27:1255–1263.

Hill, D.S. 1983. Agricultural Insect Pests of the Tropics and Their Control, 2nd ed. Cambridge, MA: Cambridge University Press.

Hill, D.S. 1987. Agricultural Insect Pests of Temperate Regions and Their Control. Cambridge, MA: Cambridge University Press.

Hoess, R.H., and K. Abremski. 1990. The Cre-*lox* recombination system. Pp. 99–109 in Nucleic Acids and Molecular Biology, Vol. 4, F. Eckstein and D.M.J. Lilley, eds. Berlin: Springer-Verlag.

Holden, L.R., J.A. Graham, R.W. Whitmore, W.J. Alexander, and R.W. Pratt. 1992. Results of the national alachlor well water survey. Environmental Science and Technology 26:935–943.

Holloway, J. 1991. Biodiversity and tropical agriculture: A biogeographic view. Outlook on Agriculture 20:9–13.

Holm, L.G., D.L. Plucknett, J.V. Pancho, and J.P. Herberger. 1977. The world's worst weeds: Distribution and biology. Honolulu: University Press of Hawaii.

Holm, L., J.V. Pancho, J.P. Herbarger, and D.L. Plucknett. 1979. A Geographical Atlas of World Weeds. New York: John Wiley & Sons.

Holt, J.S., and A.B. Boose. 2000. Potential for spread of *Abutilon theophrasti* in California. Weed Science 48: 43–52.

Holt, R.D., and M.E. Hochberg. 2001. Indirect interactions, community modules and biological control: A theoretical perspective. Pp.13–37 in Evaluating indirect effects of biological control, E. Wajnberg, J.K. Scott, and P.C. Quimby, eds. Wallingford, UK: CAB International.

Hood, E.E., D.R. Witcher, S. Maddock, T. Meyer, C. Baszczynski, M. Bailey, P. Flynn, J. Register, L. Marshall, D. Bond, E. Kulisek, A. Kusnadi, R. Evangelista, Z. Nikolov, C. Wooge, R.J. Mehigh, R. Herman, W.K. Kappel, D. Ritland, C.P. Li, and J.A. Howard. 1997. Commercial production of avidin from transgenic maize: Characterization of transformant, production, processing, extraction, and purification. Molecular Breeding 3:291–306.

Houseman, J.G., F. Campos, N.M.R. Thie, B.J.R. Philogene, J. Atkinson, P. Morand, J.T. Arnason. 1992. Effects of the maize-derived compounds DIMBOA and MBOA on growth and digestive processes of European corn borer (Lepidoptera: Pyralidae). Journal of Economic Entomology 85(3):669–674.

Hoy, C.W., J. Feldman, F. Gould, G.G. Kennedy, G. Reed, and J.A. Wyman. 1998. Naturally occurring biological control in genetically engineered crops. In Conservation Biological Control, P. Barbosa, ed. New York: Academic Press.

Huxel, G.R. 1999. Rapid displacement of native species by invasive species: Effect of hybridization. Biological Conservation 89:143–152.

Hynes, M.J. 1996. Genetic transformation of filamentous fungi. Journal of Genetics 75: 297–311.

IRRI (International Rice Research Institute). 1995. Water: A Looming Crisis. Manila, IRRI.

Ishida, Y., H. Saite, S. Ohta, Y. Hiei, T. Komari, and T. Kumashiro. 1996. High efficiency transformation of maize (*Zea mays L.*) mediated by *Agrobacterium tumefaciens*. Nature Biotechnology 14:745–750.

Islam, A.K.M.R., and K.W. Shepherd. 1991. Recombination between wheat and barley chromosomes. Barley Genetics VI:68–70.

Isshiki, M., K. Morino, M. Nakajima, R.J. Okagaki, S.R. Wessler, T. Izawa, and K. Shimamoto. 1998. A naturally occurring functional allele of the rice waxy locus has a GT to TT mutation at the 5= splice site of the first intron. Plant Journal 15:133–138.

Jackson, S.A., P. Zhang, W.P. Chen, R.L. Phillips, B. Friebe, S. Muthukrishnan, and B.S. Gill. 2001. High-resolution structural analysis of biolistic transgene integration into the genome of wheat. Theoretical and Applied Genetics 103:56–62.

Jackson, T.A. 1997. Long-range atmospheric transport of mercury to ecosystems and the importance of anthropogenic emissions: A critical review and evaluation of the published evidence. Environmental Reviews 5(2):99–120.

Jain, S., and P. Martins. 1999. Ecological genetics of the colonizing ability of rose clover (*Trifolium hirtum* All.). American Journal of Botany 66:361–366.

James, C. 1998. Global Review of Commercialized Transgenic Crops. ISAAA Briefs No. 8. Ithaca, NY: International Service for the Acquisition of Agri-Biotech Applications (ISAAA).

James, C. 2000. Global status of commercialized transgenic crops: 2000. ISAAA Briefs no. 21: preview. Ithaca, N.Y.: International Service for the Acquisition of Agri-Biotech Applications (ISAAA).

Jasanoff, S. 1986. Risk Management and Political Culture: A Comparative Study of Science in the Policy Context. New York: Sage.

Jefferies, R.L. 2000. Allochthonous inputs: Integrating population changes and food-web dynamics. Trends in Ecology and Evolution 15:19–22.

Jeffries, R.L. and J.L. Maron. 1997. The embarrassment of riches: Atmospheric deposition of nitrogen and community and ecosystem processes. Trends in Ecology & Evolution 12:74–78.

Jepson, P.C., B.A. Croft, and G.E. Pratt. 1994. Test systems to determine the ecological risks posed by toxin release from *Bacillus thuringiensis* genes in crop plants. Molecular Ecology 3:81–89.

Jesse, L.C.H., and J.J. Obrycki. 2000. Field deposition of *Bt* transgenic corn pollen: Lethal effects on the monarch butterfly. Oecologia 125(2):241–248.

Johnson, D.G. 1999. The growth of demand will limit output growth for food over the next quarter century. Proceedings of the National Academy of Sciences of the United States of America 96:915–920.

Johnson, N., C. Revenga, and J. Echeverria. 2001. Managing water for people and nature. Science 292:1071–1072.

Jung, R. 2001. Pioneer Hi-Bred. Personal communication.

Kais, J. 2001. Words (and axes) fly over transgenic trees. Science 292:34–36.

Kalaitzandonakes, N.G. 1999. Agrobiotechnology in the developing world. AgBioForum 2(3&4):147–149.

Kalendar, R., J. Tanskanen, S. Immonen, E. Nevo, and A.H. Schulman. 2000. Genome evolution of wild barley (*Hordeum spontaneum*) by BARE-1 retrotransposon dynamics in response to sharp microclimatic divergence. Proceedings of the National Academy of Sciences of the United States of America 97(12):6603–6607.

Kalter, R.J. 1985. The new biotech agriculture: Unforeseen economic consequences. Issues in Science and Technology 13:25–133.

Kareiva, P., W. Morris, and C.M. Jacobi. 1994. Studying and managing the risks of cross-fertilization between transgenic crops and wild relatives. Molecular Ecology 3:15–21.

Kareiva, P., I.M. Parker, and M. Pascual. 1996. Can we use experiments and models in predicting the invasiveness of genetically engineered organisms? Ecology 77:1670–1675.

Kasperson, R. 1992. The social amplification of risk: Progress in developing an integrative framework. Pp.153–178 in Social Theories of Risk, S. Krimsky and D. Golding, eds. New York: Praeger Press.

Kaul, S., et al. (Arabidopsis Genome Initiative). 2000. Analysis of the genome sequence of the flowering plant *Arabidopsis thaliana*. Nature 408:796–815.

Keeler, K.H. 1989. Can genetically engineered crops become weeds? Bio/Technology 7:1134–1139.

Kenney, M. 1986. Biotechnology: The University-Industrial Complex. New Haven: Yale University Press.

Kempin, S.A., S.J. Liljegren, L.M. Block, S.D. Rounsley, M.F. Yanofsky, and E. Lam. 1977. Targeted disruption in *Arabidopsis*. Nature 389:802–803.

Kerbes, R.H., P.M. Kotanen, and R.L. Jeffries. 1990. Destruction of wetland habitats by lesser snow geese: A keystone species on the west coast of Hudson Bay. Journal of Animal Ecology 27:242–248.

Khan, M.S., and P. Maliga. 1999. Fluorescent antibiotic resistance marker for tracking plastid transformation in higher plants. Nature Biotechnology 17:910–915.

Khush, G.S. 1995. Modern varieties—their real contribution to food supply and equity. GeoJournal 35:275–284.

Kiang, Y.T., J. Antonovics, and L. Wu. 1979. The extinction of wild rice (*Oryza perennis-formosana*) in Taiwan. Journal of Asian Ecology 1:1–9.

Kim, K.C. 1983. How to detect and combat exotic pests. Pp. 262–319 in Exotic Plant Pests and North American Agriculture, C.L. Wilson and C.L. Graham, eds. New York: Academic Press.

Kinney, A.J. 1994. Genetic modification of the storage lipids of plants. Current Opinion in Biotechnology 5:144–151.

Kirkpatrick, K.J., and H.D. Wilson. 1988. Interspecific gene flow in Cucurbita: *C. texana* vs. *C. pepo*. American Journal of Botany 75:519–527.

Kishore, G.M., and C. Shewmaker. 1999. Biotechnology: Enhancing human nutrition in developing and developed worlds. Proceedings of the National Academy of Sciences of the United States of America 96:5968–5972.

Kitamura, K. 1995. Genetic improvement of nutritional and food processing quality in soybean. Japan Agricultural Restoration Quarterly (JARQ) 29:1–8.

Kloppenburg, J., Jr. 1988. First the Seed: The Political Economy of Plant Technology: 1492–2000. Cambridge: Cambridge University Press.

Knowles, L. In press. Attitudes to biotechnology: A U.S.-European comparison. The Hastings Report.

Knowlton, F.F., E.M. Gese, and M.M. Jaeger. 1999. Coyote depredation control: An interface between biology and management. Journal of Range Management 52:398–412.

Kodama, H., T. Hamada, G. Horiguchi, M. Nishimura, and K. Iba. 1994. Genetic enhancement of cold tolerance by expression of a gene for chloroplast omega-3 fatty acid desaturase in transgenic tobacco. Plant Physiology 105:601–605.

Kodama, H., G. Horiguchi, T. Nishiuchi, M. Nishimura, and K. Iba. 1995. Fatty acid desaturation during chilling acclimation is one of the factors involved in conferring low temperature tolerance to young tobacco leaves. Plant Physiology 107:1177–1185.

Kohli, A., M. Leech, P. Vain, D.A. Laurie, and P. Christou. 1998. Transgene integration in rice engineered through direct DNA transfer supports a two-phase integration mechanism mediated by the establishment of integration hot spots. Proceedings of the National Academy of Sciences of the United States of America 95:7203–7208.

Kohli, A., D. Gahakwa, P. Vain, D.A. Laurie, and P. Christou. 1999. Transgene expression in rice engineered through particle bombardment: Molecular factors controlling stable expression and transgene silencing. Planta 208:88–97.

Koprek, T., S. Range, D. McElroy, J.D. Louwerse, R.E. Williams-Carrier, and P.G. Lemaux. 2001. Transposon-mediated single-copy gene delivery leads to increased transgene expression stability in barley. Plant Physiology 125:1534–1362.

Kramer, K. J. 2000. Avidin, an egg-citing insecticidal protein in corn. Agricultural Research 48(8):8–9.

Kramer, K.J., T.D. Morgan, J.E. Throne, F.E. Dowell, M. Bailey, and J.A. Howard. 2000. Transgenic maize expressing avidin is resistant to storage insect pests. Nature Biotechnology 18(6):670–674.

Kramer, M.G., and K. Redenbaugh. 1994. Commercialization of a tomato with an antisense polygalacturonase gene—The Flavr Savr™ tomato story. Euphytica 79:293–297.

Krimsky, S. 2000. Risk assessment and regulation of bioengineered food products. International Journal of Biotechnology 2:231–238.

Krimsky, S., and R. Wrubel. 1996. Agricultural Biotechnology and the Environment: Science, Policy and Social Issues. Urbana, IL: University of Illinois Press.

Kurzer, M.S., and X. Xu. 1997. Dietary phytoestrogens. Annual Review of Nutrition 17:353–381.

Kwon, S.H., K.H. Im, and J.R. Kim. 1972. Studies on diversity of seed weight in the Korean soybean land races and wild soybean. Korean Journal of Breeding 4:70–74.

Lacher, T.E., Jr., R.D. Slack, L.M. Coburn, and M.I. Goldstein. 1999. The role of agroecosystems in wildlife biodiversity. Pp. 147–165 in Biodiversity in Agroecosystems, W.W. Collins and C.O. Qualset, eds. Boca Raton, FL: CRC Press.

Lander, E.S. (International Human Genome Sequencing Consortium). 2001. Initial sequencing and analysis of the human genome. Nature 409(6822):860–921.

Landres, P.B. 1992. Ecological indicators: Panacea or liability? Pp. 1295–1318 in Ecological Indicators, vol. 2, D.H. McKenzie, D.E. Hyatt, and V.J. McDonald, eds. London: Elsevier Applied Science.

Langeland, T. 1983. A clinical and immunological study of allergy to hen's egg white. IV. Specific IgE antibodies to individual allergens in hen's egg white related to clinical and immunological parameters in egg-allergic patients. Allergy 38:493–500.

Langridge, W.H.R. 2000. Edible vaccines. Scientific American. September. Available online at *www.sciam.com/2000/0900issue/0900langridge.html*.

Laurila, J., I. Lasko, J.P.T. Valkonen, R. Hiltunen, and E. Pehu. 1996. Formation of parental type and novel glycoalkaloids in somatic hybrids between *Solanum brevidens* and *S. tuberosum*. Plant Science 118:145–155.

Letourneau, D.K., J.A. Hagen, and G.S. Robinson. 2001. *Bt* crops: Evaluating benefits under cultivation and risks from escaped transgenes in the wild. Pp. 33–98 in Genetically Engineered Organisms: Assessing Environmental and Human Health Effects, D.K. Letourneau and B.E. Burrows, eds. Boca Raton, FL: CRC Press.

Levin, D.A. 1981. Dispersal versus gene flow in plants. Annals of the Missouri Botanical Garden 68:233–253.

Levin, D.A., and H.W. Kerster. 1974. Gene flow in seed plants. Evolutionary Biology 7: 39–220.

Levin, D.A., J. Francisco-Ortega, and R.K. Jansen. 1996. Hybridization and the extinction of rare plant species. Conservation Biology 10:10–16.

Levinson, H.Z., and E.D. Bergmann. 1959. Vitamin deficiencies in the housefly produced by antivitamins. Journal of Insect Physiology 3:293–305.

Levinson, H.Z., J. Barelkovsky, and A.R. Bar Ilan. 1967. Nutritional effects of vitamin omission and antivitamin administration on development and longevity of the hide beetle *Dermestes maculatus* (Coleoptera: Dermestidae). Journal of Stored Products Research 3:345–352.

Levinson, H.Z., A.R. Levison, and M. Offenberger. 1992. Effect of dietary antagonists and corresponding nutrients on growth and reproduction of the flour mite (*Acaris siro* L.). Experientia 48:721–729.

Lewis, H.W. 1980. The safety of fission reactors. Scientific American 242:53–65.

Lewontin, R.C. 2000. P. 38 in The Triple Helix: Gene, Organism, and Environment. Cambridge, MA: Harvard University Press.

Lipton, M., and R. Longhurst. 1989. New Seeds for Poor People. Baltimore: Johns Hopkins University Press.

Livnah, O., E. Bayer, M. Wilchek, and J. Sussman. 1993. Three-dimensional structures of avidin and the avidin-biotin complex. Proceedings of the National Academy of Sciences of the United States of America 90:5076–5080.

Lloyd, E., and W. Tye. 1982. Systematic Safety. London: Civil Aviation Authority.

Longden, P.C. 1993. Weed beet: A review. Aspects of Applied Biology 35:185–194.

Lonsdale, W.M. 1999. Concepts and synthesis: Global patterns of plant invasions, and the concept of invasibility. Ecology 80:1522–1536.

Losey, J.E., L.S. Rayor, and M.E. Carter. 1999. Transgenic pollen harms monarch butterflies. Nature 399:214.

Losey, J.E., J.J. Obrycki, and R.A. Hufbauer. 2001. Impacts of Genetically-Engineered Crops on Non-Target Herbivores: *Bt*-Corn and Monarch Butterflies as a Case Study. In Genetically Engineered Organisms: Assessing Environmental and Human Health Effects, D.K. Letourneau and B.E. Burrows, eds. Boca Raton, FL: CRC Press.

Ludwig, D., B. Walker, and C.S. Holling. 1997. Sustainability, stability, and resilience. Conservation Ecology 1(1):8. Available online at *www.consecol.org*.

Luttrell, R.G. 1993. Strategies for resistance management. Pp. 15–19 in Proceedings of the Beltwide Cotton Conference. Memphis, Tenn.: National Cotton Council of America.

Luttrell, R.G., G.P. Fitt, F.S. Ramalho, and E.S. Sugonyaev. 1994. Cotton pest-management: A worldwide perspective. Annual Review of Entomology 39:517–526.

MacDonald, J.F. 1993. Agricultural Biotechnology: A Public Conversation About Risk. Ithaca, NY: National Agricultural Biotechnology Council.

Macdonald, R.W., L.A. Barrie, T.F. Bidleman, M.L. Diamond, D.J. Gregor, R.G. Semkin, W.M.J. Strachan, Y.F. Li, F. Wania, M. Alaee, L.B. Alexeeva, S.M. Backus, R. Bailey, J.M. Bewers, C. Gobeil, C.J. Halsall, T. Harner, J.T. Hoff, L.M.M. Jantunen, W.L. Lockhart, D. Mackay, D.C.G. Muir, J. Pudykiewicz, K.J. Reimer, J.N. Smith, G.A. Stern, W.H. Schroeder, R. Wagemann, and M.B. Yunker. 2000. Contaminants in the Canadian Arctic: 5 years of progress in understanding sources, occurrence and pathways. Science of the Total Environment 254:93–234.

MacIntosh, S.C., T.B. Stone, S.R. Sims, P.L. Hunst, J.T. Greenplate, P.G. Marrone, F.J. Perlak, D.A. Fischoff, and R.L. Fuchs. 1990. Specificity and efficacy of purified *Bacillus thuringiensis* proteins against agronomically important insects. Journal of Invertebrate Pathology 56:258–266.

Mack, R.N. 1996. Predicting the identity and fate of plant invaders: Emergent and emerging approaches. Biological Conservation 78:107–121.

Mack, R.N., D. Simberloff, W.M. Lonsdale, H. Evans, M. Clout, and F.A. Bazzaz. 2000. Biotic invasions: Causes, epidemiology, global consequences, and control. Ecological Applications 10:689–710.

Malcolm, S.B., and M.P. Zalucki. 1993. Biology and conservation of the monarch butterfly. Los Angeles, CA: Natural History Museum of Los Angeles County.

Malinowski, M.J. 2000. Biotechnology in the USA: Responsive regulation in the life science industry. International Journal of Biotechnology 2:16–25.

Mange, A.P., and E.J. Mange. 1990. Genetics: Human aspects, 2nd ed. Sunderland, MA: Sinauer.

Mariani, C., V. Gossele, M. Debeuckeleer, M. Deblock, R.B. Goldberg, W. Degreef, and J. Leemans. 1992. Nature 357:384–387.

Markwick, N.P., J.T. Christeller, L.C. Dochterty, and C.M. Lilley. 2001. Insecticidal activity of avidin and streptavidin against four species of pest Lepidoptera. Entomologia Experimentalis et Applicata 98:59–66.

Marroquin, L.D., D. Elyassnia, J.S. Griffitts, J.S. Feitelson, and R.V. Aroian. 2000. *Bacillus thuringiensis* (*Bt*) toxin susceptibility and isolation of resistance mutants in the nematode *Caenorhabditis elegans*. *Genetics* 155:1693–1699.

Marvier, M. 2001. Ecology of transgenic crops. American Scientist 89:160–167.

Marvier, M., and P. Kareiva. 1999. Extrapolating from field experiments that remove herbivores to population-level effects of herbivore resistance transgenes. Pp. 57–64 in Workshop on Ecological Effects of Pest Resistance Genes in Managed Ecosystems, P.L. Traynor and J.H. Westwood, eds. Blacksburg, VA: Information Systems for Biotechnology.

Matzke, M.A., A.J.M. Matzke, and M.F. Mette. 2000. Transgene silencing by the host genome defense: Implications for the evolution of epigenetic control mechanisms in plants and vertebrates. Plant Molecular Biology 43:401–415.

Matson, P.A., W.J. Parton, A.G. Power, and M.J. Swift. 1997. Agricultural intensification and ecosystem properties. Science 277:504–509.

Matteson, P. 2000. Insect pest management in tropical Asian irrigated rice. Annual Review of Entomology 545:549–574.

Matthee M., and D. Vermersch. 2000. Are the precautionary principle and the international trade of genetically engineered organisms reconcilable? Journal of Agricultural & Environmental Ethics 12:59–70.

Matthews, R.E.F. 1991. Plant Virology, 3rd ed. San Diego: Academic Press.

Mazur, B., E. Krebbers, and S. Tingey. 1999. Gene discovery and product development for grain quality traits. Science 285:372–375.

McCann, K.S. 2000. The diversity-stability debate. Nature 405:228–233.

McCormick, L.L. 1977. Weed survey–southern states. Southern Weed Science Society Research Report 30:184–215.

McGloughlin, M. 1999. Ten reasons why biotechnology will be important to the developing world. AgBioForum, 2(3&4):163–174.

Meyer, D.L., J. Schultz, Y. Lin, A. Henry, J. Sanderson, J.M. Jackson, S. Goshorn, A.R. Rees, and S.S. Graves. 2001. Reduced antibody response to streptavidin through site-directed mutagenesis. Protein Science 10:491–503.

M'Gonigle, R.M. 1998. The political economy of precaution. Pp. 123–147 in Protecting Public Health & the Environment: Implementing the Precautionary Principle, C. Raffensperger and J. Tickner, eds. Washington, DC: Island Press.

Miller, H., and G. Conko. 2001. Precaution without principle. Nature Biotechnology 19:302–303.

Minnich, R.A. 1998. Historical decline of coastal sage scrub in the Riverside-Perris Plain, California. Western Birds 29:366–391.

Mishima, S. 2001. Recent trends of nitrogen flow associated with agricultural production in Japan. Soil Science & Plant Nutrition 47:157–166.

Mississippi State University Extension Service. 1999. Managing Insects Attacking Corn. Available online at *http://msucares.com/pubs/pub899.htm*.

Montalvo, A.M., P.A. McMillan, and E.B. Allen. In press. The relative importance of seeding method, soil ripping, and soil variables on seeding success. Restoration Ecology.

Morgan, T.D., B. Oppert, T.H. Czapla, and K.J. Kramer. 1993. Avidin and streptavidin as insecticidal and growth inhibiting dietary proteins. Entomologia Experimentalis et Applicata 69:97–103.

Mourrain, P., C. Béclin, T. Elmayan, F. Feuerbach, C. Godin, J-B. Morel, D. Jouette, A-M. Lacombe, S. Nikic, N. Picault, K. Rémoué, M. Sanial, T-A. Vo, and H. Vaucheret. 2000. Arabidopsis SGS2 and SGS3 genes are required for posttranscriptional gene silencing and natural visus resistance. Cell 101:533–542.

Mücher, T., P. Hesse, M. Pohl-Orf, N.C. Ellstrand, and D. Bartsch. 2000. Characterization of weed beet in Germany and Italy. Journal of Sugar Beet Research 37(3):19–38.

Mysterud, A., N.C. Stenseth, N.G. Yoccoz, R. Langvatn, and G. Steinheim. 2001. Nonlinear effects of large-scale climatic variability on wild and domestic herbivores. Nature 410:1096–1099.

Nabhan, G.P. 1993. Songbirds, Truffles, and Wolves: An American Naturalist in Italy. New York: Penguin.

NRC (National Research Council). 1972. Genetic Vulnerability of Major Farm Crops. Washington, DC: National Academy Press.

NRC (National Research Council). 1983. Risk Assessment in the Federal Government: Managing the Process. Washington, DC: National Academy Press.

NRC (National Research Council). 1984. Genetic Engineering of Plants: Agricultural Research Opportunities and Policy Concerns. Washington, DC: National Academy Press.

NRC (National Research Council). 1987. Agricultural Biotechnology Strategies for National Competitiveness. Washington, DC: National Academy Press.

NRC (National Research Council). 1989. Field Testing Genetically Modified Organisms: Framework for Decisions. Washington, DC: National Academy Press.

NRC (National Research Council). 1994. Assessing Genetic Risks: Implications for Health and Social Policy. Washington, DC: National Academy Press.

NRC (National Research Council). 1996. Understanding Risk: Informing Decisions in a Democratic Society. Washington, DC: National Academy Press.

NRC (National Research Council). 2000a. Ecological Indicators for the Nation. Washington, DC: National Academy Press.

NRC (National Research Council). 2000b. The Future Role of Pesticides in U.S. Agriculture. Washington, DC: National Academy Press.

NRC (National Research Council). 2000c. Genetically Modified Pest-Protected Plants: Science and Regulation. Washington, DC: National Academy Press.

NRC (National Research Council). 2000d. Improving the Collection, Management, and Use of Marine Fisheries Data. Washington, DC: National Academy Press.

Naveira, H., and A. Barbadilla. 1992. The theoretical distribution of lengths of intact chromosome segments around a locus held heterozygous with backcrossing in a diploid species. Genetics 130:205–209.

Nepsted, D.C. et al. 1999. Large-scale impoverishment of Amazonian forests by logging and fire. Nature 398:505–508.

Nestle, M. 2001. Genetically engineered "golden" rice unlikely to overcome Vitamin A deficiency. Letter to the Editor, Jurnal of the American Dietetic Association 101: 289–290.

Nickson, T.E. 2001. The role of communication in ecological assessments of genetically modified crops. In Methods for Risk Assessment of Transgenic Plants, vol. IV. K. Ammann, Y. Jacot, G. Simonsen, G. Kjellsson, eds. Basel, Switzerland: Birkhauser.

Nielsen, K.M., J.D. van Elsas, and K. Smalla. 2000. Transformation of *Acinetobacter* sp. strain BD413 (pFG4Δ*nptII*) with transgenic plant DNA in soil microcosms and effects of kanamycin on selection of transformants. Applied Environmental Microbiology 66:1237–1242.

Nilsen, E.T., and W.H. Muller. 1980a. The relative naturalization ability of two Schinus species in Southern California. I: Seed germination. Bulletin of the Torrey Botanical Club 107:51–56.

Nilsen, E.T., and W.H. Muller. 1980b. The relative naturalization ability of two Schinus species in Southern California. II: Seedling establishment. Bulletin of the Torrey Botanical Club 107:232–237.

Nohara, S., and T. Iwakuma. 1996. Residual pesticides and their toxicity to freshwater shrimp in the littoral and pelagic zones of Lake Kasumigaura, Japan. Chemosphere 33:1417–1424.

Nordlee, J.A., S.L. Taylor, J.A. Townsend, L.A. Thomas, and R.K. Bush. 1996. Identification of a brazil-nut allergen in transgenic soybeans. New England Journal of Medicine 334:688–692.

Nuffield Council on Bioethics. 1999. Genetically Modified Crops: The Ethical and Social Issues. London: Nuffield Council on Bioethics.

Nusser, S.M., and J.J. Goebel. 1997. The national resources inventory: A long-term multi-resource monitoring programme. Environmental and Ecological Statistics 4:181–204.

Oberhauser, K.S., M.D. Prysby, H.R. Mattila, D.E. Stanley-Horn, M.K. Sears, G. Dively, E. Olson, J.M. Pleasants, W.F. Lam, and R. Hellmich. 2001. Temporal and spatial overlap between monarch larvae and corn pollen. Proceedings of the National Academy of Sciences of the United States of America 98(21):11913–11918.

Oehlert, G.W. 2000. A First Analysis in Design and Analysis of Experiments. Houndmills, England: Palgrave Publishers Ltd.

OSTP (Office of Science and Technology Policy). 1986. Coordinated Framework for the Regulation of Biotechnology. Federal Register 51:23303–23350.

OSTP (Office of Science and Technology Policy). 1992. Exercise of federal oversight within scope of statutory authority: Planned introductions of biotechnology products into the environment. Federal Register 57:6753–6762.

OSTP/CEQ (Office of Science and Technology Policy/Council on Environmental Quality). 2001. CEQ/OSTP Assessment: Case Studies of Environmental Regulation for Biotechnology. January. Available online at *www.ostp.gov.html/012201.html*.

Ogawa, T., H. Tsun, N. Bando, K. Kitamura, Y-L. Zhu, H. Hirano, and K. Nishikawa. 1993. Identification of the soybean allergenic protein, *Gly m* Bd 30K, with the soybean seed 34-kDa oil-body-associated protein. Bioscience, Biotechnology, Biochemistry 57:1030–1033.

Olsen, A.R., J. Sedransk, D. Edwards, C.A. Gotway, W. Liggett, S. Rathburn, K.H. Reckhow, and L.J. Young. 1999. Statistical issues for monitoring ecological and natural resources in the United States. Environmental Monitoring Assessment 54:1–45.

Orr, D.B., and D.A. Landis. 1997. Oviposition of European corn borer (Lepidoptera: Pyralidae) and impact of natural enemy populations in transgenic versus isogenic corn. Journal of Economic Entomology 90(4):905–909.

Orr, R.L., S.D. Cohen, and R.L. Griffin. 1993. Generic non-indigenous pest risk assessment process (for estimating pest risk associated with the introduction of non-indigenous organisms). Washington, DC: U.S. Department of Agriculture, Animal and Plant Health Inspection Service.

Ostlie, K., and W. Hutchinson. 1997. *Bt* Corn & European Corn Borer: Long-Term Success Through Resistance Management. University of Minnesota Extension Service. Available online at *www.extension.umn.edu/distribution/cropsystems/DC7055.html*.

Ostlie, K.R., W.D. Hutchison, and R.L. Hellmich, eds. 1997. *Bt* corn and European corn borer. NCR publication 602. St. Paul: University of Minnesota.

Otway, H. 1992. Public wisdom, expert fallibility: Toward a contextual theory of risk. Pp. 215–228 in Social Theories of Risk, S. Krimsky and D. Golding, eds. New York: Praeger Press.

Ozanoff, D. 1998. The precautionary principle as a screening device. Pp. 100–105 in Protecting Public Health & the Environment: Implementing the Precautionary Principle, C. Raffensperger and J. Tickner, eds. Washington, DC: Island Press.

Pair, S.D., J.R. Raulston, D.R. Rummel, J.K. Westbrook, W.W. Wolf, A.N. Sparks, and M.F. Schuster. 1987. Development and production of corn earworm and fall armyworm in the Texas high plains: Evidence for reverse fall migration. Southwestern Entomology 12:89–99.

Palmer, J.D., K.L. Adams, Y. Cho, C. L. Parkinson, Y-L. Qiu, and K. Song. 2000. Dynamic evolution of plant mitochondrial genomes: Mobile genes and introns and highly variable mutation rates. Pp. 35–57 in Variation and Evolution in Plants and Microorganisms: Toward a New Synthesis 50 Years After Stebbins, F.J. Ayala, W.M. Fitch, and M.T. Clegg, eds. Washington, DC: National Academy Press.

Parker, I.M., D. Simberloff, W.M. Lonsdale, K. Goodell, M. Wonham, P.M. Kareiva, M. H. Williamson, B. Von Holle, P.B. Moyle, J.E. Byers, and L. Goldwasser. 1999. Impact: Toward a framework for understanding the ecological effects of invaders. Biological Invasions 1:3–19.

Paterson, A.J., K.F. Schertz, Y-R. Lin, S-C. Liu, and Y-L. Chang. 1995. The weediness of wild plants: Molecular analysis of genes influencing dispersal and persistence of johnsongrass, *Sorghum halepense* (L.) Proceedings of the National Academy of Sciences of the United States of America 92:6127–6131.

Pawlowski, W.P., and D.A. Somers. 1998. Transgenic DNA integrated into the oat genome is frequently disrupted by host DNA. Proceedings of the National Academy of Sciences of the United States of America 95:12106–12110.

Peck, S.L., B. McQuaid, C. Lee Campbell. 1998. Using ant species (Hymenoptera: Formicidae) as a biological indicator of agroecosystem condition. Environmental Entomology 27(5):1102–1110.

Perfecto, I., R.A. Rice, R. Greenberg, and M. Van der Voort. 1996. Shade coffee: A disappearing refuge for biodiversity. Bioscience 46:598–608.

Perkins, J.H. 1982. Insects, Experts, and the Insecticide Crisis: The Quest for New Pest Management Strategies. New York: Plenum Press.

Perrins, J., M. Williamson, and A. Fitter. 1992. Do annual weeds have predictable characters? Acta Oecologica-International Journal of Ecology 13:517.

Perry, D.A. 1995. Self-organizing systems across scales. Trends in Ecology and Evolution 10:241–244.

Phillips, P.J., G.J. Wall, and C.M. Ryan. 2000. Pesticides in wells in agricultural and urban areas of the Hudson River basin. Northeastern Geology and Environmental Sciences 22:1–9.

Phillips, R.L. 1999. Unconventional sources of genetic diversity: De novo variation and elevated epistasis. Pp. 103–131 in Plant Breeding in the Turn of the Millennium. Brazil: Federal University of Vicosa. March. 2–3.

Pilon-Smits, E., and M. Pilon. 2000. Breeding mercury-breathing plants for environmental cleanup. Trends in Plant Science 5:235–236.

Pimentel, D. 1987. Is silent spring behind us? In Silent Spring Revisited, G.J. Marco, R.M. Hollingsworth, and W. Durham, eds. Washington, DC: American Chemical Society.

Pimentel, D., U. Stachow, D.A. Takacs, H.W. Brubaker, A.R. Dumas, J.J. Meaney, J.A.S. O'Neil, D.E. Onsi, and D.B. Corzilius. 1992. Conserving biological diversity in agricultural/forestry systems. BioScience 42:354–362.

Pimentel, D., L. Lach, R. Zuniga, and D. Morrison. 2000. Environmental and economic costs of nonindigenous species in the United States. BioScience 50:53–65.

Pleasants, J.M., R.L. Hellmich, G.P. Dively, M.K. Sears, D.E. Stanley-Horn, H.R. Mattila, J.E. Foster, T.L. Clark, and G.D. Jones. 2001. Corn pollen deposition on milkweeds in and near cornfields. Proceedings of the National Academy of Sciences of the United States of America 98(21):11919–11924.

Poirier, Y., D.E. Dennis, K. Klomparens, and C. Somerville. 1992. Polyhydroxybutyrate, a biodegradable thermoplastic, produced in transgenic plants. Science 256:520–523.

Polis, G.A., and D. Strong. 1996. Food web complexity and community dynamics. The American Naturalist 147:813–846.

Polis, G.A., W.B. Anderson, and R.D. Holt. 1997. Toward an integration of landscape and food web ecology: The dynamics of spatially-subsidized food webs. Annual Review of Ecological Systems 28:289–316.

Ponnamperuma, F.N. 1979. Soil problems in the IRRI farm. IRRI Thursday Seminar, Nov. 8. Mimeo paper. Manila, Philippines: International Rice Research Institute.

Powell-Abel, P., R.S. Nelson, B. De, N. Hoffmann, S.G. Rogers, R.T. Fraley, and R.N. Beachy. 1986. Delay of disease development in transgenic plants that express the tobacco mosaic virus CP gene. Science 232:738–743.

Prather, T.S., J.M. Ditomaso, and J.S. Holt. 2000. Herbicide resistance: Definition and management strategies. University of California Division of Agriculture and Natural Resources Publication #8012. Available online at *http://anrcatalog.ucdavis.edc/pdf/8012.pdf.*

Pretty, J. 1995. Regenerating Agriculture: Policies and Practices for Sustainability and Self-Reliance. London: Earthscan.

Provvidenti, R., R.W. Robinson, and H.M. Munger. 1978. Resistance in feral species to six viruses infecting *Cucurbita*. Plant Distribution Reporter 62:326–329.

Quideng, Q., and R.A. Jorgensen. 1998. Homology-based control of gene expression patterns in transgenic petunia flowers. Developmental Genetics 22:100–109.

Rabalais, N.N., R.E. Turner, D. Justic´, Q. Dortch, and W.J. Wiseman, Jr. 1999. Characterization of Hypoxia: Topic 1 Report for the Integrated Assessment on Hypoxia in the Gulf of Mexico. NOAA Coastal Ocean Program Decision Analysis Series No. 15. Silver Spring, MD: NOAA Coastal Ocean Program.

Raboy, V. 1997. Accumulation and storage of phosphate and minerals. Pp. 441–477 in Cellular and Molecular Biology of Plant Seed Development, B.A. Larkins and I.K. Vasil, eds. Netherlands: Kluwer.

Raffensperger, C., and J. Tickner, eds. 1998. Protecting Public Health & the Environment: Implementing the Precautionary Principle. Washington, DC: Island Press.

Raskin, I. 1996. Plant genetic engineering may help with environmental cleanup. Proceedings of the National Academy of Sciences of the United States of America 93:3164–3166.

Rasmusson, D.C., and R.L. Phillips. 1997. Plant breeding progress and genetic diversity from de novo variation and elevated epistasis. Crop Science 37:303–310.

Raulston, J.R., S.D. Pair, A.N. Sparks, J.G. Loera, F.A.M. Pedraza, A.T. Palamon, A. Ortega, J.M. Sanchez Ruiz, P.C. Marquez, H.A. Rueles, J.M. Perez, R.R. Rodriguez, H.R. Carrillo, R.R. Archundia, and R. Herrera Francisco. 1986. Fall armyworm distribution and population dynamics in the Texas-Mexico gulf-coast area. Florida Entomologist 69 (3):455–468.

Regal, P. 1986. Models of genetically-engineered organisms and their ecological impact. In Ecology of Biological Invasions of North America and Hawaii, Harold Mooney, ed. New York: Springer-Verlag.

Renaud, M.L. 1986. Detecting and avoiding oxygen-deficient sea water by brown shrimp, *Penaeus aztecus (Ives)*, and white shrimp *Penaeus setiferus (Linnaeus)*. Journal of Experimental Marine Biology and Ecology 98:283–292.

Rhymer, J.M., and D. Simberloff. 1996. Extinction by hybridization and introgression. Annual Review of Ecology and Systematics 27:83–109.

Riera-Lizarazu, O., H.W. Rines, and R.L. Phillips. 1996. Cytological and molecular characterization of oat x maize partial hybrids. Theoretical and Applied Genetics 93:123–135.

Risser, P. 1995. The status of the science examining ecotones. BioScience 45:318–325.

Rissler, J., and M. Mellon. 1996. The Ecological Risks of Engineered Crops. Cambridge, MA: The MIT Press.

Robertson, G.P. 1997. Nitrogen use efficiency in row-crop agriculture: Crop nitrogen use and soil nitrogen loss. Pp. 347–66 in Ecology in Agriculture, L.E. Jackson, ed. San Diego, CA: Academic Press.

Robinson, G.S. 1999. HOSTS—A database of the hostplants of the world's Lepidoptera. Nota Lepidopterologica 22:35.

Robinson, G.S., P.R. Ackery, I.J. Kitching, G.W. Beccaloni, and L.M. Hernández. 2000. HOSTS—The Natural History Museum's Database of Caterpillar Hostplants and Foods, Public Intranet Database. Available online at *http://flood.nhm.ac.uk/cgi-bin/perth/hosts/index.dsml*.

Rosenberg, C.E. 1961. No Other Gods: On Science and American Social Thought. Baltimore: Johns Hopkins University Press.

Roush, R.T. 1997. Managing resistance to transgenic crops. Pp. 271–294 in Advances in Insect Control: The Role of Transgenic Plants, N. Carozzi and M. Koziel, eds. London: Taylor and Francis.

Rowe, W.D. 1977. An Anatomy of Risk. New York: Wiley.

Rundle, L. 2000. The producer's role—serf or partner in the biobased economy? Pp. 93–96 in The Biobased Economy of the Twenty-First Century: Agriculture Expanding into Health, Energy, Chemicals and Materials, A. Eaglesham, W.F. Brown, and R.W.F. Hardy, eds. Ithaca, NY: National Agricultural Biotechnology Council.

Sánchez González, J.J., and J.A. Ruiz Corral. 1997. Teosinte distribution in Mexico. Pp. 18–36 in Gene Flow Among Maize Landraces, Improved Maize Varieties and Teosinte: Implications for Transgenic Maize, J.A. Serratos, M.C. Willcox, and F. Castillo, eds. Mexico: Center for Research on Wheat and Maize (CIMMYT).

Sandman, P.M., N.D. Weinstein, and M.L. Klotz. 1987. Public response to the risk from geological radon. Journal of Communication 37:93–108.

Sanford, J.C., F.D. Smith, and J.A. Russell. 1993. Optimizing the biolistic process for different biological applications. Methods in Enzymology 217:483–509.

Saxena, D., and G. Stotzky. 2000. Insecticidal toxin is released from roots of transgenic *Bt* corn *in vitro* and *in situ*. FEMS Microbiology Ecology 33:35–39.

Saxena, D., S. Flores, and G. Stotzky. 1999. Insecticidal toxins in root exudates from *Bt* corn. Nature 402:480.

Schroeder, J.I., J.M. Kwak, and G.J. Allen. 2001. Guard cell abscisic acid signalling and engineering drought hardiness in plants. Nature 410(6826):327–330.

Schultheis, J.R., and S.A. Walters. 1998. Yield and virus resistance of summer squash cultivars and breeding lines in North Carolina. HortTechnology 8:31–39.

Schnepf, E., N. Crickmore, J. Van Rie, D. Lereclus, J. Baum, J. Feitelson, D.R. Zeigler, and D.H. Dean. 1998. *Bacillus thuringiensis* and its pesticidal crystal proteins. Microbiology and Molecular Biology Reviews 62:775–806.

Scientists Working Group on Biosafety. 1998. Manual for Assessing Ecological and Human Health Effects of Genetically Engineered Organisms. Part One: Introductory Materials and Supporting Text for Flowcharts. Part Two: Flowcharts and Worksheets. Edmonds: The Edmonds Institute. Available online at *www.edmonds-institute.org/manual.html*.

Sears, M.K., R.L. Hellmich, D.E. Stanley-Horn, K.S. Oberhauser, J.M. Pleasants, H.R. Mattila, B.D. Siegfried, and G.P. Dively. 2001. Impact of *Bt* corn pollen on monarch butterfly populations: A risk assessment. Proceedings of the National Academy of Sciences of the United States of America 98(21):11937–11942.

Sedlacek, J.D., S.R. Komaravalli, A.M. Hanley, B.D. Price, and P.M. Davis. 2001. Life history attributes of Indian meal moth (Lepidoptera: Pyralidae) and Angoumois grain moth (Lepidoptera: Gelechiidae) reared on transgenic corn kernels. Journal of Economic Entomology 94:586–592.

Sen, A.K. 1982. Poverty and Famines: An Essay on Entitlement and Deprivation. Oxford: Oxford University Press.

Shaeffer, S.E. 1998. Recruitment models for mallards in eastern North America. Auk 115:988–997.

Shaeffer, S.E., and R.A. Malecki. 1996. Predicting breeding success of Atlantic population Canada geese from meteorological variables. Journal of Wildlife Management 60:882–890.

Sheng, J., and V. Citovsky. 1996. Agrobacterium-plant cell DNA transport: Have virulence proteins, will travel. Plant Cell 8:1699–1710.

Shintani, D., and D. DellaPenna 1998. Elevating the vitamin E content of plants through metabolic engineering. Science 282:2098–2100.

Sillett, T.S., R.T. Holmes, and T.W. Sherry. 2000. Impacts of global climate cycle on population dynamics of a migratory songbird. Science 288:2040–2042.

Simberloff, D., and P. Stiling. 1996. How risky is biological control? Ecology 77:1965–1974.

Simmonds, N.W., and J. Smartt. 1999. Principles of Crop Improvement. 2nd Ed., London: Blackwell Science.

Simpson, B.B., and M.C. Ogorzaly. 1995. Economic Botany: Plants in Our World, 2nd ed. New York: McGraw-Hill.

Singh, R.J., and T. Hymowitz. 1999. Soybean genetic resources and crop improvement. Genome 42:605–613.

Skinner, M.W., and B.M. Pavlik. 1994. Inventory of Rare and Endangered Vascular Plants of California. Sacramento, CA: California Native Plant Society.

Slatkin, M. 1987. Gene flow and the geographic structure of natural populations. Science 236:787–792.

Smalla, K., L.S. van Overbeek, R. Pukall, and J.D. van Elsas. 1993. Prevalence of *nptII* and *Tn5* in kanamycin resistant bacteria from different environments. Microbiology Ecology 13:47–58.

Smartt, J., and N.W. Simmonds. 1995. Evolution of Crop Plants, 2nd ed. Harlow: Longman.

Smith, K.J., and W. Huyser. 1987. World distribution and significance of soybean. Pp. 1–22 in Soybeans: Improvement, Production, and Uses, 2nd ed., J.R. Wilcox, ed. Madison, WI: American Society of Agronomists.

Snow, A.A., and P. Moran-Palma. 1997. Commercialization of transgenic plants: potential ecological risks. BioScience 47:86–96.

Sogawa, K. 2000. Chinese National Rice Research Institute. Personal communication.

Somers, D. 2001. University of Minnesota. Personal communication.

Somerville, C.R., and Bonetta. 2001. Plants as factories for technical materials. Plant Physiology 125:168–171.

Sonti, R.V., M. Chiurazz, D. Wong, C.S. Davies, G.R. Harlow, D.W. Mount, and E.R. Signer. 1995. Arabidopsis mutants deficient in T-DNA integration. Proceedings of the National Academy of Sciences of the United States of America 92:11786–11790.

Soule E. 2000. Assessing the precautionary principle. Public Affairs Quarterly 14:309–328.

Soule, J., D. Carre, and W. Jackson. 1990. Ecological impact of modern agriculture. Pp. 165–188 in Agroecology, Carroll, C.R., J.H. Vandermeer, and P.M. Rosset, eds. New York: McGraw-Hill.

Srivastava, V., O.D. Anderson, and D.W. Ow. 1999. Single-copy transgenic wheat generated through the resolution of complex integration patters. Proceedings of the National Academy of Sciences of the United States of America 96:1117–11121.

Stanley-Horn, D.E., G.P. Dively, R.L. Hellmich, H.R. Mattila, M.K. Sears, R. Rose, L. C.H. Jesse, J.E. Losey, J. J. Obrycki, and L. Lewis. 2001. Assessing the impact of Cry1Ab-expressing corn pollen on monarch butterfly larvae in field studies. Proceedings of the National Academy of Sciences of the United States of America 98(21):11931–11936.

Steffey, K., M.E. Rice, J. All, D.A. Andow, M.E. Gray, and J.W. Van Duyn, eds. 1999. Handbook of Corn Insects. Lanham, MD: Entomological Society of America.

Stotzky, G. 2000. Persistence and biological activity in soil of insecticidal proteins from *Bacillus thuringiensis* and of bacterial DNA bound on clays and humic acids. Journal of Environmental Quality 29:691.

Stotzky, G. 2001. Release, persistence, and biological activity in soil of insecticidal proteins from *Bacillus thuringiensis*. In Genetically Engineered Organisms: Assessing Environmental and Human Health Effects, D.K. Letourneau and B.E. Burrows, eds. Boca Raton, FL: CRC Press.

Straub, R.W., and B. Emmett. 1992. Pests of monocotyledon crops: Sweetcorn. Pp. 213–235 in Vegetable Crop Pests, R.G. McKinlay, ed. Boca Raton, FL: CRC Press.

Strong, D.R. 1974. Rapid asymptotic accumulation in phytophagous insect communities: The pests of cacao. Science 185:1064–1066.

Strong, D.R., E.D. McCoy, and J.R. Rey. 1977. Time and number of herbivore species: The pests of sugarcane. Ecology 58:167–175.

Subramanian, N., and P.R. Adiga. 1997. Mapping the common antigenic determinants in avidin and streptavidin. Biochemistry and Molecular Biology International 43:375–382.

Sukopp, H., and U. Sukopp. 1993. Ecological long-term effects of cultigens becoming feral and of naturalization of non-native species. Experientia 49:210–218.

Sun, M., and H. Corke. 1992. Population genetics of colonizing success of weedy rye in Northern California. Theoretical and Applied Genetics 83:321–329.

Suneson, C.A., K.O. Rachie, and G.S. Khush. 1969. A dynamic population of weedy rye. Crop Science 9:121–124.

Susskind, L.E., and L. Dunlap. 1981. The importance of nonobjective judgments in environmental impact assessments. Environmental Impact Assessment Review 2:335–366.

Svitashev, S.K., and D.A. Somers. 2001. Genomic interspersions determine the size and complexity of transgene loci in transgenic plants produced by microprojectile bombardment. Genome 44:691–697.

Swanson, B.E. 2000. Organic Corn. Illinois Specialty Farm Products. Available online at *http://web.aces.uiuc.edu/value/factsheets/fact-organic-corn.htm*.

Swift, M.J., and J.M. Anderson. 1993. Biodiversity and ecosystem function in agricultural systems. Pp. 15–41 in Biodiversity and Ecosystem Function, E.D. Schultz and H.A. Mooney, eds. Berlin: Springer-Verlag.

Swift, M.J., J. Vandermeer, P.S. Ramakrishnan, J.M. Anderson, C.K. Ong, and B.A. Hawkins. 1995. Biodiversity and agroecosystem function. Pp. 261–298 in Functional Roles of Biodiversity: A Global Perspective, H.A. Mooney, J.H. Cushman, E. Medina, O.E. Sala, and E-D. Schulze. New York: John Wiley & Sons.

Tabashnik, B.E. 1994. Evolution of Resistance to *Bacillus thuringiensis*. Annual review of Entomology 39:47–79.

Takano, M., H. Egawa, J.E. Ikeda, and K. Wadasa. 1997. Structure of integration sites in transgenic rice. Plant Journal 11:353–361.

Talalay, P., and Y. Zhang. 1996. Chemoprotection against cancer by isothiocyanates and glucosinolates. Biochemistry Society Transfer 24:806–810.

Taliaferro, C.M. 1995. Diversity and vulnerability of bermuda turfgrass species. Crop Science 35:327–332.

Taubes, G. 2001. The soft science of dietary fat. Science 291:2536–2545.

Thompson, J.D. 1991. The biology of an invasive plant. BioScience 41:393–401.
Thompson, P.B. 1997a. Food Biotechnology in Ethical Perspective. London: Chapman and Hall.
Thompson, P.B. 1997b. Science policy and moral purity: The case of animal biotechnology. Agriculture and Human Values 14:11–27.
Thompson, P.B. 2000. Approaches to Risk and Risk Assessment. Pp. 238–245 in Incorporating Science, Economics, and Sociology in Developing Sanitary and Phytosanitary Standards in International Trade. Washington, DC: National Academy Press.
Tiedje, J.M., R.K. Colwell, Y.L. Grossman, R.E. Hodson, R.E. Lenski, R.N. Mack, and P.J. Regal. 1989. The planned introduction of genetically engineered organisms: Ecological considerations and recommendations. Ecology 70:298–315.
Tilman, D.G. 1996. Biodiversity: Population versus ecosystem stability. Ecology 77:350–363.
Tilman, D.G. 1999a. The ecological consequences of changes in biodiversity: A search for general principles. Ecology 80:1455–1474.
Tilman, D.G. 1999b. Global environmental impacts of agricultural expansion: The need for sustainable and efficient practices. Proceedings of the National Academy of Sciences of the United States of America 96:5995–6000.
Tilman, D.G., and J.A. Downing. 1994. Biodiversity and stability in grasslands. Nature 36:363–365.
Traber, M.G., and H. Sies. 1996. Vitamin E in humans: Demand and delivery. Annual Review of Nutrition 16:321–347.
Traynor, P.L., and J.H. Westwood. 1999. Ecological effects of pest resistance genes in managed ecosystems. Blacksburg, VA: Information Systems for Biotechnology.
Trzcinski, M.K., L. Fahrig, and G. Merriam. 1999. Independent effects of forest cover and fragmentation on the distribution of forest breeding birds. Ecological Applications 9: 586–593.
Tzfira, T., A. Zuker, and A. Altman. 1998. Forest-tree biotechnology: Genetic transformation and its application to future forests. Trends in Biotechnology 16:439–446.
University of Illinois. 2001. Illinois Specialty Farm Products: Organic Corn. Available online at *http://web.aces.uiuc.edu/value/factsheets/fact-organic-corn.htm*.
U.S. Congress. 5 USC sec. 551 et seq. 1946. Federal Administrative Procedures Act. Washington, DC: Government Printing Office.
U.S. Congress, Office of Technology Assessment. 1993. Harmful, Non-indigenous Species in the United States. Washington, DC: U.S. Government Printing Office.
USDA (U .S. Department of Agriculture). 1994a. 7 CFR § 201.76. Federal Seed Act Regulations; Minimum Land, Isolation, Field, and Seed Standards. Federal Register 59:64516.
USDA (U.S. Department of Agriculture). 1994b. Response to the Upjohn Company/Asgrow Seed Company Petititon 92-204-01 for Determination of Nonregulated Status for ZW-20 Squash. Washington, DC: USDA.
USDA (U.S. Department of Agriculture). 1995. Response to Ciba Seeds Petition for Determination of Nonregulated Status for Insect-Resistant Event 176 Corn. Washington, DC: USDA.
USDA (U.S. Department of Agriculture). 1996. Response to the Asgrow Seed Company Petition 95-352-01 for Determination of Nonregulated Status for CZW-3 Squash. Washington, DC: USDA.
USDA (U.S. Department of Agriculture). 1997a. DuPont Petition 97-008-01p for Determination of Nonregulated Status for Transgenic High Oleic Acid Soybean Sublines G94-1, G94-19, and G-168: Environmental Assessment and Finding of No Significant Impact. Appendix A. Determination of Nonregulated Status or High Oleic Acid Transgenic Soybean Sublines G94-1, G94-19, and G-168 Derived from Transformation Event 260-05. Washington, DC: USDA.

USDA (U.S. Department of Agriculture). 1997b. USDA/APHIS Petition 97-013-01p for Determination of Nonregulated Status for Events 31807 and 31808 Cotton Environmental Assessment and Finding of No Significant Impact. Washington, DC: USDA.

USDA (U.S. Department of Agriculture). 1998. Determination of Nonregulated Status for *Bt* Cry9C Insect Resistant and Glufosinate Tolerant Corn Transformation Event CBH-351. Washington, DC: USDA. Available online at *ftp://www.aphis.usda.gov/pub/bbep/Determinations/ascii/9726501p_det.txt.*

USDA (U.S. Department of Agriculture). 1999. Crop lines no longer regulated by USDA. Washington, DC: USDA. Available online at *www.aphis.usda.gov/biotech/not_reg.html.*

USDA (U.S. Department of Agriculture). 2000. USDA Forest Service Strategic Plan. Integrity and Accountability: A Framework for Natural Resource Management. Washington, DC: USDA.

USDA (U.S. Department of Agriculture). 2001a. Biotechnology Risk Assessment Research Grants Program. Available online at *www.reeusda.gov/crgam/biotechrisk/biotech.htm.*

USDA (U.S. Department of Agriculture). 2001b. Initiative for Future Agriculture and Food Systems. Fiscal Year 2000: Abstracts of awards. Available online at *www.reeusda.gov/1700/programs/IFAFS/ifafssum3.htm.*

USDA-NASS (National Agricultural Statistics Service). 2000. Prospective Plantings. Available online at *http://usda.mannlib.cornell.edu/reports/nassr/field/pcp-bbp/pspl0300.txt.*

USDA-NASS (National Agricultural Statistics Service). 2001. Prospective Plantings. Available online at *http://usda.mannlib.cornell.edu/reports/nassr/field/pcp-bbp/pspl0301.pdf.*

van Frankenhuyzen, K., and C. Nystrom. 1989. Residual toxicity of a high-potency formulation of *Bacillus thuringiensis* to spruce budworm (lepidoptera, tortricidae) Journal of Economic Entomology 82(3):868–872.

van Frankenhuyzen, K., and C. Nystrom. 1999. The *Bacillus thuringiensis* toxin specificity database. Available online at *www.glfc.forestry.ca/Bacillus/Bt_HomePage/netintro99.htm.*

Vasil, I.K. 1998. Biotechnology and food security for the 21st century: A real world perspective. Nature Biotechnology 16:399–400.

Venkateswerlu, G., and G. Stotzky. 1992. Binding of the protoxin and toxin proteins of *Bacillus thuringiensis* subsp. *kurstaki* on clay materials. Current Microbiology 25:225–233.

Venter, J.C., M.D. Adams, E.W. Myers, P.W. Li, et al. 2001. The sequence of the human genome. Science 291(5507):1304–1351.

Verboom, J., and R. van Apeldoorn. 1990. Effects of habitat fragmentation on the red squirrel, *Sciurus vulgaris* L. Landscape Ecology 4:171–176.

Victor, D.G. 2000. Risk management and the world trading system: Regulating international trade distortions caused by national sanitary and phytosanitary policies. Pp. 118–169 in Incorporating Science, Economics, and Sociology in Developing Sanitary and Phytosanitary Standards in International Trade. Washington, DC: National Academy Press.

Vitousek, P.M., J.M. Aber, R.W. Howarth, G.E. Likens, P.A. Matson, D.W. Schindler, W.H. Schlesinger, and D.G. Tilman. 1997. Human alteration of the global nitrogen cycle: Sources and consequences. Ecological Applications 7:737–750.

Von Winterfeldt, D. 1992. Expert knowledge and public values in risk management: The role of decision analysis. Pp. 321–342 in Social Theories of Risk, S. Krimsky and D. Golding, eds. Westport: Praeger.

Wagner, D.B., and R.W. Allard. 1991. Pollen migration in predominantly self-fertilizing plants: Barley. Journal of Heredity 82:302–304.

Waggoner, P.E., and D.E. Aylor. 2000. Epidemiology: A science of patterns. Annual Review of Phytopathology 38:1–24.

Walker, B., and M. Lonsdale. 2000. Genetically modified organisms at the crossroads: Comments on "Genetically Modified Crops: Risks and Promise" by Gordon Conway. Conservation Ecology 4:11. Available online at *www.consecol.org/vol4/iss1/art12.*

Walmsley A.M., and C.J. Arntzen. 2000. Plants for delivery of edible vaccines. Current Opinions in Biotechnology 11:126–129.

Watanabe, T. 1993. Dietary biotin deficiency affects reproductive function and prenatal development in hamster. Journal of Nature 123:2101–2108.

Welch, R.M., G.F. Combs, and J.M. Duxbury. 1997. Toward a "greener" revolution. Issues in Science and Technology 14:50–58.

Whittaker, R.H. 1975. Communities and Ecosystems, 2nd ed. New York: Macmillan.

Whitaker, T.W., and W.P. Bemis. 1964. Evolution of the genus *Cucurbita.* Evolution 18:553–559.

Wiebold, W.J., C.G. Morris, H.L. Mason, D.R. Knerr, R.W. Hasty, T.G. Fritts, and E. Adams. 2000. Missouri Crop Performance 2000, Corn. Crop performance testing. Available online at *http://agebb.missouri.edu/cropperf/corn/.*

Wilcove, D.S., C.H. Mclellan, and A.P. Dobson. 1986. Habitat fragmentation in the temperate zone. Pp. 237–256 in Conservation Biology: The Science of Scarcity and Diversity, M.E. Soule, ed. Sunderland, MA: Sinauer.

Wilkes, H.G. 1997. Teosinte in Mexico: Personal retrospective and assessment. Pp. 10–17 in Geneflow Among Maize Landraces, Improved Maize Varieties and Teosinte: Implications for Transgenic Maize, J.A. Serratos, M.C. Willcox, and F. Castillo, eds. Mexico: Center for Research on Wheat and Maize (CIMMYT).

Williams, M.C. 1980. Purposefully introduced plants that have become noxious or poisonous weeds. Weed Science 28:300–305.

Williams, M.E. 1995. Genetic engineering for pollination control. Trends in Biotechnology 13:344–349.

Williamson, M. 1994. Community response to transgenic plant release: Predictions from British experience of invasive plants and feral crop plants. Molecular Ecology 3:75–79.

Williamson, M. 1996. Biological Invasions. London: Chapman and Hall.

Williamson, M., and A. Fitter. 1996. The varying success of invaders. Ecology 77:1661–1666.

Williamson, M.J. 1993. Invaders, weeds, and the risk from genetically manipulated organisms. Experientia 49:219–224.

Williamson, M.J. 1994. Community response to transgenic plant release: Predictions from British experience of invasive plants and feral crop plants. Molecular Ecology 3:75–79.

Wilson, H.D. 1993. Free-living *Cucurbita pepo* in the United States: Viral resistance, gene flow, and risk assessment. Prepared for USDA Animal and Plant Health Inspection Service. Hyattsville, MD: USDA. Available online at *www.csdl.tamu.edu/Flora/flcp/flcp1.htm.*

Winberry, J.J., and D.M. Jones. 1974. Rise and decline of the "miracle vine": Kudzu in the southern landscape. Southeastern Geography 13(2):61–70.

Wolf, D.E., N. Takebayashi, and L.H. Rieseberg. 2001. Predicting the risk of extinction through hybridization. Conservation Biology 15(4):1039–1052.

Woodward, A., K.J. Jenkins, and E.G. Schreiner. 1999. The role of ecological theory in long-term ecological monitoring: Report on a workshop. Natural Areas Journal 19:223–233.

Woolsey, J. 2000. Hydrocarbons to carbohydrates: The strategic dimension. Pp. 43–50 in The Biobased Economy of the Twenty-First Century: Agriculture Expanding into Health, Energy, Chemicals and Materials, A. Eaglesham, W.F. Brown, and Ralph W.F. Hardy, eds. Ithaca, NY: National Agricultural Biotechnology Council.

Wraight, C.L., A.R. Zangerl, M.J. Carroll, and M.R. Berenbaum. 2000. Absence of toxicity of *Bacillus thuringiensis* pollen to black swallowtails under field conditions. Proceedings of the National Academy of Sciences of the United States of America 97(14):7700–7703.

Wright, G.M., and B.H. Thompson. 1935. Fauna of the United States: Wildlife management in the national parks. Fauna Series 2. Washington, DC: U.S. Department of the Interior.

Wright, G.M., J.S. Dixon, and B.H. Thompson. 1933. Fauna of the national parks of the United States: A preliminary survey of faunal relations in the national parks. Fauna Series 1. Washington, DC: U.S. Department of the Interior.

Wright, S. 1969. Evolution and the Genetics of Populations, vol. 2. The Theory of Gene Frequencies. Chicago, IL: University of Chicago Press.

Yao, J-L., Y.H. Dong, and B.A.M. Morris. 2001. Parthenocarpic apple fruit production conferred by transposon insertion mutations in a MADS-box transcription factor. Proceedings of the National Academy of Sciences of the United States of America 98:1306–1311.

Yara, K., Y., Kunimi, and H. Iwahana. 1997. Comparative studies of growth characteristic and competitive ability in *Bacillus thuringiensis* and *Bacillus cereus* in soil. Applied Entomology and Zoology 32(4):625–634.

Ye, X., S. Al-Babili, A. Klöti, J. Zhang, P. Lucca, P. Beyer, and I. Potrykus. 2000. Engineering the provitamin A (β-carotene) biosynthetic pathway into (carotenoid-free) rice endosperm. Science 287:303–305.

Yodzis, P. 1988. The indeterminacy of ecological interactions. Ecology 69:508–515.

Young, N.D., and S.D. Tanksley. 1989. RFLP analysis of the size of chromosomal segments retained around the TM-2 locus of tomato during backcross breeding. Theoretical Applications of Genetics 77:353–359.

Zangerl, A.R., D. McKenna, C.L. Wraight, M. Carroll, P. Ficarello, R. Werner, and M.R. Berenbaum. 2001. Effects of exposure to event 176 *Bacillus thuringiensis* corn pollen on monarch and black swallowtail caterpillars under field conditions. Proceedings of the National Academy of Sciences of the United States of America 98(21):11908–11912.

Zhu, J.K. 2001. Plant salt tolerance. Trends in Plant Science 6:66–71. Available online at *www.soygrowers.wegov2.com/file_depot/0-10000000/0-10000/735/folder/4944/2000SoyStats.pdf*.

Zhu, Y., H. Chen, J. Fan, Y. Wang, Y. Li, J. Chen, J. Fan, S. Yang, L. Hu, H. Leung, T.W. Mew, P.S. Teng, Z. Wang, and C.C. Mundt. 2000. Genetic diversity and disease control in rice. Nature 406:718–722.

Zimmerman, R., J. Nance, and J. Williams. 1996. Trends in shrimp catch in the hypoxic area of the northern Gulf of Mexico. In Proceedings of the First Gulf of Mexico Hypoxia Management Conference. Washington, DC: U.S. Environmental Protection Agency.

Zitnak, A., and G.R. Johnston. 1970. Glycoalkaloid content of B5141-6 potatoes. American Potato Journal 47:256–260.

Zyriax, B.C, and E. Windler. 2000. Dietary fat in the prevention of cardiovascular disease: A review. European Journal of Lipid and Technology 102:355–365.

Appendixes

A

Workshop to Assess the Regulatory Oversight of GM Crops and the Next Generation of Genetic Modifications for Crop Plants

Friday, October 13th
National Research Council
2001 Wisconsin Avenue, NW
(Room 130 of the Cecil and Ida Green Building)

Agenda

I. Regulatory Oversight

Time	Topic
8:45	Welcome and Introduction **Fred Gould**, North Carolina State University, Chair, Committee on Environmental Impacts Associated with Commercialization of Transgenic Crops: Issues and Approaches to Monitoring
9:00	Overview and the Current Reassessment of the Coordinated Framework **Sharon Friedman**, National Science and Technology Council Representative, Office of Science and Technology Policy
9:20	Comprehensive and Commensurate Requirements **Terry Medley**, Director of Regulatory and External Affairs, Dupont Agricultural Enterprise

9:40 USDA's Legal Authority and Perspectives on Assessment of Environmental Effect
Stan Abramson, Chair—Environmental Practice Group, and Partner, Arent, Fox, Kintner, Plotkin, & Kahn
10:00 BREAK
10:20 Forward-Looking Regulation of Agricultural Biotechnology
David Adelman, Staff Attorney, International and Nuclear Programs, Natural Resources Defense Council
10:40 Panel Discussion
11:00 Committee and Audience Questions and Discussion
12:00 LUNCH

II. Next Generation of Transgenic Crops

1:15 Introduction
Brian Larkins, University of Arizona, member of Committee on Environmental Effects Associated with Commercialization of Transgenic Crops
1:25 Nutritional Enhancement
Dean DellaPenna, Professor, Department of Biochemistry and Molecular Biology, Michigan State University
1:50 The Next Decade of Plant Biotechnology Products
Ganesh Kishore (formerly with Monsanto Company)
2:15 Stress Tolerance
Michael Thomashow, Professor, Crop and Soil Sciences, Michigan State University
2:40 Prospects for Hypoallegenic Transgenic Soybeans
Herman M. Eliot, Acting Research Leader, Climate Stress Laboratory, USDA, ARS, Beltsville, Maryland
3:00 BREAK
3:15 Discussion and questions

III. Public Comments

3:45 Introduction
Kim Waddell, National Research Council
3:50 Selected Letter Responses
Faith Campbell, American Lands Alliance
Maureen K. Hinkle, National Audubon Society (retired)
4:10 Other Comments
5:30 Workshop adjourns

B

Workshop Presenters / Panelists

Friday, October 13th
National Research Council
2001 Wisconsin Avenue, NW

(Room 130 of the Cecil and Ida Green Building)

Committee on Environmental Impacts Associated with Commercialization of Transgenic Crops

WORKSHOP PRESENTERS/PANELISTS

Stanley Abramson, Partner
Arent, Fox, Kintner, Plotkin, & Kahn
1050 Connecticut Ave., NW
Washington, DC 20036-5339
PH: (202) 857-8935
Fax: (202) 857-6395
Email: abramsos@arentfox.com

David E. Adelman, Attorney
International and Nuclear Programs
Natural Resources Defense Council
1200 New York Avenue, NW—Suite 400
Washington, DC 20005
PH: (202) 289-2371
Fax: (202) 289-1060
Email: dadelman@nrdc.org

Faith Campbell
American Lands Alliance
726 7th Street, SE
Washington, DC 20003
PH: (202) 547-9120
Email: phytodoer@aol.com

Dean DellaPenna, Professor
Department of Biochemistry and Molecular Biology
Michigan State University
Biochemistry Bld. Room 215
East Lansing, MI 48824-1319
PH: (517) 432-9284
Fax: (517) 353-9334
Email: dellapen@mus.edu

Sharon Friedman, Agency Representative
Office of Science and Technology Policy
Executive Office of the President
1600 Pennsylvania Avenue, NW
Washington, DC 20502
PH: (202) 395-7347
Fax: (202) 456-2497
Email: sfriedma@ostp.eop.gov

Eliot Herman, Acting Research Leader
Climate Stress Laboratory
USDA / ARS
10300 Baltimore Avenue
Bldg. 006 Room 203
Beltsville, MD 20705
PH: (301) 504-5258
Fax: (301) 504-6626
Email: herman3@ba.ars.usda.gov

Maureen K. Hinkle
National Audubon Society (retired)
5511 Northfield Road
Bethesda, MD 20817
Email: mhinkle2@aol.com

Ganesh Kishore
(formerly with Monsanto)
Email: gkishore@aol.com

Terry L. Medley
Director of Regulatory and External Affairs
Dupont Agricultural Enterprise
Dupont Nutrition and Health
Route 141 & 48
Barley Mill Plaza (P38-1280)
Wilmington, DE 19805-0038
PH: (302) 992-4097
Fax: (302) 992-6098.
Email: terry.l.medley@USA.dupont.com

Michael F. Thomashow
Professor
Crop and Soil Sciences
Michigan State University
A291 Plant & Soil Science
East Lansing, MI 48824
PH: (517) 355-2299
Fax: (51) 353-5174
Email: thomash6@pilot.msu.edu

WORKSHOP ATTENDEES

Faith Campbell
American Lands Alliance Societies
PH: (202) 547-9120
phytodoer@aol.com

Thomas Cors
Dynamics Technology
PH: (703) 841-0990
FAX: (703) 841-8395
tcobs@dynamic.com

Maureen Hinkle
National Audubon Society (retired)
mhinkle2@aol.com

Shirley Ingebritsen
USDA/APHIS

Warren Leon
Northeast Sustainable Energy Association
PH: (413) 774-6051 x 17
FAX: (413) 774-6053
wleon@ultranet.com

Deborah Olster
National Science Foundation
PH: (703) 292-7318
FAX: (703) 292-9078
dolster@nsf.gov

Craig Roseland
Biotechnology Unit
Permits and Risk Assessment
USDA/APHIS
PH: (301) 734-7935
FAX: craig.r.roseland@usda.gov

Joan Rothenerg
Institute of Food Technoloy
PH: (202) 466-5980
FAX: (202) 466-5988

Allison Snow
Ohio State University, and
Member, NRC Standing Committee on Agricultural Biotechnology, Health, and the Environment

C

"Dear Colleague" Letter

The National Research Council Committee on Environmental Effects Associated with Commercialization of Transgenic Plants is seeking your input on identifying specific environmental issues associated with plant biotechnology. Please read the following letter from the chair of the committee explaining how you can contribute to this study by providing the committee with your perspective on this topic.

The letter is also attached as a file in several formats for your use as needed.

Karen L. Imhof
Project Assistant
Board on Agriculture and Natural Resources
Ph: (202) 334-3062
Fax: (202) 334-1978
Email: kimhof@nas.edu

THE NATIONAL ACADEMIES

Advisers to the Nation on Science, Engineering, and Medicine

National Academy of Sciences
National Academy of Engineering
Institute of Medicine
National Research Council

Board on Agriculture and Natural Resources

August 18, 2000

Dear Colleague,

The National Research Council recently established a committee to examine "Environmental Effects Associated with Commercialization of Transgenic Plants". Our committee held its first meeting on July 15th and 16th, 2000, during which we clarified our goals and approaches to meeting this charge. Our committee is now in the phase of gathering information. The final product from our deliberations will be a detailed report.

Our committee has broad expertise in environmental sciences, agriculture and other relevant areas. However, we think it is essential for us to seek input from individuals with diverse perspectives on environmental issues, so that we don't miss important insights. We are contacting you based on your previous interest in issues related to plant biotechnology.

The goals of our committee are as follows:

> We will review the scientific basis that supports the scope and adequacy of USDA's oversight of environmental issues related to current and anticipated transgenic plants and their products.
>
> In order to address these issues, the committee will:
>
> Evaluate the scientific premises and assumptions underpinning the environmental regulation and oversight of transgenic plants. This evaluation will include comparison of the processes and products of genetic engineering with those of conventional plant breeding as they pertain to environmental risks. This evaluation may result in recommendations for research relevant to environmental oversight and effects of transgenic plants.
>
> Assess the relevant scientific and regulatory literature in order to evaluate the scope and adequacy of APHIS' environmental review regarding the process of notification and determination of non-regulated status. The committee will focus on the identification of effects of transgenic plants on non-target organisms and the environmental assessment (EA) of those effects. The study will also provide guidance on the assessment of non-target effects, appropriate tests for environmental evaluation, and

assessment of cumulative effects on agricultural and non-agricultural environments.

Evaluate the need for and approaches to environmental monitoring and validation processes.

We would appreciate it if you would help us meet this charge. We are committed to getting sufficiently broad input to ensure that all important perspectives and information are considered. Please send us a concise letter describing what you see as specific environmental issues associated with plant biotechnology that need attention. We would be particularly appreciative if you would emphasize potential positive and/or negative impacts that you think are underrepresented in most discussions of plant biotechnology. Please send your letter via e-mail to kwaddell@nas.edu. Your letter may also be faxed to him at 202–334–1978.

It would be very helpful to us if you could return your letter by September 5th. We are planning to have a second meeting in October and will need some time to assess outside information such as that in your letter. If we have follow-up questions regarding your letter, we may contact you by phone or e-mail, or we may request that you present more details directly to our committee in October.

If you have a colleague or friend who is likely to have a novel perspective on these issues, we would appreciate your forwarding this letter to that person. While we are not trying to assess the weight of public opinion, we are trying to identify the diversity of perspectives in relation to our charge.

Attached you will find a committee membership list. You may also want to visit the National Academies' Current Projects website at http://www4.nas.edu/cp.nsf for further information* about our committee, and the Academy's standing committee on Biotechnology, Food and Fiber Production, and the Environment, which is examining a broader range of biotechnology related issues.

Please call Kim Waddell or Karen Imhof at 202–334–3062 if you need further information, or assistance with transmitting your letter. Your efforts will be appreciated.

On Behalf of the Committee,

Fred Gould
Committee Chair

* Search for committee information by entering title of project in the Current Projects website.

D

"Dear Colleague Letter" Recipients

"Dear Colleague Letter"—August 2000: Recipients

1.	Benbrook, Charles	Northwest Science and Environmental Policy Center
2.	Belding, John and Susan	Old Stage Farm (Organic growers)
3.	Burrows, Beth	The Edmonds Institute
4.	Campbell, Faith	American Lands Alliance
5.	Chrispeels, Martin	University of CA (Biology)
6.	Conko, Gregory	Competitive Enterprise Institute
7.	Cummins, Joseph E.	University of Western Ontario (Genetics)
8.	Duke, Steve	USDA, ARS, NPURU (University of MS)
9.	Federici, Brian	University of CA
10.	Fomous, Cathy	Council for Responsible Nutrition
11.	Gordon, Milton	University of Washington (Biochemistry)
12.	Gray, Chris	Computer Professionals for Social Responsibility (computer and cultural studies)
13.	Gressel, Jonathan	Weizmann Institute of Sciences
14.	Head, Graham	Monsanto
15.	Hardy, Ralph	National Agricultural Biotechnology Council

16.	Hinkle, Maureen Kuwamo (Retired)	National Audubon Society
17.	Jenkins, Peter	Center for Science in the Public Interest
18.	Levin, Simon	Princeton University
19.	McAfee, Kathleen	University of CA (Environmental Studies)
20.	McClung, Gwendolyn	US EPA—Office of Pollution Prevention and Toxics
21.	Moar, William	Auburn University (Entomology)
22.	Nielson, Patrick	Dole
23.	Pimentel, David	Cornell University (Entomology)
24.	Riddle, Jim	Organic Independents/Organicworks!
25.	Roseland, Craig	USDA-APHIS
26.	Schuh, Edward G.	Hubert Humphrey Institute
27.	Shaner, Dale	American Cyanamid
28.	Sommerville, Chris	Carnegie Institution
29.	Stabinsky, Doreen	Greenpeace
30.	Steward, Robert	Technology Sciences Group
31.	Stewart, Neal	University of North Carolina (Biology)
32.	Sweet, Jeremy	National Institute of Agricultural Biotechnology
33.	Walbot, Virginia	Stanford University (Biological Sciences)
34.	Vargas, Ronald	Corbana (Costa Rica—National Banana Corporation)

About the Authors

Fred L. Gould, *Chair*, is William Neal Reynolds professor of entomology at North Carolina State University. He has researched the ecological genetics of pest adaptation to chemical, biological, and cultural control tactics. His major emphasis in recent years has been on developing methodologies for delaying pest adaptation to transgenic crops that produce insecticidal proteins derived from the bacterium *Bacillus thuringiensis*. Dr. Gould participated in the National Research Council Workshop on Pesticide Resistance: Strategies and Tactics for Management (1986) and was a committee member for several NRC reports. He received his Ph.D. in ecology and evolution from the State University of New York at Stony Brook in 1977.

David A. Andow is professor of insect ecology at the University of Minnesota's Department of Entomology. His research interests include the ecology of beneficial insects for use in biological control in sweet corn, vegetational diversity and conservation of natural enemies, biotechnology science policy, ecology of insect species invasions, host plant resistance in corn, and resistance management of transgenic plants. He is currently a member of the National Research Council's Committee on Agricultural Biotechnology, Health, and the Environment and has also served on the National Science Foundation's Graduate Panel on Biological Sciences. Dr. Andow also is editor-in-chief of *Environmental Biosafety Research* and editor of *Ecological Applications*. He received his B.S. in biology from Brown University in 1977 and his Ph.D. in ecology from Cornell University in 1982.

Bernd Blossey is an assistant professor and director of the Biological Control of Non-Indigenous Plant Species Program of the Department of Natural Resources at Cornell University. His work involves the theoretical and practical aspects of biological weed control, including insect-plant interactions, factors underlying successful invasion by nonindigenous plants, and the development of field- and lab-rearing techniques for mass production of biocontrol agents. His primary efforts have centered on the well-known invasive plant, purple loosestrife. He also has recently initiated new biocontrol programs targeting *Phragmites austalis* and *Alliaria petiolata*. Dr. Blossey received his Ph.D. in zoology/ecology from the University of Kiel.

Ignacio Chapela is assistant professor in the College of Natural Resources, University of California, Berkeley. His current research includes the symbiosis between leaf-cutter ants and their cultivated fungi, the role of fungi in the balance between forest health and disease, and the role of fungi in sustainable management of the diverse and delicate environments of local communities in Southern Mexico. He recently served as faculty chair and vice-chair for the College of Natural Resources at UC Berkeley. He is founder and scientific director of the Mycological Facility in Oaxaca, Mexico. Dr. Chapela received his B.Sc. in biology from the Universidad Nacional Autónoma de México in 1984 and his Ph.D. in fungal ecology from the University of Wales in 1987.

Norman C. Ellstrand is professor of genetics in the Department of Botany and Plant Sciences at the University of California, Riverside. His research focuses on applied plant population genetics with current emphasis on the consequences of gene flow from domesticated plants to their wild relatives. Dr. Ellstrand has participated in a number of government and National Research Council meetings concerning genetically modified organisms including the NRC planning meeting on technology and intellectual property challenges associated with genetically modified seeds. He received his Ph.D. from the University of Texas, Austin, in 1978.

Nicholas Jordan is an associate professor in the Department of Agronomy and Plant Genetics at the University of Minnesota. His current studies explore weed competitive cultivars, ecology and evolution of herbicide-resistant weeds, and integrated weed management approaches focusing on interactions between weeds and soil biota. He is noted for his efforts in developing new working relationships between scientists and farmers, crop consultants, and other agriculturalists to increase the human resources available to develop more durable, effective, and more environmentally sound weed management methods. Dr. Jordan received his B.A.

in biology from Harvard College in 1979 and his Ph.D. in botany and genetics from Duke University in 1986.

Kendall R. Lamkey is a research geneticist with the U.S. Department of Agriculture's Agricultural Research Service and a professor of agronomy in the Department of Agronomy at Iowa State University. His research focuses specifically on the origin, maintenance, and utilization of genetic variation for important agronomic and grain quality traits in maize. Currently, he is technical editor for the journal *Crop Science* and associate editor for the *Journal of Heredity*. Dr. Lamkey received his B.S. in agronomy in 1980 and his M.S. in plant breeding in 1982, both from the University of Illinois, and his Ph.D. in plant breeding from Iowa State University in 1985.

Brian A. Larkins is the Porterfield professor of plant sciences, and professor of molecular and cellular biology in the plant sciences and molecular and cellular biology departments at the University of Arizona, Tucson. He has spent over 20 years researching the regulation of seed development and the synthesis of seed storage proteins. He has particularly focused on identifying the storage proteins that contribute the majority of essential amino acids and on developing genetic strategies to increase their content. In addition to being elected a member of the National Academy of Sciences in 1996, Dr. Larkins has received numerous awards, honors, and scholarships from several universities and the American Society of Plant Physiologists. He received his B.S. in biology in 1969 and his Ph.D. in botany in 1974 from the University of Nebraska.

Deborah K. Letourneau is professor of environmental studies at the University of California, Santa Cruz. She studies tritrophic interactions in a variety of contexts and communities, including tropical rain forests in Costa Rica and Papua New Guinea, subsistence farmers' fields in Africa, and tomato agroecosystems in California. Her most recent publications include issues examining the ecological and human health effects of genetically modified organisms and top-down and bottom-up forces in tropical forest communities. Prior to joining the University of California faculty in 1987, Dr. Letourneau was a member of the resource faculty and co-coordinator of the Organization for Tropical Studies at Duke University. She received her B.S. in zoology in 1976 and her M.S. in biology in 1978 from the University of Michigan and her Ph.D. in entomology from the University of California, Berkeley in 1983.

Alan McHughen is professor and senior research scientist in the Crop Development Centre, College of Agriculture, University of Saskatchewan.

He has helped develop Canada's regulations covering the environmental release of plants with novel traits. He has first-hand experience developing internationally approved commercial crop varieties using both conventional breeding and genetic engineering techniques. He also is an educator and consumer advocate, helping nonscientists understand the environmental and health impacts of both modern and traditional methods of food production. Currently, McHughen is chair of the International Advisory Committee for the International Symposia on the Biosafety of Genetically Modified Organisms. He received his B.Sc. from Dalhousie University in 1976 and his D.Phil. from Oxford University in 1979.

Ronald L. Phillips is regent's professor and McKnight presidential chair in genomics at the University of Minnesota. Throughout this career Dr. Phillips has coupled the techniques of classical cytogenetics with research advances in tissue culture and molecular biology to enhance understanding of basic biology of cereal crops and to improve these species by innovative methods. His research program at the University of Minnesota was one of the early programs in modern plant biotechnology related to agriculture. He is a founding member and former director of the Plant Molecular Genetics Institute of the University of Minnesota. Dr. Phillips served as chief scientist of the U.S. Department of Agriculture (1996–1998) in charge of the National Research Initiative Competitive Grants Program. In 1991 he was elected a member of the National Academy of Sciences. He also served as both a member and chair of the National Research Council's Scientific Council to the Plant Gene Expression Center from 1985 to 1993. Dr. Phillips currently serves as president of the Crop Science Society of America. He earned his B.S. and M.S. degrees from Purdue University and his Ph.D. from the University of Minnesota.

Paul B. Thompson is the Joyce and Edward E. Brewer distinguished professor of philosophy and director of the Center for Food Animal Productivity and Well-Being at Purdue University. He has published extensively on the ethics, policies, perceptions, and impact of agriculture and biotechnology for both the agricultural communities and the American public. Dr. Thompson received his B.A. in philosophy from Emory University in 1974 and his M.A. (1979) and Ph.D. (1980) in philosophy from the State University of New York at Stony Brook.

Board on Agriculture and Natural Resources Publications

POLICY AND RESOURCES

Agricultural Biotechnology: Strategies for National Competitiveness (1987)
Agriculture and the Undergraduate: Proceedings (1992)
Agriculture's Role in K-12 Education: A Forum on the National Science Education Standards (1998)
Alternative Agriculture (1989)
Brucellosis in the Greater Yellowstone Area (1998)
Colleges of Agriculture at the Land Grant Universities: Public Service and Public Policy (1996)
Colleges of Agriculture at the Land Grant Universities: A Profile (1995)
Designing an Agricultural Genome Program (1998)
Designing Foods: Animal Product Options in the Marketplace (1988)
Ecological Monitoring of Genetically Modified Crops (2001)
Ecologically Based Pest Management: New Solutions for a New Century (1996)
Ensuring Safe Food: From Production to Consumption (1998)
Forested Landscapes in Perspective: Prospects and Opportunities for Sustainable Management of America's Nonfederal Forests (1997)
Future Role of Pesticides in U.S. Agriculture (2000)
Genetic Engineering of Plants: Agricultural Research Opportunities and Policy Concerns (1984)
Genetically Modified Pest-Protected Plants: Science and Regulation (2000)

Incorporating Science, Economics, and Sociology in Developing Sanitary and Phytosanitary Standards in International Trade: Proceedings of a Conference (2000)
Investing in Research: A Proposal to Strengthen the Agricultural, Food, and Environmental System (1989)
Investing in the National Research Initiative: An Update of the Competitive Grants Program in the U.S. Department of Agriculture (1994)
Managing Global Genetic Resources: Agricultural Crop Issues and Policies (1993)
Managing Global Genetic Resources: Forest Trees (1991)
Managing Global Genetic Resources: Livestock (1993)
Managing Global Genetic Resources: The U.S. National Plant Germplasm System (1991)
National Research Initiative: A Vital Competitive Grants Program in Food, Fiber, and Natural-Resources Research (2000)
New Directions for Biosciences Research in Agriculture: High-Reward Opportunities (1985)
Pesticide Resistance: Strategies and Tactics for Management (1986)
Pesticides and Groundwater Quality: Issues and Problems in Four States (1986)
Pesticides in the Diets of Infants and Children (1993)
Precision Agriculture in the 21st Century: Geopspatial and Information Technologies in Crop Management (1997)
Professional Societies and Ecologically Based Pest Management (2000)
Rangeland Health: New Methods to Classify, Inventory, and Monitor Rangelands (1994)
Regulating Pesticides in Food: The Delaney Paradox (1987)
Soil and Water Quality: An Agenda for Agriculture (1993)
Soil Conservation: Assessing the National Resources Inventory, Volume 1 (1986); Volume 2 (1986)
Sustainable Agriculture and the Environment in the Humid Tropics (1993)
Sustainable Agriculture Research and Education in the Field: A Proceedings (1991)
The Role of Chromium in Animal Nutrition (1997)
Toward Sustainability: A Plan for Collaborative Research on Agriculture and Natural Resource Management (1991)
Understanding Agriculture: New Directions for Education (1988)
The Use of Drugs in Food Animals: Benefits and Risks (1999)
Water Transfers in the West: Efficiency, Equity, and the Environment (1992)
Wood in Our Future: The Role of Life Cycle Analysis (1997)

NUTRIENT REQUIREMENTS OF DOMESTIC ANIMALS SERIES AND RELATED TITLES

Building a North American Feed Information System (1995)
Metabolic Modifiers: Effects on the Nutrient Requirements of Food-Producing Animals (1994)
Nutrient Requirements of Beef Cattle, Seventh Revised Edition, Update (2000)
Nutrient Requirements of Cats, Revised Edition (1986)
Nutrient Requirements of Dairy Cattle, Seventh Revised Edition (2001)
Nutrient Requirements of Dogs, Revised Edition (1985)
Nutrient Requirements of Fish (1993)
Nutrient Requirements of Horses, Fifth Revised Edition (1989)
Nutrient Requirements of Laboratory Animals, Fourth Revised Edition (1995)
Nutrient Requirements of Poultry, Ninth Revised Edition (1994)
Nutrient Requirements of Sheep, Sixth Revised Edition (1985)
Nutrient Requirements of Swine, Tenth Revised Edition (1998)
Predicting Feed Intake of Food-Producing Animals (1986)
Role of Chromium in Animal Nutrition (1997)
Scientific Advances in Animal Nutrition: Promise for the New Century (2001)
Vitamin Tolerance of Animals (1987)

Further information, additional titles (prior to 1984), and prices are available from the National Academy Press, 2101 Constitution Avenue, NW, Washington, D.C. 20418, 202–334–3313 (information only). To order any of the titles you see above, visit the National Academy Press bookstore at *http://www.nap.edu/bookstore*.

Index

B

D

E

O

P

R

S

T

U

V

W

Y